Traditional Agricultural Practices

Traditional Agricultural Practices

Applications & Technical Implementations

by:

Dr. T. Rathakrishnan *(Professor and Head)*
Dr. N. Anandaraja *(Assistant Professor)*
Dr. M. Ramasubramanian *(Assistant Professor)*
Dr. L. Nirmala *(Assistant Professor)*

Department of Agricultural Extension & Rural Sociology
Agricultural College & Research Institute, Madurai, Tamil Nadu, India
(Tamil Nadu Agricultural University), Coimbatore-641 003, Tamil Nadu, India

&

Dr. M. Israel Thomas *(Assistant Professor)*
Department of Agricultural Extension, Kerala Agricultural University
Thrissur, Kerala

New India Publishing Agency

Pitam Pura, New Delhi- 110 088

Published by
Sumit Pal Jain *for*
New India Publishing Agency
101, Vikas Surya Plaza, CU Block, L.S.C. Mkt.,
Pitam Pura, New Delhi- 110 088, (India)
Phone: 011-27341717, Fax: 011-27341616
E-mail: newindiapublishingagency@gmail.com
Web: www.bookfactoryindia.com

ISBN : 978-93-80235-02-8

Composed and Designed by NIPA

Tamil Nadu Agricultural University
Coimbatore-641 003
Tamil Nadu, India

Dr. B. Chandrasekaran, Ph.D.
Director of Research

FOREWORD

Newer technologies for increasing agricultural production are now becoming available at a much faster rate. But there is no mechanism for transferring them to the uneducated and small producers in an effective manner. There is an urgent need for a communication network to the needy farmers in our country like the global communication network which makes the latest technologies available without delay in any particular field. It is not only knowledge that is needed, but an approach which will be able to supply the right knowledge and tools to the right people and at the right time.

This book presents, what should be the extension approaches in technology transfer and also the importance of the Indigenous Technical Knowledge (ITK) which will be useful to the farming community after refinement of the most useful indigenous technology thus pave a way to increase the adoption rate among the farming community. The authors have brought outstanding information in an exhaustive manner about Indigenous Technical Knowledge and the scientific rationality of them.

In juxtaposition, large number of books is available on Indigenous Technical Knowledge at present, but this book fills up the cleft and is a blend of ITK in agriculture, medicinal plants, ethno veterinary practices, fisheries practices and what not with scientific rationality. The book is frittered into various Chapters covering ifs and buts of ITK and rationality. The illustrations portrayed are chosen to serve as a tour de force in understanding the subject matter.

The authors have extensively worked to provide the valuable information to the students, researchers and the academics in the field of Indigenous Technical Knowledge and this can be visualized by the clear and elaborate nature of its content. I hope that, this book will be extremely useful not only to those who are in the field of agricultural extension but also to all agricultural graduates and researchers besides serving as a valuable tool providing useful insights to planners, policy makers, research and extension functionaries, as well as practioners who are concerned with agricultural development in developing countries.

I complement the authors for their strenuous effort to bring out this book and I ardently trust this will be appreciated at the national level in the field of Indigenous Technical Knowledge.

I congratulate the authors for their work in bringing this valuable book.

9-3-09

(B. CHANDRASEKARAN)

PREFACE

The term 'indigenous knowledge' is used synonymously with traditional and local knowledge to differentiate the knowledge developed by a given community from the international knowledge system. Such a system is viewed to be in dynamic equilibrium with the environment, influenced by innovations emerging from within the system as well as those adopted from other indigenous system. In Tamil Nadu, many of the worthwhile traditional agricultural knowledge are still practiced by the farmers. But the scientist in general concentrated their efforts in high input technologies. Most of the existing knowledge of traditional agriculture has not been properly documented and very limited work has been done in recent times in bringing this knowledge to light. This work can be carried out by pooling of traditional knowledge, short listing and evaluating them in the context of modern scientific and technological environment and harnessing it for sustainable agricultural growth. So for, this exercise of documentation has been done orally and recorded in the form of reports. But digital documentation with the help of digital camera and handy cam, classifying appropriately and storing it in the form of multimedia CDs for wider popularization has been rarely tried. Rightly identifying this gap the authors tried through a DS1R funded project which culminates in the documentation of 505 indigenous agricultural and allied practices from 16 districts of Tamil Nadu through farmer's participatory approach which adds different dimension to the methodology of documentation hitherto practiced.

The authors of this book feel privileged to communicate the documented practices through this publication to all stakeholders of agricultural profession. The documented practices had been classified into traditional varieties, crop protection, crop management, storage structure, medicinal plants, farm tools, ethnoveterinary practices, fisheries, postharvest techniques and other miscellaneous practices for the easy understanding of the readers and to cater to scientists, farmers of differentia! interests. The entire process has been classified into documentation of ITK and fixing the rationality for the identified practices. The book has Eleven chapters in which the first two chapters were devoted to introduction of ITK and conceptual clarification and third chapter dealt with the description of documented ITK with clear pictorial illustrations. The chapters four to eleven dealt with rationality of documented practices which is classified appropriately.

The indigenous practices documented may be a starting point for many research endeavours. The scientists who are not even thinking of a particular aspect may get a breakthrough in research which may result in feasible innovations for sustainable agriculture. The organizations and scientists who were already into the area of documentation of indigenous agricultural practices can make use of the practices documented in this project and there is a possibility of exchange of practices which may results in exhaustive list of practices in any given area. Most of the indigenous agricultural and allied practices which were collected through this project were Low External Input Sustainable Agricultural (LEISA) practices. Many IKs requires less or no cost on the part of farmers. Scientists who are working on strategies for sustainable agriculture and farm management may get benefited by this documentation. Agricultural Extension scientists will find this documentation interesting since they may upscale some of these Indigenous Agricultural Practices to the agricultural zones where the practices are in usage by farmers. The diffusion of such practices would go a long way in improving economic conditions of farmers.

The indigenous varieties documented and collected from interior parts of Tamil Nadu would be a useful germ plasm collection for plant breeders. They can go in for crop improvement using these indigenous varieties by transferring some useful genes to bring in desirable traits. For example, if a indigenous rice variety namely soolai kuruvai possess drought tolerance, breeders can think of isolating the gene from the indigenous variety and can hybridize with another existing variety.

The documentation has got many practices pertaining to plant protection. Agricultural Entomologists and Plant Pathologists can include such treatments in their research agenda to test the efficacy of ingredients spelt out by farmers, which they might have used in their field and got good results. For example, poochi viratti a combination of some herbs with insecticidal properties was found to be effective. This may be included in the research agenda of Agricultural scientists. Many intercropping methods, crop rotation techniques, mixed cropping techniques; integrated farming practices were collected by research team. These practices can serve as useful inputs for agronomists who are bestowed with the task of arriving optimal crop combinations for varied regions and conditions for achieving economic sustainability.

This documentation has a rich collection of ethno veterinary practices which will be a good source for veterinarians and farmers who found it difficult to meet out veterinary medicinal expenses. In this collection, many practices can be done by the farmers without a veterinarians help, which is a very good aspect since the availability and accessibility of experienced veterinarians is a biggest problem in many villages. Agricultural Engineering scientists found this documentation very useful since a major chunk of collection falls into the category of farm tools. The indigenous fabrication skills of the farmers can be traced by the agricultural engineers and they can blend it with their usual research protocol to arrive at cheap, affordable and economically appropriate farm tools and equipments.

One fifth of the documentation is collection of medicinal plants. The knowledge about medicinal and aromatic plants has become increasingly important among agricultural scientists since there is a huge scope for alternative medicinal system like siddha, ayurvedha which uses medicinal plants. The scientific community can test verify the medicinal properties of documented medicinal plants which were reported to possess specific medicinal values.

The digital documentation adds value to the authenticity of the information and enhances its reproducibility and storage capacity. In shortest period of time this documentation can reach wider scientific community breaking the space barrier. At any time the scientists of varied specialization can go through the digital video information and make use of for his /her research programme. The digitized soft copy of the information increases possibility of global scientific community get to know the indigenous practices of different regions and interact with their peer group about these practices. Besides, a blend of indigenous knowledge and modern technology may be most appropriate for sustainable development which can be furthered through Participatory Technology Development (PTD). So the documented ITKs through this book will definitively pave way for sustainable agricultural development.

The authors place on record their sincere thanks to Department of Scientific and Industrial Research (DSIR), Government of India, for funding this documentation of ITKs through Farmer Participatory Approach and the farmers of different districts of Tamil Nadu who have contributed by giving valuable information about the ITKs.

T.RATHAKRISHNAN
N.ANANDARAJA
M.RAMASUBRAMANIAN
L.NIRMALA
M. ISRAEL THOMAS

CONTENTS

1

Indigenous Technical Knowledge Need and Importance

India over several millenniums has been the treasure land of biological wealth, intellectual knowledge and spiritual wisdom. During both pre and post Vedic periods, arts, science and technology had flourished well in various fields. Indian agriculture to some extent is much traditional and dates back to about 10,000 years. India with its traditional crop varieties and crop production technologies was able to feed its population and also exporting some selected produces to other continents. All the Indian villages had been the self sufficient, self sustained and self reliant units. The development of modern agricultural production was achieved by creating large scale specialized farm production units, increased mechanization and use of chemical inputs.

Apart from the substantial increase in the crop yields through green revolution, these modern technologies have made the Indian farmers become poorer as they required heavy investment on costly external resources of uncertain future availability. These technologies with the principle of monoculture resulted in the rapid erosion of crop and livestock diversity, natural soil fertility and biological pest regulation, enhanced the soil erosion, salinization and environmental problems, reduced the nutrition once obtained from the wild foods on farms, increased the need for expensive and poisonous chemical fertilizers and pesticides and finally made the farmers more dependant on markets and outside agencies resulting in an imbalanced growth between rural and urban areas. The impact was more on millions of small farmers and pastoralists. Hence, a section of policy makers and agricultural scientists have now understood that continuation of modern agriculture might lead to severe ecological and economic problems. They are also convinced that modern agriculture may not be able to meet the requirements of the ever increasing population in the future. So, they are searching for alternative technologies. Several alternatives have been proposed such as low external input agriculture, sustainable agriculture, organic farming, biodynamic farming etc. However, they require sometimes little or considerable external inputs whose availability may be uncertain in future.

In developing countries, the other alternative viz., traditional methods have special advantages over modern agricultural techniques. The capital and technological skill requirements in the use of traditional technologies are generally low and their adoption often requires little restructure of the traditional societies. These traditional technologies are not other than the indigenous knowledge. By adopting such indigenous knowledge, our ancestors did not face any problem of large scale pest out break or economic crisis unlike the today's farmers.

INDIGENOUS KNOWLEDGE

The term indigenous knowledge is used synonymously with traditional and local knowledge to differentiate the knowledge developed by a given community from the international knowledge system. Indigenous knowledge is the product of centuries of trial and error, natural selection and keen observations that can form a knowledge base on which researchers and extension workers can plan their research strategy and experimental procedure. According to International Institute for Rural Reconstruction (IIRR, 1996) Indigenous knowledge (IK) is defined as the knowledge that the people in a given community have developed overtime and contribute to develop. It is based on experience, often tested over centuries of use, adapted to local culture and environment and it is dynamic and changing. Indigenous knowledge is not confines to tribal groups or the original inhabitants of an area. It is not even confined to rural people. Rather any community possess indigenous knowledge is rural and urban, settled and nomadic, original inhabitants and migrants. As viewed by Grenier (1998), indigenous knowledge is the unique, traditional, local knowledge existing with in and developed around the specific condition of women and man indigenous to a particular geographic area. Indigenous knowledge is the knowledge that people in a given community has developed over time and continues to develop.

Indigenous knowledge is the basis for self sufficiency and self determination for at least two reasons.

- People are familiar with indigenous practices and technologies. They can understand, handle and maintain them better than introduced western practices and technologies.
- Indigenous knowledge draws on local resources. People are fewer dependants on outside supplies, which can be costly, scarce and available only irregularly.

Hence, Indigenous knowledge is the product of centuries of trial and error, natural selection and keen observations that can form a knowledge based on which researchers and extension workers can develop their research strategy and experimental procedure.

Types of Indigenous Knowledge

Indigenous knowledge is now considered to be cultural knowledge in its broadest sense, including all of the social, political, economic and spiritual aspects of a local way of life. Sustainable development researchers have found the following categories of Indigenous knowledge to be of particular interest : resource management knowledge and the tools, techniques, practices and rules related to pastoralism, agriculture, agroforestry, water management and the gathering of wild food, classification systems for plants, animals, soils, water and weather, empirical knowledge about flora, fauna and inanimate resources and their practical uses and the worldview or way the local group perceives its relationship to the natural world (Emery,1996). While research may focus on a particular category or type of Indigenous knowledge, any Indigenous knowledge under investigation must be viewed in terms of the overall cultural context. Indigenous knowledge is embedded in a dynamic system in which spirituality, kinship, local politics and other factors are tied together and influence one another. Researchers should be prepared to examine any other aspects of a culture that may play an important role in shaping the Indigenous knowledge in question. For example, religion is an integral part of Indigenous knowledge and cannot necessarily be separated from technical forms of knowledge. Spiritual beliefs about nature may influence how resources are managed and how willing people are to adopt new resource management strategies (IIRR, 1996a).

Importance of Indigenous Knowledge

There are two basic important reasons for researchers to consider Indigenous knowledge when carrying out research projects. First and foremost one is incorporating Indigenous knowledge into research projects can contribute to local empowerment and development, increasing self-sufficiency and strengthening self-determination (Thrupp, 1989). Utilizing Indigenous knowledge in research projects and management plans gives it legitimacy and credibility in the eyes of both local people and outside scientists, increasing cultural pride and thus motivation to solve local problems with local ingenuity and resources. Local capacity-building is a crucial aspect of sustainable development, and researchers and development specialists should design approaches which support and strengthen appropriate indigenous knowledge and institutions. Second one is indigenous people can provide valuable input about the local environment and how to effectively manage its natural resources. Outside interest in indigenous knowledge systems has been noticed by the recent worldwide ecological crisis and the realization that its causes lie partly in the overexploitation of natural resources based on inappropriate attitudes and technologies. Scientists now recognize that indigenous people have managed the environments in which they have lived for generations, often without significantly damaging local ecologies (Emery, 1996).

Many feel that indigenous knowledge can thus provide a powerful basis from which alternative ways of managing resources can be developed. Indigenous knowledge technologies and know-how have an advantage over introduced forms in that they rely on locally available skills and materials and are thus often more cost-effective than introducing exotic technologies from outside sources (IIRR, 1996a). As well, local people are familiar with them and so do not need any specialized training. The following are some of the features of Indigenous knowledge which have relevance to conservation and sustainable development

Locally appropriate **:** Indigenous knowledge represents a way of life that has evolved with the local environment, so it is specifically adapted to the requirements of local conditions.

Restraint in resource exploitation : Production is for subsistence needs only; what is needed for immediate survival only is taken from the environment.

Diversified production systems : There is no overexploitation of a single resource; risk is often spread out by utilizing a number of subsistence strategies.

Respect for nature : A 'conservation ethic' often exists. The land is considered sacred, humans are dependent on nature for survival, and all species are interconnected.

Flexible : Indigenous knowledge is able to adapt to new conditions and incorporate outside knowledge.

Social responsibility : There are strong family and community ties, and with them feelings of obligation and responsibility to preserve the land for future generations. (*Source :* Dewalt, 1994)

Indigenous Technical Knowledge

Indigenous Technical Knowledge is a practical nature, concerned with operationalized local thinking in such fields as agriculture, fisheries, health, horticulture and forestry. Indigenous Technical Knowledge corresponds to and contrasts with institutionally organized science and technology. The view point of organized science or the culture of which it is a part, have conventionally regarded as the knowledge of other cultures as pre-logical or irrational and in so doing have either dismissed or greatly played down its validity. In seeking to redress the balance, many proponents of Indigenous Technical Knowledge have argued that it is eminently practical and utilitarian. Indigenous Technical Knowledge differed from science; it is only encompassed areas of direct practical value.

Levi-Strauss (1966) argued forcefully against such a distinction on the grounds that human societies could not, possibly have acquired the skills to make water-tight pots without a genuinely scientific attitude and a desire for knowledge for its own sake. Indigenous Technical Knowledge like scientific knowledge should be regarded in the first instance as something which became possible as a result of a more general intellectual process of creating order out of disorder, and not simply as a response to practical human needs such as sustenance and health.

Some of the knowledge arising in this way would of course have direct practical applications, and equally new knowledge about the way. The world worked might arise as the result of a process of inquiry triggered initially by the wish to solve a problem of a practical kind. An appreciation of this underlying similarity between Indigenous Technical Knowledge and science is important, if the full potential of Indigenous Technical Knowledge is to be realized. An important difference between science and Indigenous Technical Knowledge lies in the way in which phenomena are observed and ordered. The scientific mode of thought is characterized by a greater ability to break down data presented to the senses and to reassemble it in different ways. The mode of Indigenous Technical Knowledge on the other hand, is concrete and relies almost exclusively on intuition and evidence directly available to the senses.A second distinction derives from the way practitioners of the two modes of thought represent to themselves the nature of the enterprise in which they are engaged. Science is an open system whose adherents are always aware of the possibility of alternative perspectives to those adopted at any particular point of time. Indigenous Technical Knowledge, on the other hand, as a closed system, is characterized by a lack of awareness that there may be other ways of regarding the world. This is not to say that Indigenous Technical Knowledge does not change, but rather that those changes which occur are in nearly all instances comparable to the achievements of Kuhn (1962) termed normal science or to the detailed working out of relatively minor 'puzzles' within an established 'paradigm' of thought. Science, in contrast, constantly carries with it the possibility of 'revolutionary change' in which one paradigm would be destroyed and replaced by another.

Indigenous Technical Knowledge and science are comparable on the first criterion; science is generally superior on the second and markedly superior on the third. Indigenous Technical Knowledge can itself be classified in various ways, including :

- In terms of the idioms and conceptual tools through which Indigenous Technical Knowledge becomes possible. This can be separated into two clusters – the propensity to classify and the propensity to quantify;
- In terms of the objects towards which thought is directed. Possible subdivisions here include : physical/inanimate (e.g. soils, water, climate); biological (e.g. crops, weeds, pests, domestic and other animals, insects); medical; and energy related;
- In terms of knowledge about fabrication and use of artifacts;
- In terms of knowledge of the operation of the social and economic structures within which production is embedded.

Natural resource management can be defined as taking a firm decision about the future of any area of resources, applying it and monitoring the evaluation. These resources should be managed to meet social, economic, ecological, cultural and spiritual needs of present and future generations. Within natural resource management systems, Indigenous Technical Knowledge embraces knowledge of tools and techniques for the assessment, acquisition, transformation and utilization of resources, which are area specific.

Indigenous Technical Knowledge in Agriculture

Indigenous practices in agriculture are organic in nature. They do not cause any damage to the air, water and soil, sage to the human beings and are free from causing environmental pollution. These practices are dynamic because they are region specific, depending upon soil type, rainfall, topography etc. and are often modified by the local farmers. Therefore, indigenous agricultural practices can play a key role in the design of sustainable and eco-friendly agricultural systems, increasing the likelihood that the rural populations will accept, develop and maintain innovations and interventions. It is also suggested that if the modern techniques are integrated with the traditional and indigenous practices, that will alleviate the poverty and result in the prosperity of the country.

IIRR (1996) reported that the indigenous knowledge in agriculture might be classified into different types such as information, practices and technologies, beliefs, tools, materials experimentation and biological resources, human resources, education and communication.

Kumar (2001), a farmer of Sathiamangalam taluk of Tamil Nadu recently reported that he was able to obtain 3 to 5 tonnes of additional cane yield from the sugarcane fields applied with organic manures alone, as compared to the fields with chemical fertilizers.

Number of similar such findings and reports have provided adequate evidence to prove that modern agricultural practices are less effective and have adversely brought much problem to the farmers. Moreover, many of the today's modern technologies have rooted from the indigenous practices. For example, it could be understood from Vedic literature that the cultivator had knowledge on the soil fertility, selection and treatment of seeds, seasons of sowing and harvesting, crop rotation, manuring and other cultural practices. Multi-storey cropping and traditional household garden prominent in the tropical areas for centuries without major disease problems may well prove to be the useful models for the rest of the world.

Many of our indigenous practices in agriculture and allied fields have been replaced by the modern technologies and they have become obsolete, especially among the younger generations. Now these indigenous practices are endangered ones and there is a possibility for them to become extinct particularly during this era of globalization, liberalization and commercialization.

Few studies have been highlighting the potentials of indigenous knowledge practices in rural development and they emphasized mostly on agriculture, agro-forestry, livestock farming and ecological and natural resource management.

There is an immense pressure on the people of India to collect, preserve and adopt indigenous agricultural practices so as to reduce dependence on external inputs, to reduce the cost of cultivation and to propagate on eco friendly agriculture.

PARTICIPATORY TECHNOLOGY DEVELOPMENT (PTD)

PTD is an approach to learning and innovation that is used in international development as part of projects and programmes relating to sustainable agriculture. The approach involves collaboration between researchers and farmers in the analysis of agricultural problems and testing of alternative farming practices.

Origins of Participatory Technology Development

The origins of Participatory Technology Development can be found in a number of approaches, concepts and techniques that became popular in the 1970s and 1980s :

- Farming systems research and extension, an approach that introduced social-ecomomic issues into organisations previously dominated by biological and chemical scientists. FSR/E also promoted the idea on 'on-farm' trials.
- Appropriate technology, a concept that recognises the importance of matching the design of technology to the resources and culture of the user.
- Indigenous technical knowledge, a concept that recognised the importance of local expertise and traditional wisdom.
- Participatory Rural Appraisal, a set of techniques that promoted the involvement of rural people in decision-making.

The PTD Process

One of the leading authorities on this process is the Institute for Low External Input Agriculture (ILEIA) based in the Netherlands. ILEIA has described PTD as "a process between local communities and outside facilitators which involves :

- Gaining a joint understanding of the main characteristics and changes of that particular agro-ecological system
- Defining priority problems
- Experimenting locally with a variety of options derived both from indigenous knowledge ... and from formal science, and
- Enhancing farmer's experimental capacities and farmer-to-farmer communication."

PTD Facilitators

PTD facilitators are usually researchers, sometimes consisting of a team that includes both biological and social scientists. While PTD is closely linked to research, it often crosses the boundary into Agricultural Extension because it involves learning activities with farmers.

In the wake of the Farmer First workshop and publications in the mid-1980s, participatory approaches to agricultural research and extension were promoted across all continents and groups of development actors. Many of these approaches, together often known under the generic term of Participatory Technology Development (PTD), put small-scale farmers and their concerns at the center of the development agenda. The goal invariably emphasized the strengthening of farmers' capacities to develop, assess and adapt technologies. Joint learning by farmers, extension workers and researchers was the main vehicle for the development of ecologically-oriented agriculture and Natural Resource Management.

By the 1990s, an increasing number of organizations and projects had started implementing PTD in various settings. Often, partnerships were forged between non-governmental and governmental organizations involved in research and development in rural areas. Earlier attempts have captured some of the lessons learned in field applications, including Joining Farmers' Experiments : Experiences in Participatory Technology Development (1991), Farmer Participatory Research, Rhetoric or Reality (1994), and Farmers' Research in Practice (1997). Those studies were based mainly on evidence going back to the late 1980s and/or focused on a selection of well-known and well-documented pioneering cases.

The field of PTD is wide and the range of experiences diverse. This is largely so because a broader range of actors are involved. Often the nature of the activity is influenced by which agency initiates the activity (e.g. research, extension, civil society institution, etc.) or if it is a multi-stakeholder initiative. There is a need to look beyond oft-quoted cases and to learn from the experiences of projects and organizations that have tried to advance the approaches and methods of PTD by incorporating these into larger institutional frameworks. These organizations have encountered many practical constraints in implementing PTD on a larger scale. In doing this, they have modified approaches and have made methodological innovations.

Why PTD ?

In recent years, there has been growing dissatisfaction with the poor rates of adoption of agricultural technologies in resource-poor farming systems. This poor adoption has resulted partly because when agricultural technologies are developed, there is little input from farmers. Participatory Technology Development (PTD) offers a way forward, through active, decision-making involvement of farmers in every stage of technology development.

When new technology is being developed, researchers usually control all steps of the process. There is little, if any, input from farmers. The Participatory Technology Development (PTD) approach involves farmers right from the beginning, when they are asked to identify their problems.

Problem Diagnosis

The first step is Problem Diagnosis. Researchers work with a representative group of farmers to gain a greater understanding of their agricultural and livelihood systems. The farmers identify the problems that are of most concern to them. They describe what actions they have taken in the past to minimize each problem, and decide which of the problems have the highest priority. They then discuss what action they would like to take to solve these problems in future.

The job of the researcher at this point is to provide farmers with as broad a range as possible of technical solutions and technologies that may help solve the farmers' problems. Researchers must resist the temptation to offer farmers only the technology they think is the best. Farmers frequently use different criteria than researchers when they evaluate agricultural technologies.

Experimentation

Experiments can be formal or informal. Once farmers have developed some promising technologies, controlled experiments can then be conducted to validate and quantify the farmers' experiences.

Evaluation

Once farmers have begun testing technologies and have selected the most promising, there follows a period of evaluation. During this stage, farmers describe which of the technologies they like and why. They also explain which technologies they do not like, and why, and what characteristics of the preferred technologies could be improved.

Importance of Farmer Participatory Approach

Participatory approaches emerged as a way to address the technology needs of highly diverse farming systems in marginal environments, where resources are limited and conventional approaches to research and development often failed. Participatory approaches range from researcher-driven studies that simply involve farmers, to project in which farmers set the agenda, empowerment and education are key issues, and the participation of diverse stakeholders is sought. "Participation" in this effort includes notable cooperation among researchers, extension workers, private machine shop owners, credit institutions, non-government organizations, and others.

Indigenous knowledge system constitutes an alternative model for sustainable development. It is the sum total of experience and knowledge of a given group that can be used to form the basis for decision-making in the face of familiar and unfamiliar problems and challenges (Warren and Cashman : 1988). One way to narrow the gap between local communities and outside professionals is by getting closer to local people and learning from them. By getting closer is meant establishing better rapport and improved channels of communications. Farmer Participatory Approach helps in interacting with local communities, understanding them and learning from them. It helps in the process of involvement with local communities for indigenous knowledge building exercises. It is a way of learning from and with community members to investigate, analyze and evaluate constraints and opportunities and make informed and timely decisions regarding development projects.

It involves a set of principles, a process of communication and a menu of techniques for seeking villager's participation in putting forward their points of view about any issue and enabling them to do their own analysis and offer their own contribution in different forms with a view to make use of such learning. It initiates a participatory process and sustains it. Its principles and the menu of techniques help in organizing participation. It provides an alternative framework for data collection and analysis. Because of its participatory nature, it is a useful methodology to focus attention on people, their livelihoods and their inter-relationships with socio-economic and ecological factors.

Efforts in Documentation

Documentation of indigenous technologies is very important for future generation. Various sources and methods have been used for effective collection and documentation of farmer's indigenous practices.

Gupta (1990a) listed the following reasons for documentation of indigenous knowledge practices.

- To understand scientific rationale
- To accelerate technological change
- To evolve better understanding of technology development and development of newer concepts.
- To increase awareness among the younger generations and develop appreciation for the traditional system
- To revive and restore pride among the farmers themselves.

He also reported that collection of indigenous knowledge could be attempted by many ways like Farmer Participatory Appraisal (FPA) and Rapid Rural Appraisal (RRA), which can raise awareness of rural poverty and associated problems. He also stated that Manual Discriminant Analysis (MDA) is a method that draws upon farmers own innovative genius. The basic assumption is that by comparing and contrasting the farmers own practices a hypothesis can be developed. Massaqoi (1993) reported that the method of obtaining information's on the indigenous technology would depend on the particular technology and could include two or more of the following

- Observation of the technology in operation
- Interviews with users and
- Controlled trials of the traditional technology

IIRR (1996) recommended to consider the following as sources of indigenous knowledge

- Community members, especially the elders
- Folk lore, songs, poetry and theatres
- Community records like writing paints and carvings. Records can also consist of trees planted as boundaries, notched poles, bones and many other forms.
- People working with communities such as extension scientists
- Secondary sources include published and unpublished documents, databases, videos, photos, museums and exhibits.

The following methods has been suggested for recording indigenous knowledge in communities, identifying indigenous specialists, case studies, field observation, in depth interviews, participant observation, participative technology analysis, surveys, brain storming, games, group discussions, role play, SWOT analysis, village reflections, village workshop, flow chart, mapping, taxonomies, participatory video and photo / slide documentation.

Indigenous practices could be documented in the form of descriptive texts such as reports, taxonomies, inventories, maps, matrices, decision trees, audio visuals such as photos , films, videos or audio cassettes, dramas, stories , songs, drawings, seasonal pattern charts, daily calendars etc. Indigenous knowledge can also be stored in local communities, databases, card catalogues, books, journals and written documents, audio visuals, museums etc.

Rationality of Indigenous Knowledge

Any practice should have a scientific background before it is included in technology transfer programmes or disseminated to others. Testing the rationality of the selected indigenous knowledge items is essential as it has been envisaged to test the awareness and adoption of such knowledge by the farmers.

Rationality refers to practices that are explainable with scientific reasons or established based on long time experience likewise irrationality refers to practices which can not be scientifically explained or supported with long time experience.

Understanding the rationality of selected indigenous Agricultural practices is essential as it has been envisaged to test the knowledge and adoption of indigenous practices by the farmers. The classified indigenous agricultural practices were sent for testing rationality to various faculty scientists identified for this purpose with collected information of indigenous practices. Further, it will provide a better comprehension about the adoption and perceived effectiveness. Finally, the study will present a list of effective indigenous agricultural practices as perceived by the farmers. Since the indigenous agricultural practices are borne out of the creativity of farmers, the study can definitely be a pride for them and all concerned will realize their importance in the technology development process.

Farmers have been continuously undertaking a variety of experiments with nature over centuries, the result of which has accumulated as the indigenous practices. Even though they have been followed and passed on over generation s, it becomes necessary the scientific rationale underlying each of the Indigenous Agricultural Practices(IAP's) so as to validate programmes. Hence, the selected Indigenous Agricultural Practices were put to the rationality analysis.

Hiranand and Kumar (1980) concluded in a study that it becomes necessary that the scientists investigate the rationality of each one of the technical belief held by farmers so that they can clearly accept or reject a technical belief. It is essential to analyze the rationality of selected indigenous agricultural practices. For testing the rationality, the collected indigenous practices were classified into subject wise for all the crops. Subject wise classifications of all indigenous practices were given to rationality analysis for respective scientists.

Importance of Preservation to Future Generation

Compilation of indigenous knowledge is a great contribution to the history of agriculture. These pass on from generation to generation by word of mouth. With the passage of time by virtue of promoting modern technologies, indigenous technologies are gradually disappearing. Unless a concise effort is made to track and document the valuable knowledge it will be lost soon not to be regained in future at any cost (Sivanarayana, 1993).

The identification and documentation of traditional / indigenous technical knowledge of agriculture, horticulture, animal husbandry and allied enterprises from different farming situation like Wet, Garden, Dry and Coastal eco systems. The documentation work will be carried out with the help of modern electronic gadgets like Digital Camera, Digital Handy Cam, Audio Recorder, Scanners and Print Media. Using these medias, the indigenous agricultural practices are documented and preserved for future generation use.

Scope and Importance

In Tamil Nadu more than 65 per cent of the population is engaged in agriculture and state economy is largely depends on agriculture sector. The importance of indigenous technology and practice to sustainability is being brought through pooling of traditional knowledge, short listing and evaluating them in the context of modern scientific and technological environment and harnessing it for sustainable agriculture growth. For achieving sustainability in agriculture, Indigenous Technical Knowledge is essential and it should be identified, classified and documented for refining supported with scientific reasons and mixing with scientific technology through various modes for its immediate acceptance and adoption of the technology in the user system without any time lag. Since, time lag is the prime factor in the diffusion and adoption process of any technology. Use of indigenous knowledge in the development projects generally facilitates the participation of the beneficiaries. A blend of indigenous knowledge and modern technology may be most appropriate for sustainable development. So, this study will pave way for sustainable agriculture development.

In Tamil Nadu, many of the worthful traditional agricultural Knowledge are still practiced by the farmers. But the scientist in general concentrates their efforts in high input technologies. Most of the existing knowledge of traditional agriculture has not been properly documented and very limited work has been done in recent times in brining this knowledge to light. The knowledge of traditional agriculture not only be documented, but also be encouraged and helped to take rightful place in the context of modern knowledge. Hence, to achieve sustainable agriculture development, is advisable to study and understand the existing age old-indigenous agriculture and allied practices. The proposed research project will be conducted by involving farmers through Farmer Participatory Approach rather than the individual house hold survey.

Objectives

Tamil Nadu Agricultural University (TNAU) is one of the reputed University in our country successfully running with the objectives of Research, Extension and Education. It has 8 colleges, 7 regional research stations and 32 research stations representing all the seven agro climatic zones of Tamil Nadu. Every year University is releasing number of need based crop varieties and management innovations for the framing community and focusing the research and development, works on all the agricultural related aspects. Besides the latest technologies developed in TNAU, the farmers are also following traditional Indigenous Technical Knowledge (ITK) system for agriculture and allied fields. Whilst two decades ago, planning for higher agriculture output was based only on scientific knowledge in sustainable agriculture could be observed. The project is focused on following objectives.

- To identify the traditional knowledge through farmer participatory mode in agriculture and allied fields.
- To document the traditional practices with the help of modern electronic gadgets like digital camera, handy cam and print media.
- To ascertain the scientific rationale behind the identified traditional knowledge.
- To release and launch of multimedia compact disc.

Limitations of the Study

Indigenous Knowledge is sometimes accepted uncritically because of naive notions that whatever indigenous people do is naturally in harmony with the environment. There is historical and contemporary evidence that indigenous peoples have also committed environmental 'sins' through over-grazing, over-hunting, or over-cultivation of the land. It is misleading to think of IK as always being 'good,' 'right' or 'sustainable'.

Finally, an often overlooked feature of Indigenous Knowledge which needs to be taken into account is that, like scientific knowledge, sometimes the knowledge which local people rely on is wrong or even harmful (Thrupp, 1989). Practices based on, for example, mistaken beliefs, faulty experimentation, or inaccurate information can be dangerous and may even be a barrier to improving the well-being of indigenous people. However, researchers need to be careful when making such judgements. The following limitations are faced during the study

- The project covering entire Tamil Nadu, it is not possible to document indigenous practices because of varying seasonality prevailing particularly most of the areas under rain fed ecosystem.
- Lack of infrastructure facilities during documentation for example, IAP's is documented only in remote areas.
- Most of the farmers having conservative belief in agricultural practices not in scientific manner.
- Limited number of research staff for collecting Indigenous agricultural practices for this during the period of project.

2

Indigenous Technical Knowledge Meaning, Concepts and Experiences

Warren (1987) defined indigenous knowledge as local knowledge, knowledge that is unique to a given culture or society. This knowledge is the information base for the society, which facilitates communication and decision making. Indigenous knowledge is passed down from generation to generation by word of mouth.

Carter (1988) while describing the nature of indigenous knowledge indicated that IK is highly localised and restricted. Local environmental factors and cultural conditions govern the revolution of IK these vary between countries, regions and even farm to farm.

Indigenous knowledge (IK) is the sum total of knowledge and practices that are based on peoples accumulated experience in dealing with situations and problems in various aspects of life and such knowledge and practices are special for a particular culture (Wang, 1988). He also defined IK as the knowledge inbuilt in the system itself by virtue of informal documentation (or) practices evolved by the farmer clients over the years to generate technologies that have been proven sustainable.

Warren & Cashman (1988) explained local knowledge as the sum of experience and knowledge within a given group, which forms the basis for decision – making, related to familiar and unfamiliar problems and challenges.

Berkes et al. (1993) reported that traditional people who depend on their local ecosystems for their essential needs, have accumulated by trail and error a rich body of local environmental knowledge and in several cased elaborated resource management systems and development institutions appropriate for implementing these systems. Biodiversity conservation appears to be the most integral part of much traditional management system from tropical forests to coastal fisheries.

Browers (1993) defined the terms 'indigenous' and 'local knowledge' are used to refer to that knowledge which is generated and transmitted by communities, over time, in an effort to cope with their own agro ecological and socio –economic environment. The knowledge is generated and transformed through a systematic process of observing conditions, experimenting with solutions and readapting previously identified solutions to modified environmental, socio – economic and technological solutions.

Ulluwishwa (1993), IK is a people – derived science and it represents people's creativity, innovation and skills.

Fernandez (1994) indicated that the terms 'indigenous" and 'local' knowledge are used to refer to that knowledge which is generated and transmitted by communities, over time, in an effort to cope with their own agro ecological and socio – economic environments.

NATURE AND IMPORTANCE OF INDIGENOUS KNOWLEDGE

The study of indigenous knowledge is important in planning and implementing new progarammes. These are the entry points for future scientific work and development of appropriate and acceptable practice (Chitamber, 1961; Sikhana, 1994).

Biggs and Tinnermeir (1974) narrated that there is much to be learned from traditional resource management practices being utilized on small farm through out the developing world.

Andrew (1978) noted that in Japan the level of agricultural Technology has been raised steadily since the end of last century by means of indigenous research on the basis of existing plant resources and systems of production.

Garg and Hoda (1978) stated that maximization of production influences the income of the farmer to a substantial extent, but if the inputs for achieving high production are provided from outside the rural society, then a large part of the agricultural income will be drained to the urban areas. It is therefore necessary that the technology by which the various inputs are intended to be provided, should be carefully considered as to whether they come from rural society or from outside.

Gladwin (1980) provided a case in which prior ethno – scientific research would have enabled agricultural research to be more responsive to local conditions.

Srivastava (1980) opined that since traditional technologies have undergone a selective process over centuries of empirical testing they are very likely to represent optimal solutions. He also quoted the traditional tools and techniques should be studied and organized effort made by improving their efficiency and productivity.

Alcorn (1984) and Hunn (1985) opined that farmers' practical knowledge about the local ecosystem is reflected in their farming techniques and in their skill in using the natural resources to gain their livelihood. They may reveal ideas, which contain 'seeds' of adaptive value.

Altieri and Anderson (1986) suggested that incorporating indigenous knowledge in the technology development for the resource poor farmers will help in effecting moderate to high levels of food production.

Dommer (1988) observed that a growing number of individuals and organizations have come to appreciate the importance of working with and through indigenous knowledge systems in promoting sustainable agriculture instead of trying to work around them.

A research design is a plan comprising the researcher's decisions about the procedure for collection of data for a given study, which aim to fulfill the purpose of the study. This chapter describes the tools and techniques followed in the documentation process of the project. It is embedded with the methods used to gather information on ITK and analyze the data and implies how the research objectives and the problems encountered in the research have been tackled. The research methods and procedures are described under the sub heading.

a. Location of the study.
b. Data collection tools and techniques.
c. Data analysis.

Location of the Study

Tamil Nadu state is situated at the south- eastern extremity of the Indian peninsula with Karnataka and Andhra Pradesh in the North, Bay of Bengal in the east, and Indian Ocean in the south and kerala state in the west selected for the study.

Agro Climatic Zones

Based on rainfall distribution, irrigation pattern, soil characteristics, cropping pattern and other physical ecological and social characteristics, Tamil Nadu State is classified into seven distinct agro-climatic zones.

Ecosystem Classification

In Tamil Nadu, totally 32 districts are there and these districts are come under four different ecosystems like wet land, garden land, dry land and coastal tracts. Over 32 districts, the possibilities of documentation of ITK in different districts was assessed through triangulation involving key informants, progressive farmers, officers of the state department of agriculture and NGO's.

Selection of Districts

In each ecosystem, four districts were selected and totally sixteen districts *viz.*, Trichy, Erode, Kanchipuram, Krishnagiri, Salem, Villuppuram, Thiruvannamalai , Vellore, Sivagangai, Pudukkotai, Perambalur, Namakkal, Nagapattinam, Ramanathapuram, Kanyakumari and Tuticorin.

Data Collection Tools and Techniques

The research team covered sixteen districts from different ecosystems and collected the required information about indigenous traditional agricultural practices. The indigenous traditional agricultural practices are existed in very remote areas and these areas having truthful information. But identification of areas under different districts is very tedious process for unknown person. So that, the identification of villages were made on the preliminary visit and also obtained suggestion from key informants, progressive farmers, officers of the state department of agricultural and non governmental organizations.

Initially Focused Group Discussions (FGD) were conducted in the selected villages to assess the knowledge status on Indigenous Technical Knowledge (ITK). Focus Group Discussion (FGD) is a group discussion of approximately 6 - 12 persons guided by a facilitator, during which group members talk freely and spontaneously about a certain topic. It is a qualitative method and purpose is to obtain in-depth information on concepts, perceptions and ideas of a group. FGD aims to be more than a question-answer interaction. The idea is that group members discuss the topic among themselves, with guidance from the facilitator.

FGD techniques can be used to:

1. Focus research and develop relevant research hypotheses by exploring in greater depth the problem to be investigated and its possible causes.
2. Formulate appropriate questions for more structured, larger scale surveys.
3. Help understand and solve unexpected problems in interventions.
4. Develop appropriate messages for health education programmes and later evaluate the messages for clarity.
5. Explore controversial topics.

Later, farmers were organized and conducted Participatory Approaches like (by involving the farmers) Participatory Rural Appraisal tools have been used to collect the traditional knowledge related to agriculture and allied fields.

Participatory Rural Appraisal (PRA) is a methodology used for interacting with villagers, understanding them and learning for them. It involves a set of principles, a process of communication and menu of methods for seeking villagers' participation in putting forward their points of view about any issue and enabling them do their own analysis with a view to make use of such learning. PRA is a process of participation with the villages in which rapport- building paves the way for them to perform their own analysis and to express themselves whether by means of 'verbals' like narration of 'visuals' such as making a map.

Before conducting a PRA exercise, it is often possible to get some background information on the area to be covered. Such information can help in describing broad features and trends of the area concerned, covering different aspects at different points of time. Sources of such data can be both primary and secondary. The primary source constitutes materials gathered first hand and has a direct relationship with people, situations or events that are being studied such as records, letters, minutes, memoranda, notes and diaries of the institution concerned.

Secondary sources of data can be found in published materials which may include transcript or summaries of primary sources materials. Some secondary sources are surveys, books, reports, plans, official records, census records, project reports, maps, aerial photographs, journalistic articles, newspaper cuttings, research papers and data files from computers. The organizations and individuals who are usually associated with collecting such data are government, researchers, statistical agencies, local level educational and technical institutions, non-government organizations, donor agencies, remote sensing agencies etc.

There are different PRA methods involving direct participation, many of which include diagrams as an integral part. The major PRA methods involving direct participation are explained below.

Resource Map

It is a rich tool and is commonly used in PRA. It has enormous potential. Maps by villagers provide a basis for knowing different aspects of a village, for instance, the location of dwellings and buildings and land- use patterns as indicated by a social map showing the position of households in a village and their socio-economic characteristics. A village resource map can also show different kinds of natural resources and micro environmental. In a resource map the villagers draw the resource profile of the village depicting different kinds of soils, rivers, ponds, trees, crops and micro environments. The problems can be described and discussed on the basis of a resource map along with the opportunities and the constraints.

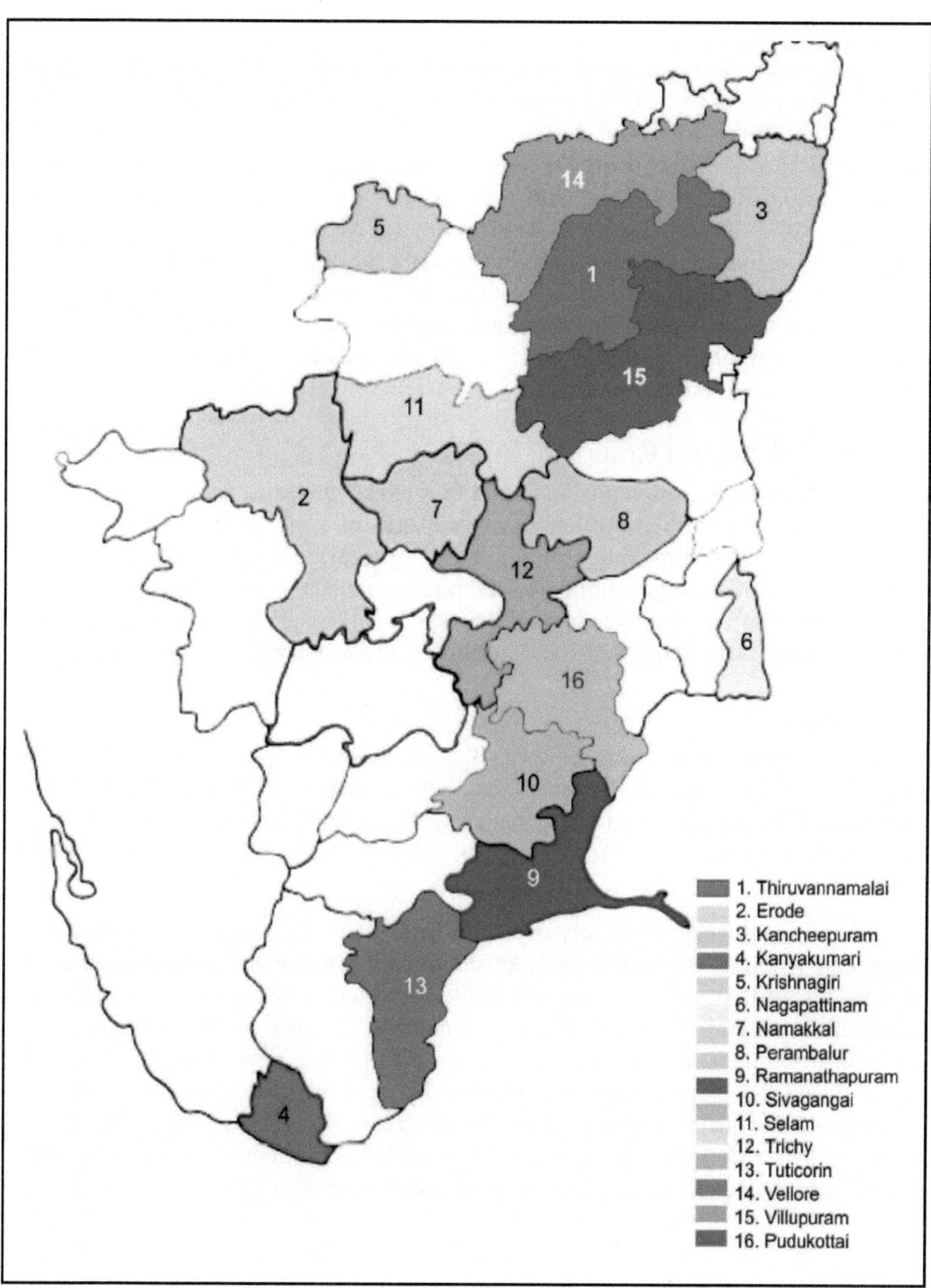

Fig. 1 : Map Showing the study area in Tamil Nadu

Transect Walk

Participatory transect walks systematically involve walking with the villagers through and area and discussing about different aspects of land –use and agro-ecological zones in the village observed during the walk. A transect is a useful method for knowing rural ecological conditions. A walk from one point of the village to another enables the outsider to discuss with rural people about soil conditions, land –use patterns, crops, livestock, micro environments, agro-ecological zones etc. the important of a transect lies not only in knowing the agro-ecological zones in rural areas but also in getting an in –depth account from the participation villagers of such zones in village, their uses, problems and opportunities. One transect walk can be supplemented by other walks so as to enable the outsider to learn more about any village and clarify doubts.

Wealth Ranking and Grouping

Wealth ranking/grouping is a method of ranking/ grouping of households on the basis of income, wealth and other local measures of well-being. Different criteria can be used for wealth ranking/grouping based of rural perceptions. Wealth ranking/ grouping is based on the presumption. That rural people have the necessary knowledge to rank/group households which implies that they have knowledge of kinds and position of households assets, other items and attributes of the households concerned.

Time Line

In the method of time line, elderly villagers narrate their life histories. Since it is often difficult to remember the exact dates of important changes, a villager can be facilitated to broadly connect such change with major events, political regimes and thus, summaries major events and changes that have taken place during his/her life time.

These approaches are highly useful to triangulate (Cross check / Validate) the documented Indigenous Traditional Knowledge. Triangulation means team composition of inter-disciplinary and can be constituted in such a manner so as to have the information generated from different perspectives. Teams can have persons from different disciplines, sex. Culture and experience so that data emerging from PRA can be discussed within a team and with local people by a team to clarify issues and select issues for further probing. The units of observation can be changed in order to cross check the information gained. One also has the option of going for deeper probing if necessary. Different groups can be approached separately of in combination so as to arrive at views from different angles.

For collection of data's, we have to formulate the interview module for collecting ITK practices. It contains

- General profile of ITK holder
- Technology details

General profile includes name and age, occupational details, educational status, wealth status and land holding details. Technology details include cropping pattern, origin, scope of the practice, description of the practice, adaptation details, cost effectiveness, advantages of the practices and constraints and limitations.

The Indigenous Technical Knowledge practices are collected from farmers as primary source and documented the practices. The documentation was not restricted with Indigenous Agricultural Practices alone. During the process, the other related valuable information's obtained on medicinal plants, ethno veterinary practices, fisheries etc. also documented. The district wise documentation of ITK's is given in table 2. Documentation of Indigenous Technical Knowledge (ITK) was carried out with the help of Digital Camera, Digital Video, and Specimen and seed sample collection.

Rationality Test

Based on each classification, a brief and concise statement over the documented ITK are prepared and given for rationality test. Rationality refers to practices that are explainable with scientific reasons or established based on long time experience likewise irrationality refers to practices which can not be scientifically explained or supported with long time experience.

Understanding the rationality of selected indigenous Agricultural practices is essential as it has been envisaged to test the knowledge and adoption of indigenous practices by the farmers. Rationality Test Module is used to test the Indigenous Traditional Knowledge practices and identified subject matter specialist in the respective area for testing the scientific rationality for practices. The Rationality Test Module is given in below.

Rationality Test Module

S.No	Documented ITK Practices	Rational		Irrational		Remarks
		Based on Scientific reason	Based on experience reason	Based on experience	Based on Scientific	

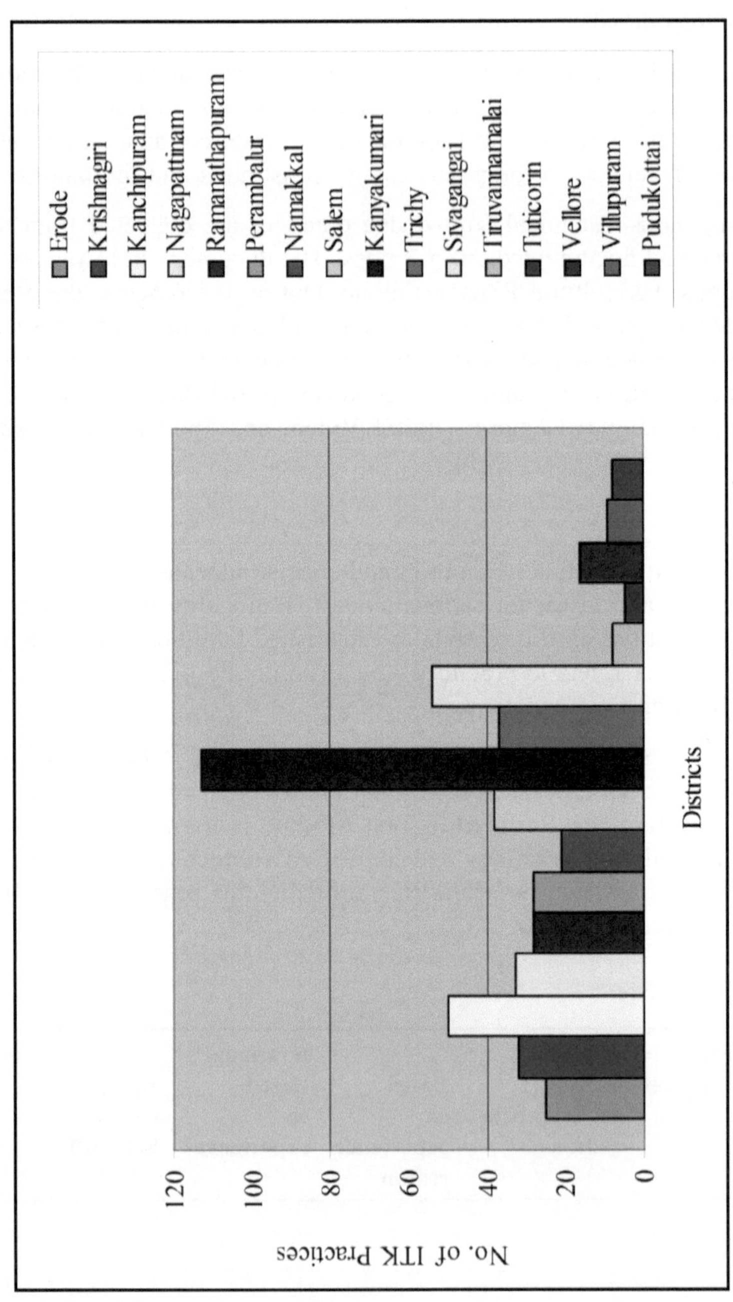

Fig. 2 : District wise Documentation of ITK Practices

Subject Wise Documented ITK Practices in Agricultural and Allied Enterprises

S. No	District	Traditional Varieties	Medi cinal plants	Crop management	Crop protection	Animal husbandry	Fisheries	Farm tools	Storage methods	Post harvest techniques	Others*	Total
1	Erode	3	4	2	1	2	0	8	5	0	0	25
2	Krishnagiri	3	11	4	1	8	0	5	0	0	0	32
3	Kanchipuram	16	10	2	14	0	0	6	2	0	0	50
4	Nagapattinam	13	0	4	3	0	3	4	4	0	2	33
5	Ramanathapuram	20	0	2	0	1	2	1	0	2	0	28
6	Perambalur	2	15	3	0	4	0	2	1	1	0	28
7	Namakkal	6	2	4	0	2	0	6	0	0	1	21
8	Salem	4	0	1	0	1	0	22	3	0	7	38
9	Kanyakumari	21	41	15	1	4	9	6	12	4	0	113
10	Trichy	2	16	6	2	9	1	0	0	1	0	37
11	Sivagangai	2	1	2	1	45	0	2	0	0	1	54
12	Tiruvannamalai	2	6	0	0	0	0	0	0	0	0	8
13	Tuticorin	1	0	1	0	0	0	3	0	0	0	5
14	Vellore	5	0	3	0	0	0	8	0	0	0	16
15	Villupuram	1	8	0	0	0	0	0	0	0	0	9
16	Pudukottai	0	0	3	2	3	0	0	0	0	0	8
	Total	101	114	52	25	79	15	73	27	8	11	505

* Others include various desi agro industry practices.

By using this test module, the subject wise documentation of ITK practices were given to respective scientists related to the field. The scientist's perceptions about ITK practices were obtained in terms of whether the practices have any scientific rationale or the practices have been followed through their personal experience. The scientists concerned were requested to substantiate if any practices possess scientific rationale. In such a way the rationality was assessed. After getting this response, the rational practices were again disseminated with a group of scientists in the same subject matter for reassessment / confirmation on the rationality. The collected rational practices were described in chapter IV. The Multimedia compact discs were prepared for selected ITK practices for launching the technologies for quick diffusion and adoption of technology.

3

Description of Indigenous Technical Knowledge

This chapter contains the description of the Indigenous Traditional Knowledge practices as subject wise manner *viz,* Traditional varieties, Crop management, Crop protection measures, Post harvest methods, Storage structures, Farm tools, Medicinal plants, Ethnoveterinary practices, Fisheries and others related to agriculture.

TRADITIONAL VARIETIES

S.No	Particulars
	RICE VARIETIES
1.	**Puzhuthi Kar Nel** grown in hilly tracks of Western Ghats in Erode district. It is suitable for rain fed and garden land conditions.
2.	**Kalurundai** in Nagapattinam district. It is resistant to drought, pest and disease as well as salinity.
3.	**Soolai kuruvai** in coastal villages of Sempodai, Periyakuthagai, and Thopputhurai at Nagai District. : It is a drought tolerant variety.

4. **Kuzhi Vedichan** in Vellapalam and Keevalur taluk of Nagapattinam district. It is resistant to salinity, drought and pest and diseases.

5. **Val Sivappu** in Vellapalam and Keevalur taluk of Nagapattinam district. Susceptible for lodging and looks tail like structure.

6. **Vellai Kuruvi Kar** in Vellapalam and Keevalur taluk of Nagapattinam district. It prefers sandy clay and saline soil condition.

7. **Seeraga Samba** in Vellapalam and Keevalur taluk of Nagapattinam district. Fine rice variety.

8. **Pichavari** in Vellapalam and Keevalur taluk of Nagapattinam district. Special feature of this variety is tolerant to flood and drought.

9. **Suran Kuruvai** in Vellapalam and Keevalur taluk of Nagapattinam district. Highly suitable for making idli and dosai.

10. **Vaei Kunda** in Keevalur taluk of Nagapattinam district. The variety is tolerant to flood and drought.

11. **Kichedi samba** in Keevalur taluk of Nagapattinam district. It is resistant to certain pest and diseases.

12. **Kuthiraival Samba** in Kampalathupatti village of Trichy district. It is resistant to lodging and also resistant to pest and disease.

13. **Chinna samba / Sada samba** in Keevalur taluk of Nagapattinam district. Susceptible for lodging.

14. **Mara nel** in Thenkanikottai taluk of Krishnagiri district. Grains of this variety have a tough seed coat.

15 **Kallimadayaan** in Iyerpalayam village of Perambalur district. Starchy watering nature influenced the grain, to keep them fresh and energetic.

16. **Karuppukavuni** in Anumandhakudi village of Sivagangai district. It is suitable for direct sowing.

17. **Chandikaar** in Anumandhakudi village of Sivagangai district. It is suitable for both direct sowing and transplanting method.

18. **Varappukkodanchan** in Chatrakudi village of Ramnad district. Suitable for rain fed condition.

19. **Kuliparichan** in Pallabachery village of Ramnad district. Suitable for coastal sandy soils and drought resistant and tall variety. Outer skin is slightly dark.

20. **Chitrakar** in Thirupullani tracts of Ramnad district. Suitable for coastal sandy soils and drought resistant

21. **Sivappu Chitraikar** red rice variety grown in Paramakudi tracts. It is drought resistant variety.

22. **Murungaikar** grown in Ramnad areas. Suitable for upland rice cultivation and drought resistant. It requires very minimal rainfall.

23. **Nootripathu** in Muthuvayal village of Ramnad district : It is drought resistant variety.

24. **Ariyaan** in Regunathapuram areas of Ramnad district. It looks horn like structure and suitable for coastal sandy soils.

25. **Sadaikkar** in Aakadavalasai village of Ramanathapuram district. It is red rice variety cultivated in rain fed condition. Its' having medicinal properties.

26. **Poongkar** in Raghunathapuram village of Ramnad district. Suitable for coastal sandy soils.

27. **Kuruvai kalanchium** in T.Karungulam, Thirupullani union of Ramnad district. Red Rice variety. The cooked rice is very good in taste even after two days.

28. **Norungan** in Raghunathapuram village of Ramnad district. Sown in rain fed condition in sandy soils.

29. **Kalurundaikaar** in Paramakudi taluk of Ramnad district. It is drought resistant.

30. **Thanga Samba** in Kanchipuram district. Ear head of this variety is very long.

31. **Neelan samba** in Sukkan kollai village of Kanchipuram district. Suitable for lactating women.

32. **Vadan Samba** in Sukkan kollai village of Kanchipuram district. Drought resistant and also resistant to pest and disease.

33 **Kaliyan samba** in Sukkan kollai village of Kanchipuram district. It grows very tall upto the height of 5.5 feet. It can be grown in water logging and low lying areas.

34. **Samba Mozsanam** in Sukkan kollai village of Kanchipuram district. Suitable to grow in Tank fed areas.

35. **Kadikazhuthan** in Sukkan kollai village of Kanchipuram district. It is Suitable for making a specialty dish called 'Puttu' of South India.

36. **Pisini** in Sukkan kollai village of Chengalpet taluk in Kanchipuram district. Highly resistant to drought. Since the ear heads are found with awn, it is important to take extra care while removing the grains.

37. **Thuyamalli** in Sukkan kollai village of Kanchipuram district. In flowering stage, the ear heads look like flowers. In Tamil, '**Thuyamalli**' literally means pure jasmine.

38. **Kalar palai** in Sukkan kollai village of Chengalpet taluk in Kanchipuram district. Suitable for kalar nilam i.e. saline soil.

39. **Kudaivazhai** in Sukkan kollai village of Chengalpet taluk in Kanchipuram district. The ear heads of this variety appear like open umbrellas and hence the name 'kudaivazhai'.

40. **Kumvalai** in Sukkan kollai village of Chengalpet taluk in Kanchipuram district. It is suitable for sandy soil and also water logging areas.

41. **Muttakar** in Edaiyur village of Thirukazhukundram taluk in Kanchipuram district. The rice has red in colour and having the coarse quality

42. **Sempalai** in Edaiyur village of Thirukazhukundram taluk in Kanchipuram district. It is highly suitable for making 'Pori' (puffed rice)

43. **Kappakar** in Thakandarapuram village of Vandavasi taluk in Thiruvannamalai district. It is suitable for growing in water logging as well as drought prone areas .Suitable for preparing Aval.

44. **Perunkar** in Thakandarapuram village of Vandavasi taluk in Thiruvannamalai district. Tolerant to stem borer and ear head bug.

MILLET VARIETIES

45. **Karikattai Ragi** in Thamarakarai village of Erode district. Suitable for hilly regions.

46. **Kattucumbu (Pearl millet)** in Kampalathupatti village of Manapparai taluk of Trichy District. It is a rain fed crop, direct sowing variety.

47 **Thattai varagu** in Veppanthatai taluk of Perambalur district. Side tillers give more number of ear heads.

48. **Puvadan Keazhvaragu (Ragi)** in Veppanthatai taluk at Perambalur district. Suitable for both direct sowing and transplanting system of cultivation.

49. **Kona Cholam (Sorghum)** in Velankadu village of Vellore district. Green and dry fodder used as cattle feed.

50. **Chittan Samai** in Alangayam, Javathu malai village of vellore district. Very good taste and flavor compared to other varieties. It is used for biscuit preparation.

51. **Sen Cholam** (Wild Sorghum) in Panaimarathupatti village of Salem district. Mainly cultivated for cattle feed.

52. **Kakacholam** (Sorghum) in Kothali village of Thoothukudi district. Straw's used for cattle feed.

OIL SEED VARIETIES

53. **Karrupu Ellu** (Black Gingelly) in Raghunathapuram village in Paramakudi taluks of Ramnad district. It has more oil content.

54. **Sivappu Ellu** (Red Gingelly) in Raghunathapuram village of Ramnad district. Higher yield and good oil content.

55. **Kothu kadalai** (Groundnut) in Vallathodu village of Kanchipuram district. It is resistant to pest and diseases.

56. **Kadalaikai** (Groundnut) in Velankadu village of Vellore district. Oil content is more than the other variety.

Vegetable Varieties

57. **Kulithakali** (Tomato) in Kanchipuram district. More tartaric acid content. It is resistant to pest and disease. Suitable for dry land cultivation.

58. **Vazhuthaalangai** (Egg plant/ Brinjal) in Vettuvenni village of Marthandum Taluk in Kanyakumari district. Tropical vegetable. It is mainly used for table purpose and its 12-15 cm height.

59. **Kannadi katthari** (Wild Brinjal) in Devoor village of Salem district.This variety possesses high pest and disease resistant.

60. **Poyyur kathari** (Brinjal) in Vellapalam village of Nagapattinam district. Fruits are consistent which gives more weight when compared to other varieties. Crop can be ratooned and maintained till optimum yield is obtained.

61. **Ramnad Gundu Milagai** (Chillies) in Raghunathapuram village of Ramnad district. It is cultivated only in less amount of rainfall.

62. **Sivappu Vendai** (Red Bhendi) in Panavilai village of Vilavancode taluk of Kanyakumari district. Rich in carotene. Main stem is multiple branched, each branch gives fruits. The variety is resistant to yellow mosaic virus.

63. **Covaikai** (Coccina) in Oolacure village of Villurupuram district. Used for vegetable purpose.

64. **Gandhari chilli** (Golkonda chilli) in Vettuvanni village in Marthandum taluk of Kanyakumari district. It helps in fat control, blood purification and used as a first aid material preventing blood clotting.

65. **Manattakkali** (Milagu Thakkali) in Malavilai village of Kanyakumari district. It cures asthma.

66. **Neela milahai** (Blue chilli) in Panavilai village of Vilavancode taluk of Kanyakumari district. Like an ornamental plants and attracts observers.

67. **Ther kothavarai** (cluster bean) in coastal areas of Nagapattinam District. It has medicinal value of curing stomach ulcer and mouth ulcer.

68. **Karuppu Avarai** (Black Beans) in Valeppur village of Namakkal district. Suited for rain fed or upland cultivation.

69. **Nattu Avarai** (Beans) in Vallathodu village of Kanchipuram district. Good taste.

70. **Nattu Pusani** (Pumpkin) in Raghunathapuram village of Ramnad district. Grown in rainfed condition.

71. **Curry leaf** in Thangachimadam areas of Ramnad district. It is suitable for coastal sandy soils and also grown as mixed crop.

72. **Vethalai** (Betel vine) in Raghunathapuram village of Ramnad district. Suitable crop for coastal sandy soils.

73. **Murungai** (Moringa) in Thangachimadam areas of Ramnad district. Suitable crop for coastal sandy soils.

74. **Mottaipayaru** (Lab lab) in Vadavathure village of Namakkal district. Rich in protein content and vegetative parts are mainly used as cattle feed.

Tuber Crop Varieties

75. **Vellai Bontha** (Tapioca) in Panaimarathupatti village of Salem district. It is mainly used for culinary purpose.

76. **Karuppu Bontha** (Tapioca) in Panaimarathupatti village of Salem district.These variety tubers have more flour content used to make sago.

77. **Kariyilai poriyan** (Tapioca) in Malavilai of kanyakumari district. Highly resistant to Tobacco Mosaic Virus and brown leaf spot.

78. **Adukku muttan** (Tapioca) in Malavilai of Kanyakumari district. Increased starch and possess good cooking quality.

79. **Nattu seeni kizhangu** (Yams) in Malavlai village of kanyakumari district. The yield is 20-25 tones / ha.

80. **Seeni kizhangu** (Sweet potato) in Irudhukottai village of Krishnagiri district .Suitable for rain fed and irrigated condition.

FLOWER CROP VARIETIES

81. **Jathi Mullai** (Jasmine) in Vettuvanam village of Vellore district. Cultivated in all types of soils and suitable for panneer preparation.

82. **Mullai** (Jasmine) in Vettuvanam village of Vellore district. Mainly used for ornamental purpose.

83. **Ramnad Gundumalli** (Jasmine) in Raghunathapuram village of Ramnad district. Grow well in sandy soils with less irrigation.

FRUIT VARIETIES

84. **Nattu vellai and senkoiya** (Guava) in Kolli hills of Namakkal district.Good taste.

85. **Annachi** (Pineapple) in Valeppur villages of Kolli hills of Namakkal district. Planted in garden land or upland crop.

86. **Naadan** (Pineapple) in Naadan Malavilai village of Kanyakumari district. It weighs upto 2 kg and bigger in size. It is mainly used as a live fence because of its sourness in taste and more number of red thorns.

87. **Ram seetha** (Anona) in Irudukottai village of Krishnagiri district. It cures asthmatic disorder.

88. **Chengavarkai (**Mango) in Erumbukadu village of Kanyakumari district. The keeping quality is very good because of thick skin.

89. **Chetti ottu** (Mango) in Erumbukadu village of Kanyakumari district. Thick skin, bigger fruits and good for consumption.

JACK VARIETIES

90. **Parukkan pala and koolam pala** (Jack) in Valeppur village of Namakkal district. Used for both consumption as well as cooking purpose.

91. **Varikkai chakkai pala** (Jack) in Vellachi vilai, Villukiri village of Kanyakumari district. It is mainly used as a table fruit.

92. **Koozhan chakkai pala** (Jack) in Vellachi vilai, Villukiri village of Kanyakumari district. It is suitable for commercial planting.

93. **Mulagumoodu** (Jack) in Mulagumoodu of Vilavancode Taluk of Kanyakumari district. Fruits are firm fleshed and medium in size weighing about 7-10 kg.

BANANA VARIETIES

94. **Palayankodan** in Vathiar villai village of Kanyakumari district. The average bunch weight is about 15 kg having 8-12 hands. In each hand with 11-18 fingers.

95. **Nendran** in Vathiar villai village of Kanyakumari district. The fruits of this variety used for preparing various dishes especially chips.

96. **Sevvazhai** in Vathiar villai village of Kanyakumari district. The bunch weight is around 20-25 kg.

97. **Matti** in Vathiar villai village of Kanyakumari district. The fruits having the medicinal value to cure dysentery.

SPICES VARIETIES

98. **Kudampuli** (Camboge) in Vettuvenni, Marthandam of Kanyakumari district. The yield ranges from 120-150 kg of pods/ear/tree. Acidity of the fruits is lesser than *Tamarindus indicus.*

99. **Manjal** (Turmeric) in Thalavadi of Erode district. High yielding variety.

100. **Nattu milagu** (Pepper) in Kolli hills of Namakkal district. Nattu milagu is small when compared to normal milagu but the spiciness is more than the later. Its quality is good. It is cultivated through twines.

101. **Jathikaai** (Nutmeg) in Malavilai village of Kanyakumari district. Seeds and aril parts are used. It cures asthma, catarrah, diarrhoea and Insomnia.

RICE VARIETIES

Puzhuthi Kar Nel – Dry Land Rice

Puzhuthi Kar Nel is being cultivated in hilly tracks of Western Ghats bounded in Erode district. It is suitable for rain fed and garden land conditions; Owing to its suitability to rain fed condition it is direct sown crop. The duration of the crop is 125-130 days and the average height of the crop is 130 cm that makes it susceptible for lodging. It yields upto 1.5 tones/ac and features of the rice are bold and red coloured, which is preferred by the hill people for consumption.Keeping quality of this dry land rice is good.

Kalurundai

It is being cultivated in Nagapattinam district. Late Samba (Purattasi – Thai i.e., Sep 15 – Feb 14) and Navarai (Margali – Masi i.e., Dec 15-Mar14) are the best suited season for cultivating this variety. It requires clay loam soil for better growth. It is direct sown crop with duration of 120 days. It grows upto a height of 120 cm (4 feet) which makes it susceptible for lodging. It is resistant to drought, pest and disease as well as salinity. During maturity stage, colour of the grain is yellow with black colour faded stripes. But the rice is bold sized with pale yellow colour. Average length of a grain is 22.5 cm and the 1000 grain weight is 32g.Due to its bold sized quality, the grain weight is comparatively more than other varieties. Straw obtained from this rice variety is used for thatching roof. It exists for more than 80 years and at present, it is cultivated in all coastal villages of Nagai District.

Soolai Kuruvai

It is cultivated in coastal villages in Sempodai, Periyakuthagai, Thopputhurai at Nagai District. Late samba (Purattasi-Thai) i.e., Sep 15-Feb 14 and Kuruvai (Mid Vaigasi – Mid Aavani i.e., June 1-Aug 31) are the suitable season for cultivating this variety. Clay loam and coastal saline soils are well suitable for cultivating this variety.

It is cultivated in both irrigated as well as rain fed conditions. In irrigated system, when water is available in plenty transplanting method is followed whereas in rain fed system, direct sowing is practiced. Since it is a drought tolerant variety, it suits even in rain fed areas. The duration of this variety is 130-140 days and grows upto a height of 110-120 cm. The feature of this variety is motta or bold and pale dirty brown color. It yields about 15 bags/ac (75kg/bag) i.e., 1125 kg/ac. This variety is characterized by the special feature like the straw yield will be more. Hence it is used as cattle feed and also used for thatching roof. As stated by Thiru. Muthusamy of Periyakuthagai (62 years) this variety in existence for more than a century.

Kuzhi Vedichan

This variety is being cultivated in Vellapalam and Keevalur taluk of Nagapattinam district. Samba (Aadi-Margazhi i.e., July 15 – Jan 14) is the apt season for cultivating kuzhi vedichan. It grows well in sandy clay soil and saline soil, so that it is very well suited for coastal areas. This can be raised by both direct sowing method as well as transplanting methods. It is resistant to salinity, drought and pest and diseases. It is a short duration crop and the cropping cycle is 110 days. On an average, this variety yields upto 950 kg/ac which is lesser than other local varieties. The average height of the crop is 2½ feet and it will leads to lodging. Size of the grain is bold in nature with pale dirty white colour.This variety is suitable for idly and dosai preparation. The specific character of this variety is induced milk secretion in lactating women. Even when the field is flooded and the crop lodges, it has an unique feature of tillering again. This is secret behind its name called "Kuzhi vedichan" in Tamil, it means "Coming out of the pit".

Val Sivappu

It is also being cultivated in Vellapalam and Keevalur taluk of Nagapattinam district. This variety is well suited for late samba (Purattasi-Thai i.e., Sep 15-Feb 14). Sandy clay and saline soils are most favourable soil type for better growth. This feature facilitates coastal farmers, especially in periyakuthagai, Thopputhurai, Sempodai villages of Nagapattinam district to cultivate val sivappu. It is saline as well as drought tolerant variety but susceptible for lodging. The duration of this crop is 145-150 days and grows upto 160cm in height. The character of the rice will be red in colour and also good taste. The grain is coarse in nature and used for idly and dosai preparation. It is a low yielding variety and yields upto 900kg/ac (12 bags/ac; 75 kg/bag). The identification character and special feature of this variety is the posterior end of each grain has a single bristle like structure which looks like a small bird's tail. So, the tail like structure and its colour (red-sivappu in Tamil) both contributed to the naming of the variety as "Val sivappu". It seems to exist even before 150 years , as mentioned by Th. Muthusamy at Periyakuthagai of Nagai District.

Vellai Kuruvi Kar

This variety is also being cultivated in Vellapalam and Keevalur taluk of Nagapattinam district .Samba (Aadi-Margazhi ie. July15-Jan 14) and late samba (Purattasi – Thai i.e., Sep 15-Feb 14) are the best suited seasons for this variety. It exists for more than 90 years. It prefers sandy clay and saline soil condition which is a special feature of this variety. Total duration of the crop is 125-135 days out of which 25-30 days are nursery period. In case of upland conditions, direct sowing can be followed. This variety grows upto 100-120 cm. Both the grains as well as straw are pale yellow to dirty white colour. During tillering stage, rat damage will be more because the stem is tastier.

Seeraga Samba

This variety is also being cultivated in Vellapalam and Keevalur taluk at Nagapattinam district .The suitable season for growing this variety is samba season. The cultivation method is transplanting system. The duration of the crop is 125-130 days. The nursery period is 30-35 days. The plant grows upto a height of 80.5 cm. It yields about 20 bags (1500 kg). The straw yield will be 1050 kg. The grain size is also very small and fine. It is highly used for making briyani on the special occasion.

Pichavari

This variety is also being cultivated in Vellapalam and Keevalur taluk of Nagapattinam district. The special feature of this variety is tolerant to flood and drought. The best season for cultivating this variety is Thaladi season. Direct sowing and transplanting system are used for cultivation of this variety. The duration is 110-115 days of which 21-25 days is nursery period. The plant grows upto a height of 100 cm. It yields about 4800 kg/ha . The straw yield is about 180 kg.

Suran Kuruvai

This variety is also being cultivated in Vellapalam and Keevalur taluk of Nagapattinam district. The suitable season for growing this variety is samba and late samba period. The cultivation method is transplantation system. The crop duration is 130-135 days and the nursery period is 30-35 days. The grain yield is about 1200 kg per acre. The straw yield will also 50 bundles (1200 kg) .The colour of the rice is dark brown and coarse in texture. This is highly suitable for making idli and dosai.

Vai Kunda

This variety is also being cultivated in Keevalur taluk of Nagapattinam district. Direct sowing and transplanting system is used for cultivating this variety. The cropping cycle is 145-150 days. The nursery period is 35-40 days. The plant grows to a height of 121 cm. It yields upto 18 bags (one bag weighs about 75 kg). The straw yield will be 180 kg. The special feature of this varity is tolerant to flood and drought.

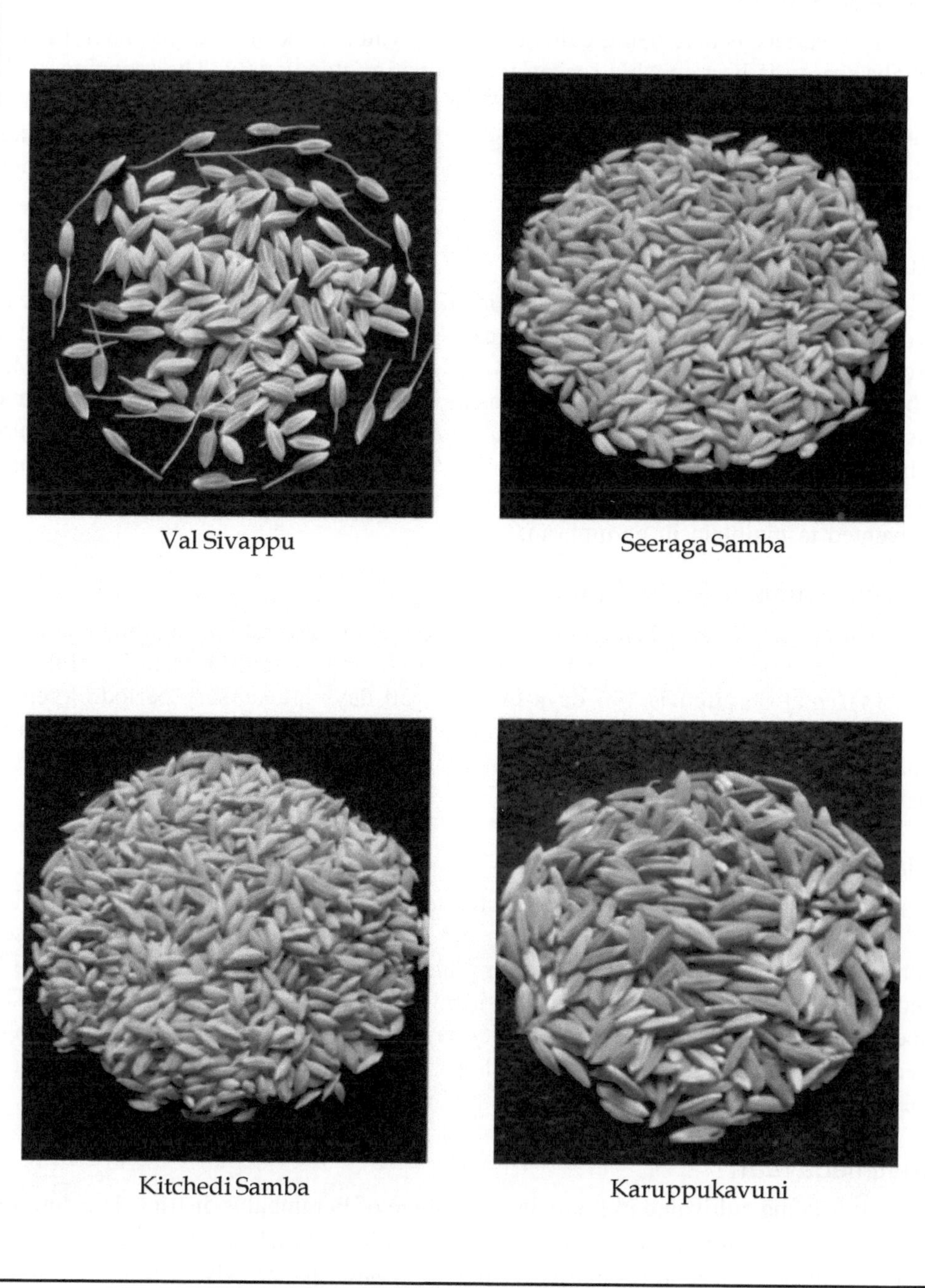

Val Sivappu

Seeraga Samba

Kitchedi Samba

Karuppukavuni

Fig. 3. Traditional Rice Varieties

Kichedi Samba

This variety is also being cultivated in Keevalur taluk of Nagapattinam district. Samba season is best suitable season for cultivating this variety. The cropping cycle is 140 days. The seed rate is 24 kg / ac. The plant grows to a height of 102 cm and it has a lodging habit. It yields upto 15 bags (1125 kg).The straw yield will be 1050 kg. The grain colour is pale white in colour. The special feature of this variety is resistant to pest and diseases.

Kuthiraival Samba

It is being cultivated in Kampalathupatti village in Trichy district. It is suitable to grow during samba season (Aadi-Margazhi i.e., July 15-Jan 14). It is a transplanted crop with a total duration of 140-150 days. Clay soil is well suited for its better growth. The height of crop is around 100-110cm (i.e. 3-3½ feet). It is resistant to lodging and also resistant to pest and disease. Grain colour is yellowish, whereas rice is pure white in colour. It yields upto 1.3 tonnes /ac. At present it is cultivated in small pockets of Trichy district which is in extinction stage that is to which should be prevented to maintain in germplasm.

Chinna Samba / Sada Samba

This variety is also being cultivated in Keevalur taluk of Nagapattinam district. Samba season is very well suitable season for sowing (Aadi-Margazhi i.e., July 15-Jan 14).Total duration is 140 days in which 30 days are nursery period. Average height of the crop is 4½ - 5 feet and susceptible for lodging. Grain is light green in colour during panicle initiation and milky stage, whereas in maturity stage it becomes pale yellow colour.

Mara Nel

It is being cultivated in Thenkanikottai taluk of Krishnagiri district. The duration of the crop is 120-125 days. Late samba and Navarai (Summer) season is best season for cultivating this variety. Transplanting method is used for cultivation.The rice is red in colour and coarse in texture. It is ideal for preparing tiffin like idly, dosai. The special feature of the grains of this variety have a tough seed coat. So the grains do not germinate even if there is heavy rainfall at the time of harvest. It has another feature i.e resistant to storage pest because of the hard seed coat.

Kallimadayaan

It is being cultivated in Iyerpalayam village of Perambalur district. The duration of the crop is six months. Plant is tall in nature with lodging habit. It has good tillering ability with a plant height ranging from 110-120cm.The panicle length is 30 cm. It yields bold grains and has the potential to give two tonnes per acre under suitable conditions prevailing during crop management. The colour of the grain is white. It has good cooking quality. It is resistant to stem borer, ear head bug and also blight disease. Farmer's opinion about this variety is the starchy watering nature influenced the grain, to keep them fresh and energetic. The seed rate is 35 kg/ac .The nursery period is 35-60 days. Now a days, few farmers are cultivating this rice variety in Iyerpalayam because most of the people don't prefer bold rice which reduces the market price of this variety. Straw obtained from this rice variety is used for thatching roof.

Karuppukavuni

It is being cultivated in Anumandhakudi village in Sivagangai district. The duration of the crop is 150-170 days. The best suitable season for growing this variety is September – January. It is suitable to direct sowing. The length of individual grain is 1cm. It needs only organic manures such as green and green leaf manure. It is quick growing crop.It does not require chemical pesticides.

Salient Features

- Non-lodging habit.
- Profuse tillering.
- The colour of the grain is black and it is only used for sweet making along with coconut milk and it is not suitable for meals preparation.
- As it has profuse tillering habit, the paddy straw yield is 150 per cent higher than any other paddy variety.
- Tillers are moist in nature and it fetches good weight.
- Grain yield is 1800 kg per acre and straw yield is 700 kg per acre.
- Price of grain is Rs.7/- per kg. But the rate for ordinary variety is Rs.3.50/- only.
- It requires lesser quantity of water comparatively and suitable for black cotton soil and red soil in dry lands.

Chandikaar

It is being cultivated in Anumandhakudi village in Sivagangai district. The best season for growing this variety is September – January. It yields about 800 kg per acre and straw yield also 1800 kg per acre. It is suitable for edible purpose in the villages and organic in nature. It is quick growing in nature. The people often cut out the stem after three months of sowing to prevent lodging. It is suitable for black cotton soil, red soil and saline soil. It is suitable for direct sowing and transplanting. Generally, it does need not require pesticides and any nitrogenous fertilizers and it is resistant to paddy blight disease. The length of individual plant is upto six feet. The variety has dense crop canopy, the field often filled with weed and it needs frequent weeding. The weed growth suppressed the plant population and it will leads to reduced grain yield.

Varappukkodanchan

It is being cultivated in Chatrakudi village in Ramnad district. This variety is suitable for rain fed condition. The outer skin is black in colour and the rice is red in colour. It is the drought resistant and tall variety. It is good in taste and cooked rice is kept upto 2- 3days. The rice cruel (kanji) of this variety is very palatable. The seed rate is 40-50 kg per acre for direct sowing .Farmers believe it gives more energy compared to other fine rice varieties for working people and the straw yield is very higher (3 tonnes /acre : 1 ton-RS.1000/-). Generally it is resistant to pests. Compared to all varieties, the cost of cultivation is less than Rs.1500. The duration of the variety is three months .It is sown normally in the middle of Puratasi (September) and harvested in the month of Tai (January).The yield is 1000kg-1200 kg/acre and it fetches higher price in Kerala state (Rs.7.50/kg).

Kuliparichan

It is being cultivated in Pallabachery village in Ramnad district. This variety is suitable for coastal sandy soils. It is drought resistant and tall variety. The outer skin is slightly dark. The inner grain is bold and round. It is good in taste and cooked rice is kept upto three days. The rice kanji of this variety is good in taste than other fine rice varieties. The seed rate is 40-50 kg per acre for direct sowing. Farmers believe, it gives more energy than other fine rice for working people and they gave higher straw yield (3 tones / acre : 1 ton-RS.1000/-). This variety is resistant to pests, generally except in very less water periods. It is affected by leaf folder in stress conditions. The cost of cultivation is less than Rs.1500.They apply only urea to the crop and one weeding is performed during season. The duration of the variety is three months .It is sown normally in the middle of Puratasi (September) and harvested in Tai month (January).The yield is 1000kg/acre and it fetches higher price in Kerala state (Rs.7.50/kg) than Tamil Nadu (Rs.4.50/kg) because they like red rice variety more than fine rice varieties.

Chitrakar

Chitrakar was the early existing red rice variety in Thirupullani tracts of Ramnad district. The rice variety was totally failure in 1964 famine. Thereafter Chitrakar was brought from nearest "Sikkal" village and are being cultivated now. Peasants generally name all these local cultivars in the name of mattai or norungan. This variety is suitable for coastal sandy soils. It is the drought resistant and tall variety.

The outer skin is slightly dark. The inner grain is bold and round. The rice kanji of this variety is good in taste. It prevents hunger for several hours. It is resistant to pests and diseases generally, except in water less times. It is affected by leaf folder in stress conditions. The cost of cultivation is less than Rs.1500. They applying only urea to the crop and only one weeding is performed during season. The seed rate is 40-50 kg per acre for direct sowing. The yield is 1000kg/acre and it fetches higher price because the local people like red rice variety more than other fine rice varieties. It is a very good straw yielder (2 tones per acre). The duration of the variety is 110 days .It is sown normally in the middle of Puratasi (September) and harvested in the month of Thai (January).

Sivappu Chitraikar

"Sivappu chitraikar" is the traditional red rice variety grown in Paramakudi tracts. The outer skin is slightly black in colour and the inner skin is red in colour. It is the drought resistant and tall variety. The duration of the crop is 110 days and it is normally sown in the month of "Aavani" and harvested in the month of "Thai". The seed rate is 40-50 kg per acre for direct sowing. The yield is1500-1800 kg per acre in normal monsoon period and in the failure of monsoon at least 600-900 kg per acre grains are harvested. So the farmers mainly rely on these traditional varieties rather than other varieties. The price of each bag is determined based on demand in market (Rs.300 to Rs.450). In the very lesser rain fed period there is small blight like formation found in the crop and ultimately lead to "Soorainookadu" (Sheath blight) disease. One or two weeding is performed in the field. "Rice kanji" of sivappu chitraikar is very tastier than other red rice varieties. It is good for preparation of dosai and idli. The seeds of this variety are stored by farmers from the previous harvest.

Murungaikar

Murungaikar is the traditional local red rice variety grown in Ramnad areas and it is suitable for even very low rainfall periods. The outer skin is slightly black in colour and the inner rice is red in colour. It is the drought resistant and tall variety. The rice kanji of this variety is very good taste in nature. The seed rate is 40 kg for direct sowing. The yield of murungaikar is 1000-1300 kg per acre in normal rain fed conditions (15-20 bags 1 bag 65 kg). Even when failure of monsoon, the crop can withstand drought conditions. The seeds of this variety are stored by the farmer in the previous year. The murungaikar variety is not affected by pests and diseases and no foliar sprays of pesticide have been applied so far. As the crop grows tall, it will lodge at the time of harvest. Comparing to "Nootripathu" variety Murungaikar has slightly lesser number of grains, but the straw yield is same (1ton per acre) as that of varappukodanchan and Nootripathu. The price of rice is fluctuating one, based on demand (Rs.350 to Rs.500). The crop is sown normally in the month of Aavani (August) and harvested in Markazhi (January). Goat manure and Urea alone are applied

Fig. 4. Traditional Rice Varieties

generally to the Murungaikar field. Only two weedings are performed in the cropping cycle but in larger land areas only one weeding is enough.

Nootripathu

Nootripathu is the traditional red rice variety grown in Ramnad district. The outer skin is white in colour and the inner coat is red in colour. It is the drought resistant and tall variety. The rice kanji of this variety is good in taste. It is sown in "Aavani" and harvested in "Thai". The seed rate is 40-50 kg per acre for direct sowing. One or two weedings are carried out during cropping period. The farmers observe "Senthalai Nookadu" (Leaf spot) disease in the paddy field in waterless condition. The total cost for cultivation of the paddy crop is Rs.2000 in rain fed condition. The grain yields about 1500-1800 kg per acre and straw yield is one ton per acre. Both black soil and red soil is found in nearby Paramakudi will lead to lesser growth of paddy, but in red soil more amount of rainfall lead to better rice growth. In some of the fields nearer to vaigai river basin, clay soil is found. The rain fed red rice varieties grow very well in clay soils rather than black and red soils. All the sivappu chitraikar, kuruvaikalnchiam and nootrypathu varieties fetch good price (Rs.6.50 per kg.) in market. The seeds of all varieties are stored by farmers from the previous harvest.

Ariyaan

There are 4 types of Ariyaan rice varieties grown in sandy soils of Regunathapuram areas at Ramnad district. Viz, "Vellai Ariyaan", Karuppu Ariyaan, Sivappu Ariyaan, and Vazhai Ariyaan. All Ariyaan varieties have awns. It grows very tall upto 5.5 to 6.5 feet. The duration of all the Ariyaan varieties is 120 days. The seed rate is 40-50 kg per acre for direct sowing. They are sown in the month of Ayppasi (October-November) and harvested in the month of Thai (January). This variety is suitable for coastal sandy soils. It is the drought resistant tall variety. The outer skin is dark. The inner grain is bold and round. The rice kanji of the variety is good in taste. It prevents hunger for several hours. It is resistant to pests and diseases, generally except leaf folder in very water less periods. It needs at least one rainfall shower in first 3 months for higher yield. The harvesting is done in three stages viz. top portion cut at first, middle portion cut at thereafter and bottom portion is cut at last. It gives higher yield of 2000-2100 kg per acre. (1 bag = 60 kg.) The total cost of cultivation for an acre is around Rs.1500 – 2000. All the Ariyaan varieties are maximum straw yielder upto 1000 kg/ acre. Among four Ariyaan varieties, karuppu ariyaan gave highest straw yield of 1-1.5 tones per acre. It is affected by leaf folder and stem borer. The seeds are stored by farmer from previous harvest.

Sadaikkar

The sadaikkaar paddy variety is available only in Aakadavalasai village in Ramanathapuram district after an unknown place in Thanjavur district. It is red rice variety cultivated in rain fed condition. It has medicinal properties. It is used to cure wound injuries in leg and hand by applying over the wounded portion. Most of the local "maatuvaidyars" (Ethno Veterinary practicers) are using this rice for its' medicinal value. This is suitable for cooking "idiaappam", "puttu" and "paniaaram" because of its sweetness. So that, the farmers are likely to grow this variety than any other red rice variety. The seed rate is 40-50 kg per acre for direct sowing and the cultivation charges are Rs.1500-2000. As the land is surrounded by sea with in 3 kilometer range in both sides, the water will drain out immediately to the subsoil. The base and stem have strong. The sowing season is September and harvested in February-March. The yield is 1200 kg per acre and it gives higher straw yield (3 tonnes per acre). As the leaves are wider, the animals prefer to eat the straw of Sadaikkar than other red rice varieties. The crop needs at least three showers in a month and it is not affected by pest and disease generally except leaf folder. It is grown in sandy soils, irrigation through well is not possible because the water will stagnate at surface level. So, the crop is grown only in rain fed condition and the crop yield is determined by the amount of rainfall received during cropping season and the number of weeding in the growing season.

Poongkar

It is being cultivated in Raghunathapuram village at Ramnad district. If the season fails in "Puratasi" (September) another traditional early duration red rice variety of Poongkar is sown in the month of "Karthigai" (October) and it is harvested in Thai (January) but the yield is lower than kuliparichan and varappukodanchan. This variety is suitable for coastal sandy soils. It is the drought resistant and tall variety. The seed rate is 40-50 kg per acre for direct sowing. The outer skin is slightly dark. The inner grain is bold and round. The rice gruel of this variety is very good taste than other fine varieties. It prevents hunger for one hour. The crop can be able to grow in very less rainfall condition and it is drought tolerant than kuliparichan and varppukodanchan. This variety is not attacked by pests generally except in very less water periods. The cost of cultivation is less than Rs.1500. They applying only urea to the crop and one weeding are performed during season. It fetches higher price than other red rice varieties (RS.10/kg).The yield is around 800 kg per ha. It is a very good straw yielder (1 tone per acre).

Kuruvai Kalanchium

"Kuruivaikalanchium" is the red rice variety grown generally in Ramnad district. The outer skin is slightly black in colour and the inner grain is red in colour. It is the drought resistant and tall variety. The rice kanji of this variety is good in taste. The duration of the variety is 110 days and it is sown normally in the month of Aavani. The seed rate is 40-50 kg per acre for direct sowing. One or two weedings are carried out during cropping season. The crop is harvested in the month of Thai. The paddy straw yield from an acre is one tonnes. There is no need of rainfall after milking stage. The total cost of cultivation for growing this variety is Rs.2000 in rain fed condition. The yield is 1500-2000 kg per acre in normal monsoon period. Rice kanji of Kuruvaikalanchiam is different than "sivappu chitraikar". The cooked rice is very good in taste even after two days. The seeds are stored by farmers from the previous harvest.

Norungan

It is one of the red rice varieties grown in Ramnad district with duration of 110 days. It is sown in rain fed condition in sandy soils. The seed rate is 40-50 kg per acre for direct sowing. The outer skin is dark. The inner grain is bold and round. The rice kanji of this variety is good in taste. It prevents hunger for several hours. It is sown in the month of Ayppasi (October) as rain fed crop and it is harvested in the month of Thai (January). The grains are bold type than other Ariyaan varieties. The crop yields about 2100 kg per acre. The straw yield from one acre is 1000 kg and it is 10 days shorter duration red rice variety than "varappu kodanchan". It is affected by leaf folder and stem borer, if there is no sufficient rain water. The crop is not affected by pest and diseases normally. The seeds are stored by farmer from previous harvest. After the end of the cropping season, gingelly or green gram is sown.

Kalurundaikaar

It is the traditional red rice variety growing in Paramakudi taluk of Ramnad district. The rice grain weight of kallurandaikaar is slightly more than kuliparichaan/ varappukodainchan varieties. It is the drought resistant and tall variety. The outer skin is slightly dark.The inner grain is bold and round. The rice kanji of this variety is good in taste.The seed rate of Kalurundaikar variety of an acre is 45 kg for direct sowing. This variety prevents hunger for more time than other fine rice. The yield is 1500-1800 kg per acre. Paddy straw yield is approximately three tons per acre. In some field, rice and kudiraivali (in border areas) system is followed. Kudiraivali yield is 600 kg per acre and it fetches higher price in market (Rs.6 per kg.). The rice variety is not generally affected by pest and diseases. The seeds are stored by farmers themselves from the previous harvest. In some fields 3 types of sorghum are grown as border crop. They are red, white and black.

Thanga Samba

It is being cultivated in Kanchipuram district. Late samba and samba period is the best season for cultivating this variety. The cultivation is by transplantation with the duration of 160-165 days out of which 40 days is nursery period. The grain is golden colour and fine in texture. The grain yields about 1500 kg per acre and the straw yield is around 60 bundles (1800 kg).

Special Features

- The ear head of this variety is very long.
- This variety of rice is extremely fine and long,it is used for the preparation of special dishes like pula.

Neelan Samba

It is being cultivated in Sukkan kollai village in Kanchipuram district. Samba season is best suitable season for cultivating this variety with duration of 175-180 days. The rice has red in colour and coarse in texture. It yields about 1500 kg /ac and the straw yield will be 1800 kg.

Special Features

- Highly suitable for water logging condition.
- Resistant to brown plant hopper and ear head bug
- Suitable for lactating women.

Vadan Samba

It is also being cultivated in Sukkan kollai village at Kanchipuram district. Samba season is best suitable season for cultivating this variety with duration of 155-160 days. Direct sown and transplanting is used for growing this variety. The rice is dull white in colour and coarse in texture. It yields about 1125 kg/ac and the straw yield will be 1200 kg.

Special Features

- Drought resistant
- Resistant to pest and disease
- Since the grains shatter, it is important to harvest them when they are still green.

Kaliyan Samba

It is also being cultivated in Sukkan kollai village at Kanchipuram district. Samba season is best suitable season for cultivating this variety with duration of 155-160 days. Out of which, 40-45 days will be nursery period. Transplanting is used for growing this variety. The seed rate is 35 kg per acre. The rice is red in colour. It yields about 1200 kg /ac and the straw yield will be 1600 kg/ac.

Special Features

- It grows very tall upto the height of 5.5 feet.
- It can be grown in water logging and low lying area.
- Highly resistant to pest and disease.

Samba Mozsanam

It is also being cultivated in Sukkan kollai village at Kanchipuram district. Samba season is best suitable season for cultivating this variety with duration of 160-165 days. Out of which, 40-45 days will be nursery period. Dry sown and transplanting methods is used for growing this variety. The rice has red in colour and coarse in texture. It yields about 1800 kg /ac and the straw yield will be 65 bundles (1950 kg/ac).

Special Features

This variety is also called Puzudikal, Erinel and Maduvu muzangi. It is suitable for growing in the vicinity of lakes. It is said that people travelled by boats and harvested the 'Samba Mosanam' in the lakes. The variety is good for preparing Aval (Flattened rice), idly and dosa.

Kadikazhuthan

It is also being cultivated in Sukkan kollai village at Kanchipuram district. Samba season is best suitable season for cultivating this variety with duration of 165 days. Out of which, 40-45 days will be nursery period. Transplanting system is used for growing this variety. The rice is white in colour and having the fine quality. It yields about 1350 kg /ac and the straw yield will be 40 bundles (1200 kg/ac).

Special Features

- It is also known as 'Garudan samba'. The appearance of the grains is like the neck of the Brahmin kite (Garudan) i.e. it has a white ring and so called this name.
- It is suitable for making a specialty dish called 'Puttu' of South India.
- It grows well in areas which have good drainage facilities.

Pisini

It is also being cultivated in Sukkan kollai village of Chengalpet taluk at Kanchipuram district. Samba season is best season for cultivating this variety with duration of 120 days. In the cropping period, 40-45 days will be nursery period. The rice has red in colour and having the coarse quality. It yields about 1500 kg /ac and the straw yield will be 50 bundles (1500 kg/ac).

Special Features

- Highly suitable for making dosa, porridge and aval.
- Highly resistant to drought. Since the ear heads are found with awn it is important to take extra care while removing the grains.

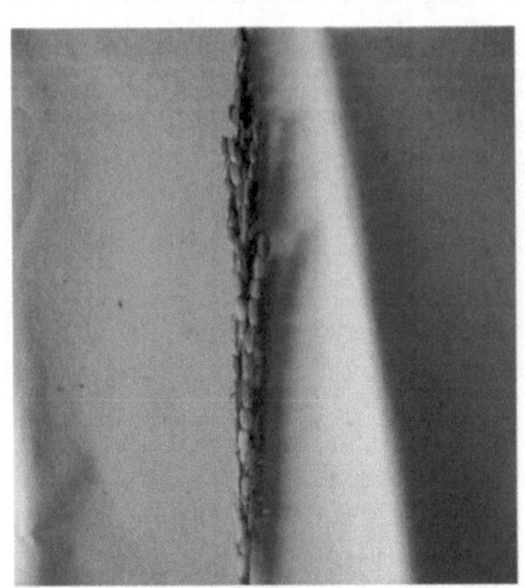

Kuruvai Kalanchium

Norungan

Samba Mosanam

Kadaikazhuthan

Fig. 5. Traditional Rice Varieties

Thuyamalli

It is also being cultivated in Sukkan kollai village at Kanchipuram district. Samba season is best suitable season for sowing this variety with duration of 135-140 days. Out of which, 30-35 days will be nursery period. Direct sown and transplanting system is used for growing this variety. The rice has white in colour and having the fine quality. It yields about 1125 kg /ac and the straw yield will be 35 bundles (1050 kg/ac).

Special Features

- During the flowering stage the ear heads look like flowers.
- Highly resistant to pest and disease. In Tamil, 'thuyamalli' literally means pure jasmine. Since the rice of this variety is white in colour like that jasmine, so that it is known by this name.

Kalar Palai

It is also being cultivated in Sukkan kollai village of Chengalpet taluk at Kanchipuram district. Navarai season is best season for cultivating this variety with duration of 120days. Transplanting system is used for growing this variety. The rice has brown in colour and having the coarse quality. It yields about 1200-1300 kg /ac and the straw yield will be 1300 kg/ac.

Special Features

- Resistant to pest and disease.
- Suitable for kalar nilam i.e. saline soil. It is the reason behind this name.

Kudaivazhai

It is also being cultivated in Sukkan kollai village of Chengalpet taluk at Kanchipuram district. Late Samba season is best season for cultivating this variety with duration of 120-125 days. In the cropping period, 25-28 days will be nursery period. Transplanting is used for growing this variety. The rice has red in colour and having the coarse quality. It yields about 1400 kg /ac and the straw yield will be 50 bundles (1500 kg/ac).

Special Features

- The ear heads of this variety appear like open umbrellas and hence the name 'kudaivazhai'.
- This variety is highly suitable for making dishes such as idly and Dosa.
- It grows upto 5 feet,the stem is 2.5 cm thick and the straw is used as roofing material.

- This variety is highly suitable for areas like water logging.
- This is a short duration variety; hence it can even be cultivated thrice a year.

Kumvalai

It is also being cultivated in Sukkan kollai village of Chengalpet taluk at Kanchipuram district. Samba season is best season for cultivating this variety with duration of 128-130 days. Direct sown and transplanting is used for growing this variety. The rice is brown in colour and having the coarse quality. It yields about 1350 kg /ac and the straw yield will be 1800 kg/ac.

Special Features

- This variety grown upto the height of 4.5 - 5 feet.
- Straw is particularly preferred as a roofing material
- It is suitable for sandy soil and also water logging area

Muttakar

It is also being cultivated in Edaiyur village of Thirukazhukundram taluk at Kanchipuram district. Navarai season is best season for cultivating this variety with duration of 120 days in which 30 days will be the nursery period. The seed rate is 45 kg per acre. Wet sown and transplanting is used for growing this variety. The plant grows upto a height of 4.5 feet and it is a lodging type.The rice has red in colour and having the coarse quality. It yields about 1350 kg /ac and the straw yield will be 40 bundles i.e. 1200 kg/ac.

Sempalai

It is also being cultivated in Edaiyur village of Thirukazhukundram taluk at Kanchipuram district. Samba season is best season for cultivating this variety with duration of 100-105 days out of which 21-25 days are nursery period. Direct sown and transplanting system is used for growing this variety. The rice is dull white in colour and coarse quality. It yields about 1125 kg /ac and the straw yield will be 45 bundles (1350 kg/ac).

Special Features

- It is highly suitable for making 'Pori' (puffed rice)

Kappakar

It is also being cultivated in Thakandarapuram village of Vandavasi taluk at Thiruvannamalai district. Samba season is best season for cultivating this variety with duration of 150-160 days in which 40 days will be the nursery period. The seed

Pisini

Kalarpalai

Kurnvazhai

Sempalai

Fig. 6. Traditional Rice Varieties

rate is 50 kg per acre. Dry sown and transplanting system is used for growing this variety. The rice is brown in colour. It yields about 1350 kg /ac and the straw yield will be 60 bundles i.e. 1800 kg/ac.

Special Features

- This variety is highly suitable for making idly and dosa.
- It gives high yield in clayey soil.
- It is suitable for growing in water logging areas as well as drought prone areas.
- It is suitable for preparing Aval.

Perunkar

It is also being cultivated in Thakandarapuram village of Vandavasi taluk at Thiruvannamalai district. The duration of the crop is 120 days. The seed rate for direct sown crop is 35 kg per acre and 40 kg per acre is used for transplanting system. It yields about 1.4 tonnes /ac and the straw yield will be 1500 kg/ac.The plant grows to a height of 4.5 feet.

Special Feature

- Tolerant to stem borer and ear head bug.
- Suitable for preparing idly and dosa

MILLET VARIETIES

Karikattai Ragi

It is Suitable for Hilly Region. It is a traditional ragi variety which is being grown in Thamarakarai village of Erode district. As per their perception, it is hardy in nature. It is suitable to prepare ethnic food called Kali, Muruku, Koozh etc. It is harvested along with straw and dried for about 15 days. Threshing can be done manually or by using cattle. Whereas the other traditional varieties only the ear heads are reaped. The duration of the crop is three months. The season for growing this variety is in the month of Aadi (July).This variety is existence over 125 years.

Kattucumbu- Pearl Millet

This traditional cumbu is cultivated in many areas of Trichy District, particularly Kampalathupatti village of Manapparai taluk, says that this traditional variety exists for more than 100 years. The apt season for cultivating this cumbu is Aadi Pattam (July-August).It is well suited for garden land and dry land eco system. Since it is a rain fed crop, direct sowing is the suitable method of sowing. The total duration of this crop is six months.When it is sown during Aadi pattam and harvesting can be done in the month of Margazhi (January). Manual harvesting is done by using sickle, where the ear head alone is cut from the stem and collected. These ear heads are sun dried for 2-3 days. It is then threshed manually by striking it with sticks or even bullocks are allowed to stamp on the ear head and thus the grains are separated. The separated grains are winnowed and cleaned. Grains are elliptical shape with pointed nose at the point of attachment with the rachis. They are faded green in colour with milky taste. Normally, it yields about 600-700 kg/ac.

After cleaning process, it is stored in traditional storage structure called 'Thaniya kuthir'. When utilized for consumption, it is pounded well into course powdery form and consumed. It can be fed to both human and cattles. It contains minerals and other nutrients. Normally, an ethnic food called "cumbu koozh" is prepared by using the coarse powder made out of cumbu grains. This ethnic food controls fever, blood pressure and even diabetes in human beings and it is very effective against Food and Mouth Disease (Kaanai) in cattles, as stated by Thiru .Kuppamuthu of the said village.

Thattai Varagu

It is also being cultivated in Veppanthatai taluk at Perambalur district. The duration of the crop is 5 ½ months. The seeds are sown directly in the field. For increasing the yield, one traditional practice is followed by the farmer since 20 years i.e. 45 days after sowing, the grazing is done in the field by allowing the sheep for one day. This practice is called "kedai aadithal" and technically it called sheep penning. For one acre 100 sheep's are allowed. Sheep's are grazed the entire field and leave only the root of the crop. After that, the side tillers are developed well. Side tillers give more number of ear heads. It gives yield upto 10-12 q / ac.

Puvadan Keazhvaragu – Ragi

It is also being cultivated in Veppanthatai taluk at Perambalur district. The durationof the crop is three months.The seed rate is 3.5 kg/ac. The apt season for cultivating this cumbu is Karthigai. Direct sowing and Transplanting system is used for growing this variety. It yields about seven qtl/ac. This crop is sown at the time of "Pogi" (Puvadan Noonbu) ie 30 th of Margahazi. This is the reason behind the name Puvadan Keazhvaragu.

Sorghum (Kona Cholam)

This sorghum is cultivated as an irrigated crop. All type of soil is suitable for this sorghum cultivation. After first ploughing, Farm Yard Manure is applied as a basal. After that 2nd ploughing is done, during the 3rd ploughing seeds are sown. Seed rate is 3 kg/acre. Season for cultivating kona cholam is Jan-Feb. Duration of this crop is 90 days. Irrigation is given once in 10 days interval. No need to done weeding. Grain yield in one acre of land is 2500kg. One kg of seed Rs. 12-15 for cultivating one acre of land Rs 3000/-is required.

Chittan Samai

Chittan samai is cultivated traditionally in Alangayam, Javathu malai village at vellore district. Samai is cultivated during the month of June-August. Land is well ploughed and seeds are sown through hand sowing. In one acre of land, 12kg seeds are required. The crop duration is 90 days. One weeding is done in DAS 30th day. For weeding, 10 labourers are required. One acre of land wages for weeding is Rs 30/ for women labour and Rs70/day for men labour. Crops are ready to harvest after 90 days. Plants are harvested by using sickle. Grains are separated by threshing. The straw yield from one acre of land is 800kg. The market rate of samai grain is Rs 10-12/kg.

Special Feature

- Very good taste and flavour compared to other varieties.
- Storage capacity is more than 4-5 years compared to other recommended varieties.
- It is used for biscuit preparation.

Sen Cholam (Wild Sorghum)

This is one of the traditional varieties in Salem district. The duration of the crop is 120 days and suitable for all the season. The color of the grain is dark reddish, ear heads are medium sized and oval in shape. The crop is mainly cultivated for cattle feed. In olden days, sorghum grains are used as a food material, but now it is not preferred. Farmers are eager to cultivate this variety while compared to other varieties. Green fodder is mainly used for milch animals to induce the milk production. Twenty kg of green fodder is fed to one animal /day (At milking time especially for cow). Dry fodders are preserved and given for all the animals. After maturation, the ear heads are harvested by using knife and it is sun dried upto 2-3 days. The grains are separated manually by using sticks. This seeds can be used for next sowing after three months.

Kakacholam (Traditional Cholam)

Traditional cholam varieties are grown in Kothali village at Thoothukudi district. It is a black seeded (Crow colour in Tamil kaka) variety. Duration of the crop is 6 months. They are cultivating this crop as rain fed and irrigated system. The seed rate is 30 kg /acre. It may be sown in the month of September- October and to be harvested in the month of January - February. The seeds are kept for next sowing by farmers themselves. This variety is suitable for livestock as fodder. It can be used as inter crop along with any other crops.

Karikattai Ragi
Eleusine coracana

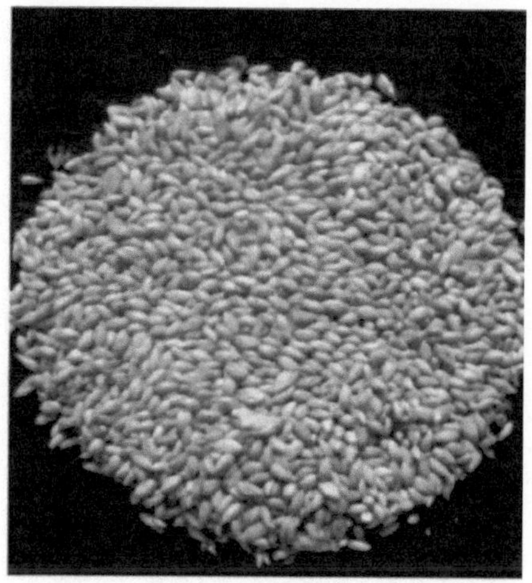

Kattucumbu
Pennisetum glacum

Thattai Varagu
Pespalum scrobiculatum

Kona Cholam
Sorghum bicolor

Fig. 7. Traditional Millet Varieties

Chittan Samai
Panicum sumatreuse

Sen Cholam
Sorghum bicolor

Red Gingelly
Sesamum indicum

Kothu Kadalai
Arachis hypogaea

Fig. 8. Traditional Millet and Oil Seed Varieties

OIL SEED VARIETIES

Black Gingelly

Traditional Gingelly crop is grown in Ramnad and Paramakudi taluks. It is black seeded variety than normal white seeded type. It has more oil content than the other varieties and the duration of the crop is longer (i.e.90 days) than the other varieties (75 days). They are cultivated as rain fed crop and the yield is 400 kg per acre and the total cost of cultivation for rain fed gingelly is Rs.500/-. The traditional gingelly has more branches and vigorous growth than any other gingelly varieties. The seeds are kept for next sowing by farmers themselves. There is no permanent season for gingelly cultivation. It may be sown in the month of Chitrai, vaikasi or Adi (April – July) based on need and demand of farmer. It is not suitable for cultivating in rainy season because it will cause certain problems.

Red Gingelly

It is being cultivated in Raghunathapuram village at Ramnad district. Gingelly is sown in month of December and harvested in March. The gingelly is red in colour and it has good oil content. The rain fed traditional gingelly yield is 320 kg/acre. It fetches good price in market (Rs.18 per kg.).The seed rate is 1.5 kg/acre and the seeds are colleted from the previous year grains. The oil content is more than other varieties. The duration (90 days) is more than other varieties (70 days). It has more branches than other varieties. The seeds are collected from the harvested grains and it is kept upto one year without addition of any chemicals.

Kothu Kadalai –Groundnut Bunch Variety

It is being cultivated in Vallathodu village at Kanchipuram district. The sowing season is Adipattam. The seed rate is 55 kg/ac. It is resistant to pest and diseases. It yields about 16 bags/ac (75 kg/ bag). This variety is in existence for more than 100 years.

Kadalaikai- Traditional Groundnut

It is being cultivated in Velankadu village at Vellore district. Land is ploughed 3-4 times before sowing by using country plough or cultivator. Five tones of farm yard manure are applied as a basal. The suitable season for growing this groundnut is June-August (it is a rain fed crop). The duration of the crop is 90 days. 50 kg of seeds are required for one acre of land. Seeds are sown behind the country plough. 30th days after planting, 1st weeding 60th day second weeding and for each weeding 20 women labourers required. For harvesting groundnut, farmers are given only produce as wages @ 5 :1 90th days after planting, ground nut are ready for harvesting. For one acre of land 20-25 women's labour is required / day.Yield is about 5000 kg/acre. One kg of ground nut is Rs10/-. The expenditure for cultivating groundnut per acre is Rs 5000/-.

VEGETABLE VARIETIES

Kulithakali (Tomato)

It is being cultivated in Kanchipuram district. The season for growing this variety is throughout the year and also the bearing season is throughout the year. The seed rate is 100g /ac and 10 plant / cent .The size is also small and more tartaric acid content. It is resistant to pest and disease.It is suitable for dry land cultivation. Root slip is enough for re-emergence (after drying of above ground portion i.e. stem and leaves). After two months, the plant will start bearing.

Vazhuthaalangai *(Solanum melongena)* - Egg Plant/ Brinjal

It is another important tropical vegetable believed to be a native of India. It is being cultivated in Vettuvenni village of Marthandum Taluk at Kanyakumari district. Local preference varies considerably from region to region and even district to district. The suitable season for growing this crop is June to October. The duration of the crop is 90-110 days. The variety is well adopted to well drain deep soil for its growth. It is mainly used for table purpose. The variety is preferred more by kanyakumari people as it is a traditional part of their daily food. Little leaf disease is a major problem caused by Mycoplasma like organisms but the variety is resistance to this disease. First harvest starts from 55-60 days after transplanting. The fruits are lengthier than normal varieties and measures 12-15 cm height. The fruit is cylindrical in shape. Harvesting can be done by once in 5 days. Yield ranges between 20-25 tones/ha.

Kannadi Katthari- Wild Brinjal

Kannadi kathari is traditionally cultivated in Devoor village at Salem district. The nursery requirement for one acre of land is 1.5 cents. The required seed rate is 200-300 grams. The nursery duration is 2 months. Seeds are collected from well ripened, pest and diseased free fruits. Seeds are coated with ash and Dithane powder to protect the seeds from stored pests. It can be stored up to 2-3 years. Cultivation starts during the month of August and January and the duration of this crop is 6-7 months. This variety is mostly preferred in Erode market and has good market value while compared to other varieties. Cow dung, goat manure and poultry waste were used as manures. Some times plant leaves are used as manure in field itself. This variety possesses high pest and disease resistant, so that farmers used to prefer to cultivate this variety. Plant height is 4-5 feet; high rainfall condition is not suitable for this crop because the root rot incidence is more in that time. After first harvest the plants are left with 1 foot height from the ground level as rattoon. The yield from this crop is upto 30-40 bags in 3 months. The farmer can get a net profit of upto Rs. 30,000- 50,000.During harvesting period, intercropping is done by using "kottai Pudalai' (Snake gourd) seeds are sown in brinjal field. The Snake gourd plant uses the wild brinjal plant as a support for its growth. The farmer generated the additional income of Rs.20, 000/- for growing kottai pudalai as inter crop.

Poyyur Kathari-Local Brinjal Variety

This local variety brinjal variety is commonly cultivated in coastal villages of Nagapattinam district.Thiru Sengutuvan of Vellapallam in Nagapattinam district cultivates this 'Poyyur kathari' for about 10-15years and still continues to cultivate the same. This traditional variety is suitable to sow during June –July. It is a transplanted crop for which a nursery is prepared with raised bed. Seeds are sown on these raised bed and 15 days after sowing, the seedlings are transplanted to the main field. While transplanting two seedlings are planted per hill. 45 days after sowing, it starts bearing. Once it starts bearing, brinjals are harvested normally by just plucking. Harvesting is done in a phased manner to have the yield through out the season.

Special Features

- It is well known and highly preferred for tasty fruit.
- The fruits are consistent which gives more weight when compared to other varieties.
- For upto 6 months, this crop can be ratooned and maintained till optimum yield is obtained.

Ramnad Gundu Chillies (Capsicum spp)

In Ramnad areas, the prevailing conditions are favoured to grow chillies well. The traditional Ramnad gundu variety is most preferably cultivated because of its yield is higher than the other samba type. When the amount of rainfall increases, the yield of chillies is drastically reduced. It is cultivated only in less amount of rainfall. Farmers get more profit in "Ramnad Gundu" chillies. The cost of cultivation of Chillies is Rs.10,000 per acre and the profit is Rs.30,000 per acre. It is sown in the month of Aavani / Karthigai and harvested in the month of Chittrai / Vaikasi. Cotton is used as intercrop in chillies field. Some of the farmers using coriander as intercrop in chillies field. The yield of chillies is 1000 kg per acre (50 bags : 1 bag = 20 kg).

Bhendi /Okra – *Abelmoschus esculentus* – Red Bhendi ("Sivappu Vendai")

This is commonly known as lady's finger. It is being cultivated in Panavilai village of Vilavancode taluk at Kanyakumari district. It is a tropical direct sown vegetable crop with duration of 100-120 days. The optimum season is June-August. Sowing can be extended upto July but harvested before October. This is mainly to prevent the crop from North East Monsoon. Deep well drained loamy soils with undulating pattern are most suitable and also available in larger area. The variety is highly useful for culinary purpose. The plant grows to a height of 4 feet. Fruits are red colored and so called red bhendi. It is rich in carotene. 20th day onwards the plants start yielding. After the emergence of fifth leaf, flowering occurs. Main stem is multiple branched, each branch gives fruits. The variety is resistant to yellow mosaic virus spread by vector white fly-*Bemisia tabaci*. Harvesting is done on alternate days, since the pods get matured. Quickly, the selection of pods with maximum size but still tender is taken. A total number of 15-20 harvests can be made. The yield varies from 12-15 tones/ha.

Covaikai (Coccinia)

The crop can be cultivated in villurupuram district. The name of the variety is Padappai. The method of propagation is by stem cutting from pistil late plants. Crop spacing is 2x2m. The crop needs a special practice like erection of pandal or trellis and also training the crop. Harvesting starts from 6 months after planting and continues upto through out the year. The yield varies from 6-8 tonnes/acre

Gandhari Chilli – *Capsicum frutescence* ("Golkonda Chilli")

Chilli is an important vegetable cum condiment called Red Pepper or hot pepper. It is being cultivated in Vettuvanni village of Marthandum taluk at kanyakumari district. The pungent principle is called 'capsicin' which is mainly present in the central placenta of the fruit, as well as seed and to some extends in the ovary wall also. Generally, the capsaicin content varies from 0.2-0.4 %. Vitamin C is present in appreciable quantities. Spicy vegetables like chilli add taste and flavour to our food. The added advantage of these spicy vegetables is their protective nature against diseases. During May to June, the nursery is raised and the duration of the crop is 100-120 days. Deep fertile loamy soils are best suited for cultivation. Gandhari chilli is more pungent than other varieties since their capsaicin content more than 0.5%. It helps in fat control, blood purification as a first aid by preventing blood clotting. The juice extracted from the ripe fruits is used for consumption. If excess of juice is consumed, it leads to ulcer. The fruits grow upwards, small in size. Yellowish or greenish colour at first stage, later on it turns to red on maturity. The fruits are also used in control of ear head bug along with garlic and neem oil as paste. The paste is diluted in 10litre of water. From it 50ml/tank is taken and sprayed. The yield varies from 2.3-2.5 tones/ha. The variety has not been popular in other places of Tamil Nadu, because of its high pungency in nature. So that, the farmers don't prefer to grow this variety.

Manattakkali *Solanum nigrum*

It is being cultivated in Malavilai village at kanyakumari district. The common name is Black Night –Shade.The whole plant is used because of its medicinal properties. It cures asthma, general disability, heart diseases and hiccough.

Blue Chilli (Neela milahai)

It is being cultivated in Panavilai village of Vilavancode taluk at Kanyakumari district. The variety resembles all the characters of gandhari chilli except the fruit size, and colour. The flower appears blue along with blue leaf. It looks like an ornamental plant and attract observers. The fruits are round shaped and turn red on maturity. Fruits grow in bunches used for cooking, also eaten with tapioca.

Ther Kothavarai-Local Cluster Bean Variety

This traditional cluster bean variety is popularly cultivated in coastal districts, especially in Nagapattinam District. This variety exists beyond 75 years. June-July is the apt season for cultivating this variety. Ridges and furrows method is followed to sow the seeds, where 3 seeds are dibbled in one hill in the ridges.30 days after sowing final harvesting can be done.It is harvested manually by plucking method. Time interval is set, once in 4 days harvesting can be done. On an average, it yields about 50kg/ week and fetch about Rs5/kg of fresh bean. Normally in the tracks, 'Ther kothavarai' is cultivated organically by using farm yard manure, vermi compost and locally available backyard manures. Thus it finds own market value. Apart from that, the taste of this variety is also good which makes it highly preferable. It is capable of curing stomach ulcer and mouth ulcer. It yields about 1800 kg/ac.

Karuppu Avarai -Black Beans

This is suited only for rain fed or upland cultivation. It is grown in Valeppur village of Namakkal district. The whole bean as well as the pod alone can be separated and used as vegetable. The method of cultivation is more similar to that of the ordinary bean cultivation. Generally, this crop can be raised thrice a year during May-June (Vaigasi) September-October (Purattasi) and January-February. It comes to bearing on third month onwards and the total duration of the crop is four months with an average yields about 2-3 kg. The major difference between black bean and white bean is the taste and energy which is comparatively higher in the earlier than the later.

Nattu Avarai (Local Bean)

It is predominantly cultivated in Vallathodu village of Kanchipuram district. The suitable season for growing this variety is Aadi pattam(July). The crop will start bearing at 45 days after sowing. After six months, it starts bearing. Harvesting can be done in weekly interval. It yields about 3-4 kg/cent/week. If it is matured plant ash is sprinkled whereas it is young one, cow dung is sprayed over the crop to control the pest and disease.

Nattu Pusani- Pumpkin

The traditional pumpkin is sown in rainfed condition and grown in Ramnad district. The matured crop weighs about 20 kg. It is sown in the month of Aadi (July and August) and harvested in the month of Thai (January). Each pit is 2.5 feet depth.

Vazhuthunangai
Solanum melongena

Ramnad Gundu Milagai
Capsicum annum

Gandhari Chillies
Capsicum annum

Ther Kothavarai
Cyamopsis tetragonoloba

Fig. 9. Traditional Vegetable Varieties

Thereafter thinning operations can be performed after growth of more seedlings. The seeds are kept by farmer themselves in previous year harvest.

Curry Leaf

The curry leaf crop is grown as mixed crop in Thangachimadam areas of Ramnad (10cents in one acre) and 2000 trees can be grown. They are picked in every months and the yield is 3000 kg/ac (The rate is Rs.5/ per kg.) and they are transported to Paramakudi and Rameshwaram markets. It is suitable for coastal sandy soils. Goat manure is used for this crop. They are resistant to pest and diseases. Normally the soil is very fine coastal sandy in nature, the crop comes very well because of the availability of water at lower depth.

Betel Vine (Vethalai)

The traditional "betel vine" variety is grown along with "Velimurungai" (Kalyana Murungai) which gives anchorage to betel vines. It is suitable crop for coastal sandy soils prevailing in Ramnad district. In the border areas, banana trees are grown and it gives higher crop intensity and also generates additional income. They are 150 banana trees surrounding around the border area of the 2 acres land and most of them are border crops in betel vine field. The field is irrigated by three wells and the water is available at 10 feet depth.

Jaffna Moringa

The traditional 40 years old Jaffna Moringa has been grown in Thangachimadam areas and the cutting was brought from Jaffna, Sri lanka during the cargo service between India and Sri lanka. It comes very well in coastal sandy soils. A 40 years old tree gives 1500 moringa sticks daily between March-June. They apply only organic manures such as goat manure to the tree. It is not usually affected by pests and diseases. The length of individual fruit is upto 12 feet and it fetches Rs.3-5/- based on demand. The rainy season of October, November and December is the off-season for moringa tree.

Motchaipayaru (Lab Lab)

It is grown in Vadavathure village of Namakkal district and the seeds are sown in December. The yield is 6 bags per acre (one bag is 50 kg). The seed rate is 1 kg/ acre. As the soil is sandy in nature, water will percolate into deeper layers. So only rain fed sowing is possible in this crop. It has twining quality, so it is not used as intercrop with any other crops. The seeds are collected from the harvested grains and it is kept upto one year without addition of any chemicals. The crop is mainly used as cattle feed.

Tuber Crop Varieties

Nattu Maravalli Kizhangu (Vellai Bontha) - Tapioca

Vellai bontha is traditionally cultivated in Salem district. The duration of this crop is 8 months (August- March). 4500- 5000 setts are required for planting in one acre. Weeding should be done once in a month, so that yield could be increased. One plant can yield about 4-5 kg of tubers. In one acre we can get 10 tonnes of tubers. The market price for 1 kg of tapioca is Rs.5/-. Intercropping is possible in the period of first three months. Black gram and groundnut are cultivated as inter crops. Coconut tree is grown as a border crop. White fly incidence is more in this variety at the end of the harvest.

Special Feature

- It is mainly used for culinary purpose.
- The colour of the skin is pink and some times white in colour and nattu variety tubers have more flour content.

Karuppu Bontha (Black Tapioca)

Karrupu bontha is traditionally cultivated in Salem district .The colour of the skin is dark brown and some times black in colour and these variety tubers have more flour content.The yield varies between 4-5 kg of tubers per plant.

Kariyilai Poriyan (Tapioca)

As the name indicates, the tubers are black in colour. The variety has duration of eight months and cultivated in the tracks of kanyakumari district. It is a good and energetic table food also used as poultry feed. It possesses less bitterness in taste than other varieties. The tuber is sun dried. After cutting into pieces and the pieces are dried in naturally.. It can be kept stored for 3 months. In pest and diseases, this is highly resistant to Tobacco Mosaic Virus and brown leaf spot respectively.

Adukku Muttan (Tapioca)

It is predominantly seen in Kanyakumari district with the cropping cycle of six months. Tubers are small with increased starch and possess good cooking quality. Processed tubers can be stored upto 3 months after dehydration (processed one). It is suitable for making food items like “kizhangu puttu”- steamed flour sweetened or spiced. After processing powder obtained from the tuber is given to children. For both the varieties described above, the best suited soil is light sandy loam. Planting can be taken up in May-June. The crop does not grow well in other districts of Tamil Nadu because the soil type, required seed material is not suitable and available at the place. So care should be taken in selection of varieties and their season. Tapioca is an

Nattu Avarai
Lab lab purpureus

Betel Vine
Piper betel

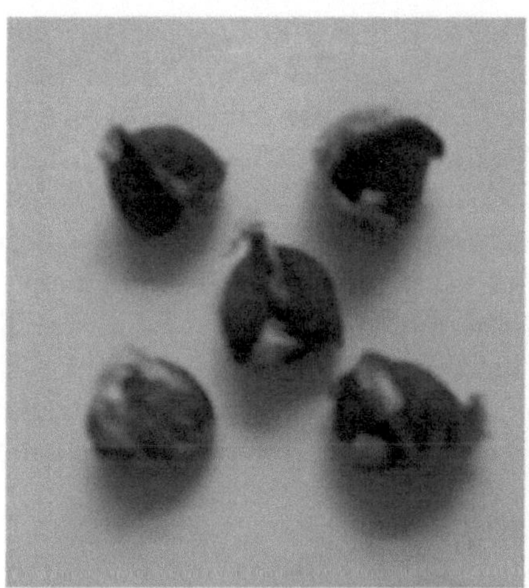

Murungai
Moringa oleifera

Sweet Potato
Ipomoea batatas

Fig. 10. Traditional Varieties

important subsidiary food and used in different forms. Yield of the crop varies from 30-35 tonnes/ha.

Nattu Seeni Kizhangu -Yams *(Dioscorea spp)*

The tubers are underground stem rich in starch and are usually single with various size and shape. Tubers are rich source of sucrose. A deep rich friable well drained sandy loam soil is ideal for development of tubers and it is prevailing in Malavlai village of kanyakumari district. The proximal end of tubers are cut into pieces and dipped in cow dung slurry allowed to dry in shade. The cropping cycle is 8-10 months. The yield varies from 20-25 tonnes/ ha. Yellowing and drying of leaves is the indication for harvesting the tubers.

Seeni Kizhangu – Local Sweet Potato Variety

It has been cultivating for more than 15 years. It is suitable for both rain fed and irrigated condition seen in Irudhukottai village of Krishnagiri district with duration of three months. Leaf cuttings are best suited method for propagating this crop. It is all season crop. Furrows are made in the field with two feet length and the leaf cuttings are planted in that furrow. For cultivating one acre, 10 cent area is used for transplanting purpose. This crop is not affected by any pest and disease. Based on the water availability, the kizhangu size will vary.

FLOWER CROP VARIETIES

Traditional Jathi Mullai

In one acre of land, 800 plants are planted. All soils are suitable for this crop *viz.*, red, clay, and clay loamy and grown in Vellore district. After field preparation, plants are planted at a spacing of 5x5 feet interval. For line planting, rope is used. Furrows are made for plant to plant. Irrigation is given to the field once in two weeks. Flowering will start six months days after planting. Pruning is done in the month of July (yearly once). Flowering is start during the month of October-February. The yield will be given upto 5 months per/day. The total yield per plant is 100 kg. The market rate is Rs.15-20/kg. Wages for picking process is Rs.12/kg. The expenditure for one acre of land is Rs 5000/- .Jathi mullai is mainly used for panner preparation.

Traditional Mullai (Jasminum Auriculatum)

It is grown in Vellore district and similar to that of traditional Jathi mullai. The market rate is Rs 20/kg. Mullai is mainly used for ornamental purpose.

Ramnad Gundumalli (Jasminum Sambac)

Ramnad gundumalli can able to grow well in sandy soils with less irrigation i.e. two irrigations per week. The name itself it is grown in Ramnad district. The crop withstands for 4 years and is sown at 2.5 x 2.5m interval. Usually marginal and small farmers are used 20 cents or 30 cents area for raising the crop. As December-January is the rainy season, no picking is done.In January, all leaves are completely removed to grow new leaves. From February to June, flower pickings are done and mainly 8 kg flowers are plucked in every day during season time and are exported to Ramnad & Madurai. The market rate is Rs.40/-. The cuttings are done in thrice a year. In one time, 15000 cuttings are cut down. (1 cutting = Rs. 0.80 to 1). Only goat manure is applied in the field. Initially cuttings are kept inside coconut leaves, but recent times cuttings are grown small greenhouse chamber (inner Plastic /outer green sheet) and irrigation is done to maintain heat and coolingness.

FRUIT VARIETIES

Guava - Nattu Vellai (White) and Senkoiya (Red Flesh)

There are two types of guava varieties. Among these, one is white coloured and the other one is pink coloured .The main difference between the varieties released and the traditional varieties with reference to guava are;

- Taste – traditional varieties are tastier than other varieties.
- Skin – traditional varieties have a light skin when compared to other varieties which have a thick skin that makes it hard to eat.

Even within the traditional varieties, especially in red guava there are two forms, one which has a pointed end with less taste and the other which is round shaped with good taste. One basket containing 10-25 kg of fruit. The cost of one basket is Rs.30-40/-.

The reason for existence of such traditional guava variety still at kolli hills is because;

- Government variety doesn't suit to the climatic condition prevailing at kolli hills.
- Even if a same tree bears fruit, due to the harder skin, it has less or no market preference. So obviously, it loses its position in cultivation.

Pineapple

Pineapple can be cultivated as two types of seed materials.

- Crown
- Bottom slip

Mostly it is planted in garden land or upland crop in Namakkal district. If crowns are used as a seed material, the duration of the crop is 1 ½ years and if bottom slips,

Jathi Malli
Jasminum grandiflorum

Mullai
Jasminum auriculatum

Ramnad Gundu Malli
Jasminum sambac

Naadan
Ananas sativus

Fig. 11. Traditional Varieties

the cropping cycle is 2 years. Nearly 10,000 slips are needed to plant in one acre area. Pits are dug with a dimension of 1'x1'x1' in rows. Each row is called 'saal' between which 3 feet distance is allowed. While digging pit, the distance between each pit is around ½ feet – ¾ feet. It needs a good drainage system, which is naturally provided by the sloppy topography of kolli hills. Generally, these traditional varieties don't need any additional external input, but weeding should be done monthly once as stated by the farmer.

The crop can be cultivated through out the year. Only 6 months particularly, Jan-June is appropriate season for good yield. Each plant bears nearly 4-5 times and even sometimes 10 slips at bottom which can be used as planting material during the next season. Previous crop is harvested and the residues are burnt, and the resultant ash is left as potash fertilizer for field.

Pineapple fruit turning into reddish or reddish yellow in colour, is the indication for its maturity. So it reaches the colour, the fruits are plucked by hand. Whereas the bottom slips are cut by using knife. After harvest, each plant yields again and again. Until 3 years from planting, the yield will be good after which it reduces.

Based on the size, each fruit costs upto Rs.2-5/-. Though pineapple is found in vazhavanthinadu areas (semmedu) and more commonly grown in Alathur, Ariyur and Valeppur villages of kolli hills. The main advantage of the traditional pineapple variety is good consumption quality. Because other varieties are good for juice making but unfit for consumption. Th.A.K.Raju states that burning the residues there is no harm to micro organisms whereas it is beneficial to the field by segmented fertilizer.

Pineapple (Annachi) - Naadan /Local - *Ananus Comosus*

It is being cultivated in kanyakumari district. A well drained loamy soil is ideal for pineapple cultivation. On 7-8 months after planting pine apple starts fruiting. It takes 3 months time to mature and marked with in a day or two. In maturity stage, the fruits are harvested in the month of April-July. Fruits obtained during peak summer (May) are faster compared to other varieties. If the fruits are harvested in rainy season, their sourness will be more. In early stage, the fruit becomes red in colour and later it turns to yellow. It weighs upto 2 kg and bigger in size. From each plant, 1-2 fruits are harvested per year. When planting is done in hill slopes, the mounds and channels should be formed across the slope and planting also should be done across the slope which will prevent soil erosion and conservation of rainwater. It is mainly used as a live fence because of its sourness in taste and more number of red thorns.

Ram seetha (*Annona reticulata*) – Local Anona Variety

Fruits are edible one and unripe fruits considered as anathematic; bark is a powerful astringent; leaves and seeds are insecticidal use. It cures asthmatic disorder.

Generally rural people use this plant as health tonic and farmers used the leaves for controlling pest. The colour of the fruit is yellowish orange. The single fruit weighs about ½ - ¾ kg. The maximum yield is 200 fruits/ year. The Cost of the single fruit is Rs. 2-3/-.

Chengavarkai (Mango- Red Shaded)

The mangoes of kanyakumari district are off season, a variety is considered to be the best as a whole. It is reddish in colour with round shape. The fruit is devoid of fibre content and possess 'karkandu' taste (candy). It weight about 150-200g a single fruit. Fruits are not used for pickles and cooking because it possesses thick pulp (flesh).The shelf life of the fruits is very good because of thick skin. Mostly the mango varieties of kanyakumari district come to harvest in summer. "chenkavarkai", sonachi to some extend it may be extended upto July. Mango can be grown on a wide range of soils, but kanyakumari possess deep well drained soils, clay loam are some places. The varieties have not been popularized in other places of Tamil Nadu because of seed scarcity, preference by other people and varied climatic conditions. The traditional varieties are now under extinction, the seed material is kept for next crop stored as germplasm at concerned research stations.

Chetti Ottu

The mangoes possess thick skin, bigger fruits, and good for consumption. The availability is extended throughout the year. But the yields are less when compared to others. The fully ripe fruits are used for preparations like jam, jelly, squash, syrups and nectars. The raw fruits before ripening can be used in preparations like pickles, chutney. Each variety has its own distinct taste, flavour, pulp consistency and yield potential. Because of this, the production of traditional varieties has been linked.

JACK VARIETIES

Jack Fruit

There are two types of jack fruit called 'Parukkan pala' and 'koolam pala'.The varieties are differentiate by two major properties. The main difference between both of these traditional varieties are the earlier one is used for both consumption as well as cooking purpose, whereas the later one is not fit for consumption but the seeds only used for culinary purpose. Another difference is the method of harvest, the Parukkan pala can be harvested only by using knife and fetches more market prices whereas the koolam pala can be plucked by hands and doesn't fetch more prices. Apart from these differences, the cultivation procedures are similar.

Normally, it starts flowering during February-March (Masi) and starts bearing during August-September (Aavani season).Mostly it is affected by pest especially a pest called ‘Nurai poochi’ which makes the fruit watery and water droplets drops from the leaves and seems as raining. In each tree on an average bearing 100 fruits and each fruit contain nearly 75-150 flakes (based on the size).If the flakes size is bigger then the number of flakes is less and vice-versa. The cost of each fruits is around Rs.10-20/-.

Varikkai Chakkai Pala

The variety thrives well in deep rich alluvial and loamy soils prevailing in Kanyakumari district. It belongs to firm flesh group and the carpel’s are thick and tightly arranged. It is mainly used as a stable fruit. Jack is a popular fruit with the house wife for making pickles. In a year, the tree bears fruits in two seasons, viz. April-June and November- December. Normally it starts fruiting from 7th – 8th year onwards. The yield ranges from 60-80 fruits/tree. The fruit weight varies from 12-18 kg containing 115-120 flakes. Flakes are pale yellow in colour. The shelf life is more when compared to other varieties.

Koozhan Chakkai Pala

The variety belongs to soft flesh group. Trees are medium tall and maximum bearing in tree trunk. Flakes are loosely arranged and poor storage life because moisture content is more when compared to other varieties. Carpels are sweet and tasty with pleasant aroma. It is suitable for commercial planting.

Mulagumoodu (Local Pala)

It’s having less moisture content, specific to the place “mulagumoodu’ of vilavancode Taluk, Kanyakumari district. It is used as a parent in PPI -1(pechiparai-1). The fruits are firm fleshed and medium in size weighing about 7-10 kg. Patch budding is done in 3-5 months old rootstock seedlings during January-February and October-November was found to be very successful. In all the varieties of jack during the pre bearing age, pulses are raised as intercrop and dried leaves are spread below the trees to serve as mulch for moisture conservation.

The type and bearing habit is called “cauliflorus”. The fruit buds should be protected from any positive damage by rubbing the body of the stray cattle such as buffaloes through spreading dried thorny bushes. This is very essential especially when there is no proper fence in the fields.

BANANA VARIETIES

Palayankodan

This is the foremost commercial variety in Tamil Nadu. The plant is tall, hard and grows vigorously under the ratooning system. The fruits are small to medium in

Ram Seetha
Annona reticulata

Chengavarkai
Mangifera indica

Chetti Ottu
Mangifera indica

Parukkan Pala
Atrocarpus heterophyllus

Fig. 12. Traditional Fruit Varieties

Varikkai Chakkai
Artocarpus heterophyllus

Koozhan Chakkai
Artocarpus heterophyllus

Sevvazhai
Musa sp

Nendran
Musa sp

Fig. 13. Traditional Fruit Varieties

size, yellow skinned, firm fleshed with a sub-acid taste and individual finger with prominent nipple. The duration varies from 11-14 months. The average bunch weight is about 15 kg having 8-12 hands.In each hand with 11-18 fingers.

Other names : Erode Poovan, Kadali, Fill basket, Adukku namarai.

Nendran : *Syn* : Ethankai & Plantain

The fruits of this variety are eater as raw and used for preparing various dishes especially chips. The pseudostem has a distinct shade of pink coloration. Leaves and young plants inclined towards ground unlike other varieties. The fruits are relatively longer and thicker than other banana varieties. The bunch is not compact like other varieties. Fruit with three prominent ridges strongly hold. Ripe fruits are yellow colored and pulp is flesh and tough. Duration of this variety is 11-12 months.

Sevvazhai : Syn : Red Banana

This variety has a free suckering habit grown in Kanyakumari district. The colour of the pseudo stem, petiole, mid rib and fruit rind is purplish red. The fruit is of good size, slightly curved with a blunt apex. The bunch weight is alone 20-25 kg with over 100 fruits. It is a long duration variety and takes about 18 months from planting to harvest.

Matti

Fruits are very small but are very good in taste. The fruits having the medicinal value to cure dysentery and suitable for giving to small children. The fruit smell is similar to that of pineapple and no flour ness, bunch is compactly arranged.

SPICES VARIETIES

Kudampuli Camboge – *Carcinia cambogia*

“Kudampuli”is a large sized, long-lived, evergreen tall tree with a spreading crown. The pulp is edible, used as souring agent for squashes, chutney in beverages, and in general cooking. Ideal soil type is deep loamy soil with adequate moisture. This condition prevails in Kanyakumari district. Flowering is seen in September-October. Fully riped fruits are come to harvest in March-April. The yield ranges from 120-150 kg of pods/ear/tree. Acidity of the fruits is lesser than *Tamarindus indicus*. The fruits are similar to orange, which has flesh in many numbers of edible juicy flakes.

Manjal - Turmeric

One such crop is turmeric, which is completely cultivated in organic method. Local variety is used with less seed rate of about 8 quintals /ac and it yields about

5000 q/ac that is the special features of this variety. The apt season for growing the crop is Puratasi, Iypasi (October-November). This variety has the advantage of withstanding in more rainfall and mitigating from root rot / wilt diseases. It needs about 20 loads/ac of farmyard manure for best result in field.

During Thai (January), the leaves will turn yellow which is the indication for time of harvest. Thereafter, during Vaikasi (May) it is harvested. An implement called 'Manjal kothi' does harvest the manjal. It is similar to hand hoe, but the handle is comparatively longer. Leaves are cut by using sickle and then, this manjal kothi is used and rhizomes are dug out. Harvested rhizomes are dried in shade for 2-3 days and stored. Such post harvest practice enhances the quality of the rhizome. This field is interrupted with onion during chithirai (May). Onion crop is grown in the furrows, and turmeric on ridges. If onion is inter cropped, then turmeric is harvested during Masi (February).

Nattu Milagu- Pepper

Nattu milagu is small when compared to normal milagu but the spiciness is more than the later. Though the yield is less, its quality is good. It is cultivated through twines in Kolli hills of Namakkal district. The twines are cut into small pieces and planted near the support (preferably a tree). Generally, it sets flower during may-June (Vaigasi-Aadi season) and starts bearing during March. Harvesting is done by plucking the berry bunches. From each tree, nearly 10 kg of pepper is obtained.

Post Harvest Process

It is the main aspect with reference to pepper. Two types of pepper are prepared after processing i.e.

i) White pepper
ii) Black pepper.

i) White Pepper

The whole berry is soaked in water for one day completely and the outer skin is peeled off carefully. After that, it is sun dried and the resultant product is called white pepper. This fetches more market prices. However, it consumes more labour and time, which are the limitation.

ii) Black Pepper

Berries are harvested and it is stamped by foot. Then they are boiled in hot water so that the black colour is obtained. In case of black pepper, the intensity of colour and the size determines the market price and care is given in selection of berries, colour processing, and while boiling. 1 kg of black pepper costs around Rs.80-100/-

Jathikaai- Nutmeg *Myristica fragrans*

It is being cultivated in kanyakumari district. Seeds and aril parts are used. It cures asthma, catarrah, diarrhoea and Insomnia.

CROP MANAGEMENT PRACTICES

S.No	Particulars
1	**Mixed Cropping (Onion+ Sorghum+ Black Gram)** in C.R.Palayam, Manachanallur taluk of Trichy district. It reduces sloppy length and consequent erosion.
2	**Kalappu Payir (Ragi, lab lab, Caster and Red gram)** in Sambanthakottai, Santhanapalli Post, Thenkanikottai taluk of Krishnagiri district. It will fixes the atmospheric nitrogen to the soil and also it will increase the soil fertility.
3	**Kalappu Payir (Black gram + Tapioca + Paddy)** in Malavilai, Velicharanai Post, Agastheeswaran taluk of Kanyakumari district. Nutrient application is less because of nitrogen fixation.
4	**Chillies +Castor +Agathi** in Kothali village of Thoothukudi district. Generate extra income.
5	**Moodu Payir (Guava + *Stylosanthus hamata* & Jack + *Stylosanthus hamata*)** in Kalluvali, Pallithambam (via), Sivagangai taluk of Sivagangai district. Suppresses the weed growth in dry land and ensure sustainable farming.
6	**Oodu payir (Pine Apple + Rubber)** in Malavilai, Velicharanai Post, Agastheeswaran taluk of Kanyakumari district. Abundant sunsine hours permit for pine apple to grow well throughout the year.
7	**Oodu Payir (Coconut + Banana)** in Vellchivilai village, Villukuri of Kanyakumari district. Optimum solar radiation provided by banana and recycling of nutrients.
8	**Rubber + Banana** in Malavilai, Velicharanai Post, Agastheeswaran taluk of Kanyakumari district. Competition of solar radiation and soil factors are less influenced.
9	**Black gram + Amaranthus** in Vathiyarvilai, Nattalam (South), Nattalam- Kallukulam taluk of Kanyakumari district. Effective usage of land and nutrient and also income generation.
10	**Banana+ Chilly** in Vathiyarvilai, Nattalam (South), Nattalam- Kallukulam taluk of Kanyakumari district Effective usage of irrigation water.
11	**Betel vine + Jaffana Moringa** in Nenmeni, Arianenthal, Paramakkudi union of Ramanadhapuram district. Which gives anchorage to the betel vines.
12	**Wild Brinjal + Snake gourd (Kottai Pudalai)** in Devoor village of Salem district. Brinjal plant stalk used as a roofing material for snake gourd.
13	**Betel vine + Agathi** in Kunnipalayam village of Namakkal district. Agathi plants are used for climbing purpose and also give favorable condition to the betel vine crop.

14 **Amla + Ramnad Jasmine** in Minnakkal village of Namakkal district. Effective utilization of land.

15 **Cotton +Red gram +Castor** in Kalathiyur village of Vellore district. Castor act as a trap crop and also generates additional income.

16 **Groundnut + Red gram** in Velankkadu village of Vellore district. Extra income generation.

17 **Groundnut + Black gram** in Volakkure village of Villuppuram district. Generate extra income.

18 **Groundnut + Amaranthus** in Kaligium village of Erode district. Greens used for house hold purposes.

19 **Border cropping (Sorghum + Kuthiraivali + Rice)** in Nenmeni, Arianenthal, and Paramakkudi union of Ramanadhapuram district. To reduces the hot air circulation.

20 **Mithi Ulunthu (Rice Fallow Pulses)** in Odugampatti village of Pudukottai district. Effectively utilize the fertilizer, water and land.

21 **Kathiriyil Ottu kattuthal (Indigenous grafting technique for Vegetable)** in Kalluvali, Pallithambam (via), Sivagangai taluk of Sivagangai district. It will increase the vegetable yield.

22 **Kasevu Neer Kuttai (Percolation Pond)** in Vellapallam village of Nagapattinam district. For effective utilization of farm pond.

23 **Snake gourd Sowing** in Vellapallam village of Nagapattinam district. Incorporation of pungam, the plant will emerge quickly and grow sturdy nature.

24 **Paddy Straw bunches** in Vellapallam village of Nagapattinam district. To avoid of soil erosion and up rooting of soils.

25 **Muthalai Poondu/Agayathamarai** in Agaraperunthottam village, Sirkazhi taluk of Nagapattinam district. Dried plant materials can be incorporated to the saline and sodic soil for reclamation.

26 **Pitcher Pot Irrigation** in Odugampatti village of Pudukottai district. It is used to irrigate the crops like Mango and Sapota.

27 **Stubble Mulching** in Odugampatti village of Pudukottai district. Effective moisture and soil conservation

28 **Kodai Uzhavu (Summer Ploughing)** in Manachanallur taluk of Trichy district. It increases the water capacity of the soil

29 **Kidai Adaithal (Sheep penning)** in C.R.Palayam village of Trichy district. Sheep manure enriches the soil fertility

30 **Padal izhuthal/ Mull izhuthal (Leveling)** in C.R.Palayam village of Trichy district. To facilitate germination

31 **Bench terracing** in C.R.Palayam village, Manachanallur taluk of Trichy district : It reduces the slope length and consequently erosion

32 **Puzhuthikal nursery (Dry land paddy Nursery)** in C.R.Palayam village, Manachanallur taluk of Trichy district. Seeds will emerge out quickly

33 **Contour Trenching** in Bogasandiram , Irudukottai post,Thenkanikottai taluk of Krishnagiri district : Water shed management practices

34 **Panai Neer Pasanam (Irrigation Method)** in Irudhu kottai post, Thenkanikottai taluk of krishnagiri district. The plants are getting water uniformly

35 **Eatram (Water lifting device)** in Sambanthakottai, Santhanapalli post, Thenkanikottai taluk of Krishnsgiri district. Uplift water from well without electricity

36 **Varappu Poosuthal (Bund Plastering)** in Iyerpalayam village, Mettur post, Veppanthattai taluk of Perambalur district. It prevents water over flow

37 **Pasunthal Uram (Green leaf manuring)** in Iyerpalayam village, Mettur post, Veppanthattai taluk of Perambalur district. Fix the atmospheric Nitrogen to the soil

38 **Natural Fruit Ripening Structure** in Iyerpalayam village, Mettur post, Veppanthattai taluk of Perambalur district. Duration for fruit ripening is 2-3 hours.

39 **Natural Dying in Cotton**

a. To get blackish red kapas in Malavilai, Vellicharanai post, Arasamoodu junction, Kulasekaram taluk of Kanyakumari district. Ingeridient : Barley, Gingelly, Turmeric, Vilasinipattai / Thamadha : To change the color of the kapas.

b. To get sky blue colour kapas in Malavilai, Vellicharanai post, Arasamoodu junction, Kulasekaram taluk of Kanyakumari district. Ingeridient : Manjatti, Gingelly, Barley, Coconut, Adapathiyam leaf, Manosilsi : To get blue color kapas.

40 **Covering Pineapple with its slips** in Marikonam, Arasumoodu junction, Kulasekaram taluk of Kanyakumari district. Highly effective to control lodging, the matured fruits create ripening and fragrant smell.

41 **Kanchi Kambu (Propping Pole)** in Marikonam, Arasumoodu junction, Kulasekaram taluk of Kanyakumari district. Highly effective to control lodging of Banana.

42 **Saalu (Irrigation Chennels)** in Vathiyaar Vilaia, Nattalam (South), Kalkulam taluk of Kanyakumari district. Water reaches the plant roots directly.

43 **Chev vaazhai- Koodai podal (Bunch covering)** in Erumbukadu, Nagarcoil of Kanyakumari district.Avoid scorching sun light falling on the fruits.

44 **Pakku Mattai Nar (Areca nut outer husk)** in Erumbukadu, Nagarcoil of Kanyakumari district. Used for moisture retention.

45 **Rubber Moodu Payir (Rubber cover cropping)** in Kaliyal village of Vilvancode taluk of Kanyakumari district. Loss of moisture and erosion by wind and water.

46 **Mazhai Neer Semippu (Rain water harvesting)** in Kaliyal village of Vilvancode taluk of Kanyakumari district. It protects the crops from drought during non rainy days.

47 **Brinjal raised in coconut ring basin** in Vettuvenni, Marthandam taluk of Kanyakumari district. Seeds are sown every coconut ring basin, arrrange the coconut husk in every brinjal seedlings.

48 **Paddy Manual Threshing** in Thiruvalluvapuram, Kaliyapetai village, Utheramerur taluk of Kanchipuram district. To avoid wastage and paddy straw damages.

49 **Alaiathi** in Thandarai, Thirukazhlukunram taluk of Kanchipuram district. Used to prevent the tsunami.

50 **Thokai Kattuthal (Shellter belt)** in kunnipalayam village of Namakkal district. To control high wind velocity, temperature and it will give cool atmosphere to the betel vine field.

51 **MaraKilai Panthal (Tree branch panthal)** in Vadavathur village of Namakkal district. For all gourds easy to stage

52 **Banana trees tying with ropes / Plastics tapes** in Kurumanthur village of Erode district. To avoid damage by wild wind.

53 **Wood / Fly ash Treatment in Banana Bunches** The new traditional technology is mixing house hold wood ash with cow urine.

Kalappu Payir - Mixed Cropping (Onion+ Sorghum+ Black Gram)

Mixed cropping is nothing but growing or raising two or more crops in a same piece of land, in a same season .Onion, Sorghum and Black gram are raised in a same piece of land as mixed crop. In this, onion is the main crop; Sorghum is the border crop, which is raised along the sides of the bunds. In between onion and sorghum black gram is raised. Black gram is a leguminous crop which fixes atmospheric nitrogen into the soil. It increases the soil fertility level and gives additional nutrients to the associated crops. So, excess dosage of nitrogenous fertilizer can be avoided, which needs extra investment in cost of cultivation. Nitrogenous fertilizer invites pests; it will also be avoided, due to lesser nitrogen fertilizer application.

Sorghum grows upto the height of 5-6 feet. It prevents heavy wind, especially northern wind which carries lot of disease causing micro organisms. So, main crop (Onion) will be safeguarded from diseases. It is used as cattle feed and it acts as live fencing to avoid cattle grazing. Sorghum stalks are used for preparation of "Pattarai"- *in situ* onion storage structure. Aadi (July-August) is the suitable season for raising these crops. Seed rate for onion is 3-4 kg/ac; it yields about 250-300kg/ac.

Advantage of Mixed Cropping

- Better use of growth resources including nutrients and water
- Suppression of weeds.
- Reduced pests and diseases.
- Yield stability.
- Ecological stability – i.e. improvement of soil health and agro eco system.

General Profile of the ITK Holder

Thiru Natarajan, C.R.Palayam, Manachanallur (TK), Trichy District

Kalappu Payir - Mixed cropping (Ragi + lab lab + Caster + Red gram)

Mixed cropping is a common practice followed by the farmers of Thenkanikottai Taluk in Krishnagiri district of Tamil Nadu. It is the practice of raising more than one crop in the same piece of land simultaneously in Rainfed farming systems. Crops grown in mixed cropping system are ragi, lablab, castor, red gram. Ragi is a main crop rests are sown in one row, as mixed up with the spacing of 1m. Seeds used for mixed cropping are traditional/local variety. In which red gram (Mara thuvarai) is a perennial crop. After the harvest of ragi rest of the crops are allowed to grow in the same field itself. Mara thuvarai will give yield upto 3 years. In this mixed cropping, lab lab and red gram are other leguminous crop which fixes the atmospheric nitrogen in to the soil. It increases the soil fertility level.

Details of the local variety are given below

S.No	Crop	Season	Seed rate kg/ac	Yield kg/ac
1.	Ragi	June-July	2 kg/ac(Transplanted crop) 20 kg/ac (Direct sown)	1000 kg/ac 1500 kg/ac
2.	Lablab	June-July	8 kg/ac(Inter crop) 16 kg /ac (Main crop)	400 kg/ac 600 kg/ac
3.	Red gram	June-July	6 kg/ac	200 kg/ac
4.	Castor	June-July	½ kg/ac	100 kg/ac

Advantages

- The available soil nutrients in different layers of the soil are tapped evenly by the different crops grown at the same time by sowing a mixed crop of cereal and legume.
- The cereal crop gets the advantage of atmospheric nitrogen fixed by the associated legume crop. Due to favourable association of crops, the total yield from a unit land area is increased.
- There is decrease in cultivation expenses by raising two or more crops simultaneously. With the same land area the cultivator gets different kind of produces in the same season which is advantageous from the point of view of total income per unit area of land.
- The fodder value in terms of quality and quantity will be higher by raising cereal-legume mixtures.

- In mixed cropping the inter row space of crops, especially in widely spaced crops, are better utilized.
- When the crops of different durations are mixed, the advantages of both the crops are derived in years of well distributed rainfall or in the event of failure of one of the monsoons at least one crop will give income avoiding total loss.

In pastures, a mixture of grass and legume provides balanced forage for animals and at the same time soil fertility is enhanced.

Kalappu Payir (Black Gram + Tapioca + Paddy)

Mixed cropping is the method of growing more than two crops either simultaneously or in sequence per unit area (same piece of land) and per unit time (one year). The intensification of cropping may be over space or time dimension or in both dimensions.

Method

Black gram is a pulse crop with duration of 90days, which is sown after 10days of tapioca planting. This combination is raised in main field whereas in 'saalu'-irrigaton channels paddy is cultivated. Paddy is with duration of 115days and so it is taken up sowing at last. Black gram is harvested by just cutting the terminal portion of the plant. Paddy gets benefited from the preceding crop black gram.

Benefits

- Legume effect- legumes fixes atmospheric nitrogen.
- The succeeding crop paddy gets benefited by nitrogen fixation of legumes.
- It has been observed that 15-20% nitrogen can be saved in the nitrogen budget of the succeeding crop.
- Carry over effect of nutrients/manures to the succeeding crop.
- Nutrient recycling-crop residue of previous crop viz., paddy straw, and stubbles can be incorporated in the soil and can be used for succeeding crop as organic manure.
- Nutrient pumping from deeper layer. This is based on thc principlc that uptake pattern of nutrients will depend upon the rooting pattern of individual crops.

Kalappu Payir - Mixed Cropping (Chillies + Castor + Agathi)

Chillies, castor and Agathi are raised in same piece of land as mixed crop. In this, chillies are the main crop. Castor is the border crop, which is raised along the side of the bunds. In between chillies and castor, Agathi is raised. Agathi is leguminous tree fodder which is a protein supplement to small ruminants and also used as green leaf manure .Agathi growth is limited upto 10 feet height for easy management. The castor crop is used for castor oil extraction and the castor leaves are used as green fodder to live stock.

Moodu Payir- Mulch Crop (Guava + *Stylosanthus hamata* & Jack + *Stylosanthus hamata*)

Normally the orchard method of crop cultivation often faces weed menace and it requires intercultural operations frequently. It will reduce the net income of farmer and renders him bankruptcy. It converts the farming unproductive and if it is dry land the damage is too high. To prevent those melodies and ensure sustainable farming Mr. Abraham has been cultivating *Stylosanthus hamata* along with guava and jack fruit trees.

Salient Features

- *Stylosanthus hamata* is a legume forage crop.
- It is highly palatable for goat and sheep.
- It often suppresses the weed growth in dry land as it is spreading in nature.
- The roots are deeper and prevent the emergence of all type of weeds.
- It requires less sunlight and grows well under fruit trees.
- It has deeper roots to conserve soil moisture and suitable for dry land.
- It is one of the components in mixed farming.
- Duration of the crop is 180-200 days.
- It increases milk yield in goat.

Oodu Payir (Pine Apple + Rubber) - Inter Cropping

Intercropping is the practice of growing of two or more crops simultaneously in the same field, intensification over time and space. Contour is an imaginary line connecting slopes of different elevation. The system followed for pineapple and rubber is row intercropping. Two crops are grown simultaneously in 2 distinct rows.

Method

At first, rubber seedlings are planted in June-July with pits of $1m^3$ and spacing of 3x2/5x5m. The slopy areas should be first cleaned and contours are made. Rubber starts yielding only after 7years. During the first 7years, pineapple gives yield every year. Pine apple is raised in rows first, with 90cm spacing between two rows or trenches and row to row spacing is 60cm with 30cm between plants. The season of planting is July-September. Duration of the crop is 18-24month. Yield ranges from 45-50tonnes/ha.

Advantages

Abundant sunshine hours permit pineapple to grow well throughout the year. Farmers get additional income; more over the land is effectively utilized. It also minimizes and controls soil erosion.

Address of the ITK Holder

Mr. Henry, Malavilai, Vellicharanai post, Agastheeswaram taluk, Kanyakumari district.

Oodu Payir (Coconut + Banana) - Inter Cropping

Banana is the intercrop planted in between the rows of main or base crop coconut with a view to obtain some extra intercrop yield without sacrificing the main or base crop yield.Coconut is planted in pits of $1m^3$ size, spacing of 8x8m between rows and 6m between plants. Coconut yields 80nuts/tree/year. Banana should be planted only after first five years of coconut. Because, the coconut comes to certain height after 5years only. Banana is raised in January-February or November-December, with duration of 12-15months. The yield range of banana is 40-60t/ha/year.

Advantages

- Higher returns from a unit area
- Optimum solar radiation provided for banana
- Nutrient recycling is provided for coconut by residues of banana.

Address of the ITK Holder

Mr. Dhas, Mano illam, Vellachi vilai, Villukuri (P.O)-629168, Ph : 04652-258064.

Rubber+ Banana – Inter Cropping

Rubber and banana are raised together in lowlands of Kanyakumari district in pechiparai and kuzhithurai blocks. Rubber is planted in June-July in pits of $1m^3$ size. Adopted spacing for rubber plantations is 5x5m. Banana is planted in January-February or during November-December. The spacing followed for banana is 2x2m with duration of 12-15months. Rubber starts to yield only after 7years.so; banana is planted till 6th year. Banana yields 40-60tonnes/ha/year. Here, competition for solar radiation and soil factors are less influenced. Because, rubber needs less solar radiation and lesser nutrients than banana till it starts yielding. After 6th year, rubber alone is grown and gives yield from 7th year onwards.

Black Gram + Amaranthus – Inter Cropping

Black gram is an important pulse crop sown in June- July and harvested in September-October after 90days. Amaranthus is also a short duration crop raised together with black gram. It usually comes well in drained loamy soils. Sowing of amaranthus can be done through out the year. Irrigation channels/'saalu' are left as such with out any crop. Black gram raised near to 'saalu' and amaranthus raised at the inner side of the field and harvested once in 3-4months. A single crop can be continued to remain till 5years and more. Yield ranges from 5tonnes/ha/year.

Banana + Chillies – Inter Cropping

Chilli is the intercrop planted in between the rows of main or base crop banana with a view to obtain some extra intercrop yield without sacrificing the main or base crop yield.Banana is planted in January-February or during November-December. The spacing followed for banana is 2x2m with duration of 12-15months.Banana yields 40-60 tonnes/ha/year. In the irrigation channels/saalu water gets stagnated. To overcome this, chilli is sown as a border cum inter crop at sides of irrigation channels. Chilli is usually sown in January-February, June -July. The yield ranges from 2-3tonnes/ha.

Betel Vine + Jaffna Moringa – Inter Cropping

The traditional "betel vine" variety is grown along with "Velimurungai" which gives anchorage to betel vines. It is a suitable crop for coastal sandy soils. In the border areas banana trees are grown and it gives higher crop intensity leading to additional income. There are 150 banana trees surrounding around the border area of the 2 acres land and most of them are border crops in betel vine field. The field is irrigated by 3 wells and the water is available at 10 feet depth.

The traditional 40 years old Jaffna Moringa has been grown in Thangachimadam areas of Ramnad district and the cutting was brought from Jaffna, Sri lanka during the cargo service between India and Sri lanka. It comes very well in coastal sandy soils. The 40 years old tree gives 1500 moringa sticks daily between March-June. They apply only organic manures such as goat manure to the tree. It is not usually affected by pests and diseases. The length of individual fruit is up to 12 feet and it fetches Rs.3-5/- based on demand. The rainy season of October, November and December is the off-season for moringa tree.

Wild Brinjal + Snake Gourd - Inter Cropping

Wild brinjal is traditionally cultivated in Devoor village at Salem district. Cultivation starts during the month of August and January and the duration of this crop is 6-7 months. This variety is mostly preferred in Erode market and has good market value while compared to other varieties. This variety possesses high pest and disease resistant, so that the farmers used to prefer to cultivate this variety. Cow dung, goat manure and poultry waste are used as manures. Some times plant leaves are used as manure in filed it self. Plant height is 4-5 feet; high rainfall is not suitable for this crop because the root rot incidence is more in that time. After first harvest the plants are left with 1 foot height from the ground level, so that other harvest can be done. Normally the yield obtain is up to 30-40 bags in 3 months. The rate fixed in the market is Rs.15 per kg. So that, the farmer fetches net profit of upto Rs. 30,000- 50,000/.-. Inter cropping is also possible in this crop, at the time of harvest "kottai Pudalai' seeds are sowed. The pudalai plant uses the wild brinjal plant as a support for its growth. From the intercrop upto Rs.20, 000 additional incomes are gained by the farmers.

The nursery requirement for one acre of land is 1.5 cents. The required seed rate is 200-300 grams. The nursery duration is 2 months. Seeds are collected from well ripened, pest and disease free fruits. Seeds are coated with ash and Dithane powder to protect the seeds from storage pests. It can be stored upto 2-3 years.

Vellaikodi (Betel Vine) + Agathi (*Sesbania grandifora*) – Inter Cropping

It is followed in kunnipalayam village at Namakkal district.

Varieties : vellaikodi

Season : January- February

Sets rate : 30000 sets/ acre.

Soil and Climate

Well drained fertile clay loams are suitable. It does not tolerate saline and alkaline conditions. Betel vines require a cool shade considerable humidity and regular supply of moisture in the soil.

Method of Propagation

The vines are propagated by terminal stem cutting or sets about 30-45cm long. Sets obtains from the top portions of the vines are easy to root and hence best for planting. Sets with vigorous apical buds and nodal adventitious roots are selected and planted at the base of the live supports, which are to be planted 4-5 months earlier.

Preparation of Field

The field is prepared to a fine tilth and beds of 2m wide are formed to a convenient length. Provide drainage trenches of 0.5m width 0.5m depth in between two adjoining beds. Plant the seeds of the live supports. *Sesbania grandiflora* (agathi) in long rows. About 300 banana suckers are planted at the edges of the beds which are used, for tying the vines on the live support and for packing the betel leaf. When the agathi support grows to a height of about 4cm they are topped. The crop is planted in two rows in beds of 180 cm width on agathi plants with spacing 45cm between plants in the row.

Irrigation

Irrigate the field immediately after planting and afterwards once in a week.

After Cultivation

Training of the Live Standards

Before the establishment of vines the side branches of agathi trees upto a height of 2cm are removed for early creeping of the vines.

Training of the Vines

Training is done by fixing the vine at intervals of 15 to 20cm along the standards loosely with the help of banana fibre training is done every 12-20 days depending upon the growth of vines.

Manuring

Apply 75 kg nitrogen 40kg phosphate and 10kg of potash in three split doses first at 15 days after lifting the vines and second and third dose at 40-45 days intervals.

Harvest

It depends up on the growth of the vines and market condition. Once harvesting starts it continues almost every day. Hundred leaves is one "kowli" hundred "kowli" is one bundle (sumai) one bundle rate in market is Rs. 1000/- .

Yield

30 to 35 lakhs leaves/ac/year.

Amla + Ramnad Jasmine – Inter Cropping

Amla : (Phyllanthus Emlica), Euphorbiaceae+Ramnad

Jasmine : *Jasminum sambac* Oleaceae

District : Namakkal

Varities : Amla NA-7, jasmine Ramnad jasmine,

Spacing : 20x20ft

Season : Amla june-july, jasmine june- july

Soil and Climate

Amla is a subtropical plant and prefers dry climate. It can be grown in variable soil condition. It is a hardy plant and tolerates salinity and alkalinity condition. Well drained fertile soil is suitable for jasmine.

Irrigation

Irrigate the plants initially for establishment. No irrigation is required during rainy and winter season.

Manuring

FYM 10kg /plant, and fertilizers 200 :500 :200 g NPK /plants. To be given immediately after pruning.

Inter Crop

Ramnad jasmine is cultivated as intercrop with Amla (Na-7). Amla planted in 20X20 feet spacing. Jasmine generates additional income and also land utilization effectively. The jasmine plants are purchased from nursery, one plant cost is Rs.1.50/-. In one acre 400 plants are planted.

Season of Flowering and Harvesting

Season of flowering commences on March- April. Fully developed, unopened flower buds should be picked in the early morning and sale for market. Extra care should be taken for jasmine plants. Only farmyard manure is applied at the time of intercultural operation. Neem oil spraying is done for controlling pest and disease. Income from amla trees is Rs.50, 000 to 60,000/-. Income from jasmine flower is Rs. 5,000-6,000/- per year. Total investment for these two crops is Rs.8, 000 to 10,000/- and net profit is Rs. 50,000/-.

Cotton + Red Gram + Castor – Inter Cropping

It is followed in Kalathiyur at Vellore district. In a cotton field red gram and castor crops are cultivated as an intercrop, during the month of July. Cotton seeds are sown at the same time. Red gram seeds are sown in line 12ft spacing of 12ft interval. Castor seeds are sown as a border crop around the cotton field. Seed rate required for red gram is 2kg/ac. Castor seeds ½ kg /ac. Cotton is harvested with in a period of 5 months. Red gram is harvested in the 6 months. Duration of the castor crop is 10 month. Yield from red gram is 400kg/ac. The cost of 1kg red gram is Rs. 50/-. Yield from castor seeds 120-150kg/ac.

Ground Nut + Red Gram – Inter Cropping

In ground nut field farmers normally grow red gram as an intercrop. It will give extra income to the farmers. It will also be used for household purpose. In one acre of land 2 kg of red gram seeds are sown. Ground nut is harvested on the120[th] day. Red gram is harvested in180th day. The yield from red gram is 400-450 kg/ac. The price for One kg of red gram is Rs 50/-. This system is traditionally followed in northern district of Tamil Nadu.

Groundnut + Black Gram – Inter Cropping

Intercropping of black gram in groundnut is practiced in Viliupuram Dt. Black gram was sowing at an interval of 2-3 m in between ground nut. It is an additional income before harvest of the groundnut

Groundnut + Amaranthus – Inter Cropping

Ground nut cum amaranths inter cropping is practiced in Erode district. Amaranthus is harvested within one month. Ground nut harvested at end of three months period. Amaranths crops will give additional income to needy farmers by selling amaranthus green leaves daily in the market.

Border Cropping (Sorghum + *Echinocloa calona* + Rice)

It is the traditional red rice variety growing in Paramakudi taluk of Ramnad district. The rice grain weight of kallurandaikaar is slightly more than kuliparichaan / varappukodainchan varieties. It is the drought resistant and tall variety. The outer skin is slightly dark. The inner grain is bold and round. The rice gruel water of the variety is good in taste. The seed rate of Kalurundaikar variety of an acre is 45 kg for direct sowing. This variety prevents hunger for more time than fine rice. The yield is 1500-1800kg per acre. Paddy straw yield is approximtely 3 tons per acre. In some field **rice + kudiraivali** (in border areas) system is followed. Kudiraivali yield is 600 kg per acre and it fetches higher price in market. (Rs.6 per kg.). The rice variety is not generally affected by pest and diseases. The seeds are stored by farmers themselves from the previous harvest. In some fields 3 types of sorghum are grown as border crop. They are red, white and black. These sorghum varieties differ in colour. They are grown for cattle feed purpose. The traditional sorghum varieties grow very tall in nature. The seeds are stored by farmers from the previous harvest.

Mithi Ulunthu (Rice Fallow Pulses)

10 days before harvesting the paddy, black gram seeds are sown in field (when field is in waxy condition) while paddy harvesting the black gram seeds struck in the stalks will be well stamped and pressed in the field. This induces the germination of those seeds. This will effectively utilize the fertilizer, water and land. Except rat control mechanism, all other technologies are still under use.

Kathiriyil Ottu Kattuthal (Indigenous Grafting Technique in Brinjal)

Mr. Abraham is technical way in nature and wants to get high yield with less number of crops and space. Thus, he is trying to harvest brinjal in a tree. Brinjal

Coconut + Banana

Rubber + Banana

Betel Vine + Agathi

Amla + Ramnad Jasmine

Castor + Red Gram + Cotton

Ground Nut + Red Gram

Fig. 14. Crop Management Practices

(*Solanum melangenum*) is a widely used crop for edible purpose and has high market value compared to "sundai". Based on the experience he understand that both crops are belonging to same genus and come to a conclusion to generate brinjal tree by crossing with "sundai".

Procedure

The technique adopted is epicotyl grafting. The rootstock and scion are "sundai" and brinjal respectively. The stem of required young tender brinjal variety is to be collected and grafted with "sundai' tree and tied with a cloth. The field is to be irrigated regularly. After a period of 15 days, the "sundai" tree bears brinjal branches and produce brinjal flowers.

Advantages

- This will increase yield 300% than the individual brinjal plant.
- As "sundai" is disease free in nature the brinjal crop generated from "sundai" plant is also disease free.
- It requires less water and can survive in black, red and saline soils.
- It requires less fertilizer compared to brinjal plant.
- Intercropping is possible.

Name and address

Mr. Abraham, Kalluvali, Pallithambam via. Sivagangai taluk, Sivagangai dt, Ph : 204 508

Kasivu Neer Kuttai (Percolation Pond)

In his farming different types of irrigation method is followed i.e. since it is a coastal village, sandy soil allows easy excavation of farm pond. Thus with a width of 20 feet (approximately) and depth of 10-15 feet a farm pond is excavated, where water oozes out even at 10 feet. For effective utilization such farm ponds are dug in centre of the field. Since this coastal garden land has no other source of irrigation, these farm ponds are the only possible source.This kind of farm pond can be seen in almost every farm in the coastal villages especially Vettaikaraniruppu, Vellapallam, Nalluvethapathy, Vizhuthamavadi, Periyakuthagai, Sempodai of Nagapattinam district.

Snake Gourd Sowing

Pit size of 1.5 feet depth and 2 feet width is dug in the pit ,a bunch of fresh Pungam (*Pongamia pinnata*) leaves are incorporated ,this will acts as green manure as well as preventing termites and ant damage. Due to the incorporation of pungam, the plant will emerge quickly and grow sturdy. After that, pungam is covered with thin layer of soil, 2.5 kg well decomposed FYM is applied over it and then it is covered with 5cm of soil. The layer of soil and FYM should be mixed well and pit is rilled at centre, in which 3 seeds are sown. Sharp edge of the seed is dibbled inside the pit and then covered with soil.

Finally the entire pit is covered with leaf litters in order to avoid excess evaporation which in turn avoids damage due to rodents. Immediately after sowing, the pit should be watered. The seedlings will emerge out seveen Days after sowing. After three weeks, the pandal is erected using Palmyra fibre/coconut coir pith rope/ Karung kodi-a twiner.

Karung kodi (Black vine) is a twine found in all coastal villages. Matured vine is used for preparation of pandal. Instead of other ropes, this is used, due to its hardy nature, after drying.From 50-60 days after sowing, it starts bearing. Naturally snake gourd grows in an irregular manner which is least preferred in the market. In order to avoid this economic loss, coconut pith is tied at the posterior end of the fruit. There are two types of local snake gourd varieties that are prevailing. One is lengthy snake gourd (Neetu pudal),which grows upto a length of 6 feet and it weighs about 2kg.Second is short and stout snake gourd (Pani pudal) which weighs about ½ kg-1kg and it grows upto 1-1.5 feet length. Suitable season for sowing snake gourd is July-August. Existence of the local variety is for 75 years.

Paddy Straw Bunches

Paddy straw bunch is kept in one corner of the nursery to avoid soil erosion and up rooting of plants.While irrigating vegetable nursery, care should be taken so that soil is not eroded, which may up root the seedlings. To avoid such damage, a small pit is dug in the corner of the plot and filled with sufficient paddy straw. Thus while pouring water upon it seedlings are watered but dose not damage it.

Muthalai Poondu/Agasathamarai - *Pistia Stratiotes*

Agasathamarai is a common weed found in all water bodies. In the month of January, whole plants are removed from the water bodies and allowed for drying. The fully dried plant materials are incorporated to the saline and sodic soil for reclamation. It can be used as Green Leaf Manure, in wet land condition. Before puddling, whole green plants are harvested from the water bodies and used as GLM. It has to be puddled well in the wet land and allowed for decomposition. Otherwise it will emerge as weed in wet land field. After incorporation of Muthalai poondu, discoloration of water i.e. red colour indicates that the salt is dissolved and removed from soil. No specific dosage in agasathamarai incorporation. Dosage depends on the quantity of availability of the weed.

Pitcher Pot Irrigation

Pitcher pot irrigation is a method of drip irrigation in which mud pot is used for irrigation. Capacity of pot is 10 litre which costs about Rs10/.Pitcher pot irrigation is a suitable method of irrigation for the area where water availability is less. Mostly, it is used to irrigate the tree crops like Mango; Sapota.One foot away from tree's trunk a pit with size of 1 feet depth and 2 feet width is dug. Small size hole is made at the bottom of the pot. Pot is buried in the pit, in such a way that it's rim is slightly above ground level. Water is poured into the buried pot and the pot is closed with mud lid, one pot of water is enough to irrigate a tree for about 4-5 days.

In this method water droplet is applied directly to the root zone of the crop. Water wastage will be avoided and water loss due to evaporation will be minimized. So, water can be effectively saved through this method of irrigation.

Stubble Mulching

Tillage with leaves to the surface of the soil, cloddy should be mulched with crop residues for an effective moisture and soil conservation. Properly mulched soil surface protects soil particles from runoff or drift. Thus erosion does not take place. The traditional methods of tillage developed for soil (Water) moisture conservation. Locally available materials such as straw and stalks could be used for mulching. Soil loss and runoff from cropped area can be reduced. Mulches improves infiltration rate, increasing termite and biological activity. Mulches have been used to reduce evaporation, increase infiltration, prevent the soil from blowing and washing away, control weeds, improve soil structure and to increase crops yields.

Mulching is a traditional agronomic practice, in which crop waste and leaf litter are used to cover the soil. Mulching is done under the tree ie; around the tree trunk. Through this method soil moisture is conserved effectively, evaporation loss is also minimized and checks the weed growth.

Kodai Uzhavu (Summer Ploughing)

Plough the field during summer and two feet height earthern bund is made around field which checks run off of the rain water.It increases water capacity of the soil. It destroys weeds and pupal case of the pests.

Kidai Adaithal (Sheep Penning)

During non cropped season or after harvest sheep penning is done with 100 sheep's in the field for about 4-5 days. Every day morning, the place of penning can be changed. So that, the entire field will be covered. At the time of penning sheep manure enriches the soil fertility. Wages for sheep penning is Rs. 80/night.Existence : .75 years.

Padal Izhuthal/ Mull Izhuthal (Levelling)

After sowing, proper levelling is highly required for germination. Hence, levelling is done by farmers using wooden plank of size 5 feet length and 1 feet width.The size of which varies from place to place. The wooden plank when pulled by farmer covers the seeds with soil. A rope is tied on the two sides of the plank and for increased efficiency on leveling : stone is kept on the plank. One member can level one acre a day. In the place of wooden planks, some farmers are using Prosopis (*Prosopis julifera*) branches for leveling and this is called as padal izhuthal or mull izhuthal. Even spade is used for leveling but it is labour intensive. Leveling gives even /proper covering of seed, which facilitates germination.

Bench Terracing

On steeply sloping and undulated land, intensive farming is possible only with bench terracing. It is usually practiced on slopes. Bench terracing consists of principally transforming relatively steep land into a series of level strips or platform across the slope of the land. It reduces the slope length and consequently erosion. The field is made into a series of benches by excavating the soil from upper part of the terrace and filling the lower part. A good soil depth is required to avoid exposure of unproductive soil during leveling.

Puzhuthikal Nursery (Dry Land Paddy Nursery)

Entire nursery for an acre is divided into 10 plots. 18 hours pre soaked seeds are sown in that nursery. (In this method, seeds are not kept in the dark room).After sowing, nursery is covered with crop waste. This is called as "Mootam Poduthal"(Mulching). Five days after sowing seeds will emerge out.

Contour Trenching (Watershed Management Practices)

Watershed is a natural hydrologic land unit. It is an area of internal drainage above a common point of outlet. Its size and shape are determined by surface topography. It is bound on all sides by a divide or ridge line.

Advantages

- Control the runoff which in turn influences the rate of soil loss, moisture availability to plants, sedimentation, ground water discharge, water quality etc.
- Useful to manage rain water falling within the unit for various purposes including agriculture without damage to the natural resources like soil.

Contour Trenching

The farmer establishes watershed structure i.e. contour trenching in his 5 acre area of the field. He is cultivating Mango, Castor, Sapota and Stylo grass in the field. Trenches are dug on contours at suitable vertical intervals depending on the slope of the land. The size and number of trench varies based on the spacing of the crop.

Pitcher Pot Irrigation
Stubble Mulching
Contour Trenching
Water Lifting Device
Green leaf Manuring
Pineapple Covering with Slip

Fig. 15. Crop Management Practices

Trenching reduces the velocity of runoff. The water collected in the trenches or pits percolated through the soil and improve soil moisture conditions. Number of these structures in a given land helps to improve the recharge of ground water and reduces the runoff.

Panai Neer Pasanam (Picture Pot Irrigation Method)

Locally available waste plastic bottles are used to irrigate the Sapota and Mango crops. A plastic bottle has the capacity of 3 litres of water is buried into the soil. . Fifteen days once the water is filled in to the bottle. At the bottom side of the bottle a small hole is made. The water is coming from that hole only. Due to this method, the water wastage is restricted and it reduces the evaporation loss also. The plants are getting the water uniformly throughout the day.

Eatram (Water Lifting Device)

Using the traditional water lifting device is called as Eatram. This is used to lift the water from well.

Components of Eatram

Eatram panai (Iron bucket)

Eatram panai made up of iron, its capacity is 10 litres. Two iron rings are fixed on two rims of the eatram panai in which the bamboo pole is tied with the help of rope.

Mun Kazhi

Dried bamboo poles are used as mun kazhi. This mun kazhi (Bamboo pole) is tied in the two rings of eatram panai. Length of mun kazhi will vary which depends on the water level/depth of water in the well.

Vasam (Lifting Bamboo has See Saw Movement)

Another bamboo pole/vasam is attached to the mun kazhi, its length is 8-10 feet. Rock stones are tied on the other end of the vasam which facilitates to lift the water up.

Supporting Pole

'Y' shaped banyan tree's branch is used as supporting pole for the entire structure (Eatram panai, Mun kazhi and vasam).

Stone Structure and Tamarind Poles

Masonry structure is built on 2 feet above the well with the help of rock stones and cement. Two tamarind poles are inserted in to that structure, which touches the rim of the well.

Operating Mechanism

A member can keep two legs on two tamarind poles and hold the mun kazhi by using hands. While pulling the mun kazhi simultaneously eatram panai will take the water from well and eatram panai with water will be uplifted and discharge the water automatically due to the weight of the rock stones on the vasam.

Most of the farmers in that hilly track are using eatram for irrigation still now. This type of traditional irrigation method can be used for Rice, Sorghum, Horse gram and Gingelly.

Varappu Poosuthal (Bund Plastering)

Raising the bund height, strengthening the bund and plastering the bunds are the common traditional practices followed by wet land farmers.

- Raising the height of the bund prevents water flow from rice field.
- Strengthening the bund and plastering with available clay in the rice field will check the water wastage due to seepage and leakage.

Pasunthal Uram (Green Leaf Manuring)

It is the application to land of green leaf, gathered from shrubs and trees' growing in waste lands. Green leaf is gathered from all sources by farmers during the rice planting season. Different kinds of shrubs growing on tank bunds, waste lands, field bunds, porombokes, etc are used. In addition, lopping from trees on the road side and forest area also gathered for use as green leaf manures. Common shrubs and trees useful for this purpose are, Neem, Teak, Pungam and Erukku leaves (Calotropis).

Advantages of Green Leaf Manuring

- Leguminous green leaf manure crops fix atmospheric nitrogen which enriches the land.
- When the green manure crop is ploughed in the soil, it decomposes easily, without leaving much of residue in the soil.
- It improves the structure of the soil. It acts as a binding material in light soils and increases their water holding capacity.
- Crumb formation is induced in clayey soils, leading to an improvement in the absorbtive capacity of the soil and percolation.
- The green manure added to the soil promotes the activity of soil micro-organisms, carbon dioxide is liberated, the chemical reactions in the soil are speeded up and more plant nutrient is made available.
- Organic acids produced during the decomposition of the green manures help to release phosphate and make them available for the use of the next crop.

Nutrient content of the green leaf manure Percentage of dry weight basis

	N	P	K
Neem	2.8	0.3	0.4
Calatropis	2.1	0.7	3.6
Pungam	3.2	0.3	1.3

Incorporation of chopped roots, stubble, green-stalks, tree leaves (neem, teak, pungam) etc., it increases the organic matter and bacterial population which in their turn to improve the soil structure.

Natural Fruit Ripening Structure

Normally, carbide stone is used in artificial chamber for fruit ripening. This is used for mango, sapota and jack. But it fetches poor market price and make the fruits into bitter in taste. To overcome these problems Mr. Abraham of Sivagangai has developed a natural fruit ripening structure by using a local carpenter.

Description of the Structure

The size of the structure is 1x1x1 m. It is made by using clay soil and covered by a wooden shield. The entire structure is smeared with cattle dung. The mini stove is attached with a hole tube so that to supply a smoke by firing farm waste. The capacity of the container is 15 kg.By burning wood ash and natural waste, the smoke is supplied into the hole. Thus, the smoke curing technique ripens the fruits.

Advantages

- The duration for fruit ripening is 2-3 hours.
- This method does not ripen the tender fruits and prevent poor quality, low price in market.
- The fruits ripened by this technique are sweet in taste and free from foul odour normally occurred during inorganic ripening method.
- The cost of preparation of the structure is less than Rs.200/-. But the stony method of fruit ripening requires Rs.200/- for single attempt.

It is a permanent structure and is not affected by termites since is smeared by cow dung.

Natural Dying in Cotton

a. To get blackish red kapas/fibre in cotton

Barley, Gingelly, Turmeric, Vilasinipattai / Thamadha are taken in equal amount (100g) and powdered. The powder is applied at the root portion of cotton and changes are noted. Kapas/fibre obtained is blackish red colour. Similarly, with the same ingredients solution is made and sprayed on plants. The kapas/fibre looks blackish red colour. It looks similar to fire.

b. To get sky blue colour kapas/fibre

Manjatti, gingelly, barley, coconut, adapathiyam leaf, manosilsi are taken in equal amount (50g), cow/goat's milk (250ml) are taken together and boiled well. Allow it to settle for some time, then apply the solution or spray in the field. After maturity, the kapas/fibre appears in sky blue colour.

Covering Pineapple with its Slips

Pineapple is a tropical and subtropical fruit found in humid hill slopes of kanyakumari district especially, Paraliar, Thadikarankonam. Crops grow about 1-1½ years and harvested in 12th month onwards. Slight fruit colour at base of the fruit indicates maturity. The matured fruits create fragrant smell which invites crows and other birds to taste it off. To avoid this, the fruits are covered with the slips around them with a rope. It also prevents the breakage of overweighed matured fruits from the plant. The operation is carried out after 1-1½months of fruit setting.

Benefits

- Prevents the major economic loss
- Quality of the fruit is maintained

Kanchi Kambu (Propping Pole for Banana)

Propping is a process by which the kanichu kambu –a pole is kept at an angle of 45^0 to banana trees. This is mainly to prevent the crop from lodging on heavy rains, winds also the over weight of the bunch. The pole is also used to make holes in the fields.

Merits

- Crop yield maintained
- Plant is made safer from natural disasters
- Highly effective to control lodging

Simple to follow in farmer's own field.

Saalu (Irrigation Channels)

In general irrigation is done through channel and water reaches the plant roots directly. In kanyakumari district, large trench like structures called 'saalu' are created. It is mainly followed for banana and paddy. Saalu is created between the rows with a breadth of 1m and depth of 1.5m. The main reason for this structure is soils of kanyakumari has proportionate amount of clay in paddy and to some extent in banana fields. To mitigate this problem and avoid water stagnation in the root zone of plants with regular supply of soil moisture 'saalu' is made.

Advantages

- Number of irrigations minimized
- Optimum amount of water is effectively used for irrigation

Border crops can also be raised on the edges of saalu.

Chev Vaazhai- Koodai Podal (Bunch Covering in Red Banana)

Dried coconut leaf are collected and made as coverage i.e., basket like structures to protect the scorching sunlight falling on the fruits. This is mainly practiced in chevaazhai and naendran varieties. Banana is a crop known for value addition and better price in the market. For this, the cover should be made and tied on the fruits 15-20 days before harvest. This results in uniform coloring and ripening through out the bunches.

Advantages

- Easily made with the farm available dried coconut leaves.
- Simple to follow.
- No chemical is involved.
- Better price in the market.

Pakku Mattai Nar (Arecanut Outer Husk) for Pot Plants

Dried outer husk of arecanut are collected and finely made in to small pieces. In home gardens and nurseries, these dried outer husks are spread on the mud pots kept for ornamental plants, flowers. Irrigate the pots, the arecanut husk absorbs moisture and retains with in it. So, the arecanut husk makes plants avail of moisture for longer time than normal one. This result in minimizing irrigation, effective utility of waste products, eco friendly and no making cost is incurred.

Rubber Moodu Payir (Rubber Cover Cropping) *Pueraria phaseoloides*

Cover crops are grown primarily to cover the soil and reduce the loss of moisture and erosion by wind and water. *Pueraria* provides 8 t/ha of nitrogen. They are sown after six months of rubber establishment. Flowering is noticed yearly once. After seed setting, fall on the ground and leaves cover the full field and automatically get decomposed.

Mazhai Neer Semippu (Rain Water Harvesting)

In rubber plantations, pits of 1m^3 size are dug seedlings are planted. The spacing between the plants is 3x2/5x5m. Planting is usually done in June- July. In between two rows of planting pits of 6'x2'x2' are dug, this is mainly to store the rainwater harvested in plantations. When heavy rains occur, water is stored in it and used at the time of scarcity. It protects the crop from drought during non rainy days. The dried leafs also fall in to the pit and gets decomposed over a period. They are collected and applied as manures.

Brinjal Raised in Coconut Ring Basin

Coconut is grown in larger tracts of Eetamozhi, Thengai patnam and Marthandam of kanyakumari district. Red sandy loam, laterite and alluvial soils are suitable to raise coconut. Planting season is June-July and December-January. Normally spacing of 25'x25' between rows and 20' between plants is adopted. Irrigation is done through ring basin method. Basins of 1m length are made around the coconut trees. In it, sow the seeds of brinjal at its periphery. Spacing between brinjal plants is 30cm. The seeds germinate in 2-3days. Once the seedlings are emerged, arrange the coconut husk in every brinjal seedlings and form a inner periphery with it. Once the coconut is irrigated, the coconut husk absorbs moisture and retains with it. The brinjal gets moisture from the coconut husks. Brinjal starts giving yield in 3months. Effective land utilization, minimize water loss, benefits the farmer with additional income.

Paddy Manual Threshing

Harvested paddy bunches are threshed manually i.e stricking on the stones. Two members are needed for manual threshing (one female labour and one male labour). Female labour is helping the male labour for taking harvested paddy bunches. Pidikayaru is used to hold the bunches for threshing.Coverage/ efficiency – 33 cents/day

Wages : Rs. 90 / day.

Advantages

- Germination percentage will be more
- No damage or less damage to grains
- Less number of broken grains

Aalayathi/ Mangrooves

- It is grown only in salt water condition.
- It is used for prevent the tsunami.
- This is planted mostly is tsunami affected area.
- The root length is more than 20 feet. Due to this habit, it is binded with soil very strongly.

Thokai Kattuthal (Shellter Belt)

Banana leaves are tied around the betel vine field up to 5-6 feet height. This method is useful to control the high wind velocity and temperature. It will give cool atmosphere to the betel vine field. More wind and temperature can be affecting the betel vine leaf. It helps maintaining the moisture condition inside the field. It is low cost method followed in Namakkal district.

Mara Kilai Panthal (Tree Branch Panthal)

It is followed in Vadavathure village at Namakkal district. Vathanarayanan and oonchai tree branches are used as a pandhal. It is easily available in field and is easy to stag. There is no need of labour to stag the tree branches. After the crop period, it is easy to dislocate from the field. Then we can go for further cultivation, it is useful for the growth of a variety of vegetables crops namely, lab lab and bitter gourds. This pandhal used by many as it is economic.

Banana Trees Tying with Ropes / Plastics Tapes

A plastic rope is tied 2/3 from the bottom of the banana tree to prevent the lodging and bunch damage while high flow of wind during the rainy season. The end of plastic ropes from 4 banana trees is tied with one feet height wooden plug which is firmly attached with earth. The cost of this indigenous technique is Rs. 4 per tree which includes the labour charges. The duration of this plastic ropes will exists for three year, which is after harvest of banana this plastic ropes can be used in subsequent seasons.

Wood / Fly Ash Treatment in Banana Bunches

The new traditional technology is mixing house hold wood ash with cow urine in a polythene bag and tying it in the bunches after cutting the last flower,In order to get the uniform growth of bunches and more no of fingers in banana. The combination to be tried is.

- Urea as such tied in the bunches.
- Wood ash + cow urine
- Wood ash + panchakavia
- Brick chamber, wood ash is more effective when it is tried in the field.

The combinations to get uniform growth of bunches have to be tried in all levels and means to find out the best combinations and then to spread the combination or technology to all the banana farmers. 250 ml of wood ash + cow urine and panchakavia is put as such in the polythene bag and tied in the bunches, the result proved good in all the banana varieties like red banana, rasthali and robusta. In earlier days panchkavia

Saalu Irrigation

Banana Bunch Covering

Rubber Cover Cropping

Shelter Belt in Banana

Tree Branch Panthal

Banana Tying with Ropes

Fig. 16. Crop Management Practices

was tried in banana and the result was good. One tea cup of panchakavia was put in the polythene bag for one tree and tied in the bunches after cutting the last flower.The technology have to be proved with wood ash, cow urine, panchakavia or other combinations with scientific background and scientific statistics. The indicators of good yield were.

- The fruit size has to be uniform and bigger in size
- 6 to 7 months the bunches have to be emerged
- The fruit must be greenish in colour
- The spacing between the tress has to be adequate.
- The sucker must be in a uniform size.

CROP PROTECTION MEASURES

S.No	Particulars
1	**Cow urine + Neem leaf extract** in Thiruvalluvapuram, Kaliyapettai, Uthiramerur taluk of Kanchipuram district. To control the insects.
2	**Poochi Viratti (Pest Repellent)** in Thiruvalluvarpuram, Kaliyapettai, Uthiramerur taluk of Kanchipuram district : Spray will control the pests and also promote the plant growth.
3	**Thennai Eriophid Sylanthi Kattupaduthuthal (Coconut Eriophid mite control)** in Marikonam, Arasamoodu junction, Kulasekaram of Kanyakumari district. Nethili Karuvaadu + Salt mixture spray to control the Eriophid mite.
4	**Illuppai seed and oil cake** in Vellapallam village of Nagapattinam district. Effectively control the pest and disease with locally available material.
5	**Pest control in Rice field**
	a. Cow Dung Application in Thiruvalluvarpuram, Kaliyapettai, Uthiramerur taluk of Kanchipuram district. It will act as plant growth promoters.
	b. Cow Dung Smoking in Thiruvalluvapuram, Kaliyapettai, Uthiramerur taluk of Kanchipuram district. It will control the pest and disease in Rice field.
	c. Kerosine + Chilly Powder Spray in Vallathodu viilage, Uthramerur taluk of Kanchipuram district. It will control pest and disease in Rice field.
	d. Cow Urine Tonic in Vallathodu viilage, Uthramerur taluk of Kanchipuram district. Controls pest and disease in rice field.
6	**Kunabam (Plant growth promoters)** in Madhoor, Uthiramerur taluk of Kanchipuram district. It will act as a plant growth promoter.
7	**Ginger + Garlic + Chillies + Cow urine** in Madhoor, Uthiramerur taluk of Kanchipuram district. It will control the boll worms in cotton field.
8	**Rice bran + Neem oil** in Madhoor, Uthiramerur taluk of Kanchipuram district. Used to control the Brown Plant Hopper in Rice.

9	**Pig Manure Smoking** in Kolathancherry, Uthiramerur taluk of Kanchipuram district. Effectively controlled in pest and disease in vegetable crops.
10	**Graveyard Ash Sprinkling** in Kolathancherry, Uthiramerur taluk of Kanchipuram district. It will control pest in vegetable crops.
11	**Payir Valarchi Ookki (Plant growth promoters)** in Kolathancherry, Uthiramerur taluk of Kanchipuram district. It will promote the plant growth.
12	**Paravai Thanki (Bird Perch)** in Agaraperunthottam, Sirkazhli taluk of Nagapattinam district. It is used to control the insects in rice field.
13	**Eli viratti (Rat Control)** in Agaraperunthottam, Sirkazhli taluk of Nagapattinam district. It will scare the rat away.
14	**Eli Kattuppaduthuthal (Rat Control)** in C. R.Palayam Manachanallure taluk of Trichy district : It is used to trap and kill the rat.
15	**Neekal Poduthal (Rat control in Rice Field)** in Thiruvalluvarpuram, Kaliyapettai, Uthiramerur taluk of kanchipuram district. Prevents rat entry into the field.
16	**Paravai Viratti (Bird scaring)** in C. R.Palayam Manachanallure taluk of Trichy district. Birds are scaring with the help of creating sound.
17	**Pasu komiam vithai nerthi (Seed treatment with cow urine)** in Odugampatti village of Pudukottai district. To control the storage pest moth.
18	**Manjal and Mor Vithai Nerthi (Seed treatment with turmeric and butter milk)** in Odugampatti village of Pudukottai district. To control the storage pest moth.
19	**Mann pzhlu uram (Vermi compost)** in Bogasandiramirudhu kottai post, Thenkanikottai taluk of Krishnagiri district. It reduces the pest and diseases incidence.
20	**Lanther Pori (Light trap)** in Vallathodu viilage, Uthramerur taluk of Kanchipuram district. Pests are attracted by the light and controlled.
21	**Poochi Pori (Pest Trap)** in Gettavadi (Post), Thalavadi (Via), Sathy taluk of Erode district. Used in pest trap particularly for Rhinoceros beetle and root grub.
22	**Mooligai Poochi Viratti (Herbal Pest Repellent)** in kalluvali, Pallithambam (via), Sivagangai taluk of sivagangai district. Used to control all types of pest in fruit trees.

Cow Urine + Neem Leaf Extract

Ingredients

- Neem – 10 kg
- Cow urine – 8-10 litres

Preparation

Neem leaves are pounded well by using ural and ullakai and extract is taken. This extract is mixed with 10 lit of cow urine. Neem leaves bunch is taken for applying the extract.

Dosage : 10 lit / ac

Time of application : 30-45 (DAT)

Existence : > 150 years]

General Profile of the ITK Holder

Th.Mohan, Thiruvalluvarpuram, Kaliyapettai, Uthramerur Tk, Kanchipuram Dt

Poochi Viratti (Pest Repellent)

Ingredients

Eruku-3Kg, Pungam-3kg,Umatham-3kg, Notchi-3kg , Kadal palai-3kg, Neem Seed-1/2kg Pungam Seed-1/2kg, Neem Cake-5kg, Cowurine-10lit, Water -30lit.

Preparation

Eruku, Pungam, Umatham, Notchi, Kadalpalai leaves are pounded well. Neem seeds, Pungam seeds are pounded separately. Fifty litre capacity mud pots are buried in the ground, because soil will give favourable microclimate for fermentation. Pounded materials are put into the pot. With this neem cake 5kg, cow urine10 litre and water 30 litres, are added. After that the mouth of the mud pot is tied with gunny bag or waste cloth. It is kept as such for about 10-15 days. Every day morning it has to be stirred well to ensure, well and quick fermentation. After 10-15 days, fermented solution is filtered and ready for spraying.

Dosage : 5lit/ac (500 ml/10 lit water).

Additional information : as stated by Mr. Marimuthu, Poochivirati spraying will control the pests and also promote the plant growth. Cow urine has nitrogenous compound which will promote the plant growth.

Coconut Eriophid Mite Kattupaduthuthal (Eriophid Mite Control)

Nethili karuvaadu Dry fish (*Stolephorus indicus*) is dried and made as powder, mixed with salt in the ratio of 1 :1 and applied as such or spray (250gm/litre).

Illupai-Seeds and Oil Cake

Illupai (*Madhuca latifolia*) belonging to Sapotaceae family. Nut of Illupai has various uses. Thiru.Sengutuvan who is residing in Vellapallam village in Keevalur taluk of Nagai district explained about its uses, as follows. Each illupai fruit has 4 brown coloured seeds with hard seed coat. Seeds are elliptical in shape with pointed edges.

Health Aspects

The oil extracted from the nuts reduces heat and cools the human body. For the extraction of oil, the outer skin is peeled off and the meso carp is removed from each seed, by using ural (stone grinders) and ullakai long (wodden pestle). This mesocarp is sun dried for 4-5 days. Now the mesocarp becomes ready for oil extraction by using oil expeller.

Cooking

Oil extracted from illupai nuts is good for cooking/culinary purpose, because of its cholesterol controlling property.

Agriculture

Oil cake obtained from illupai nuts is also advantageous. Illupai oil cake is applied at the rate of 50 kg/ac. Before transplanting or at the time of transplanting to the main field will effectively control the pest and disease incidence in the main field.

Even though this is an effective method, due to the non-availability of illupai oil cake, it rarely exists only in few areas.

Apart from this, it is also believed that illupai trees have the capacity to absorb rain bearing cloud. So it is said that wherever illupai trees are found, rainfall is received and it also absorbs carbon-di-oxide, as stated by Thiru.Sengutuvan.

Pest Control in Rice Field

a. Neem cake application

Five kg neem cake is taken in gunny bag and kept in the main channel of the rice field. During irrigation, water passes through neem cake and takes essence from neem cake which will control the root borne diseases and pest. It can be done after transplanting.

Existence : > 25 years

b. Cow dung smoking

Smoking with the help of dried cow dung in one end of the field will reduce the pests in field.

c. Kerosine + chilly powder spray

One kg of chilly powder is mixed with one litre of kerosene. This solution is sprayed in rice field. It is done only at the early morning.

d. Cow urine tonic

Ingredients

Cow urine-10 lit

Neem oil-2 lit

Preparation

Cow urine (10 lit) is taken in a mud pot or any container and boiled well until 10 litres cow urine comes to 5 litres i.e. half of the quantity. With this neem oil 2 litres and cow urine 5 litre are added. Again the above said contents are boiled well until 12 lit solutions come to 6 litres. After that the boiled solution is cooled.

Dosage : 1 lit/ac (100ml/10 litre water).

Cost : cost incurred for the preparation is Rs. 50/ac.

Kunabum- Plant Growth Promoter

Ingredients

Meat (Bone+ flesh)-1 kg

Beef (Bone+ flesh)-1 kg

Pork (Bone+ flesh)-1 kg

Cow urine-10 litres

Preparation

Meat (1 kg), Beef (1 kg) and Pork (1 kg) are boiled well. After boiling, it has to be cooled and cow urine (10 litres) has to be added into it. These contents are kept for fermentation about 10 days. It has to be stirred well daily. After 10 days, the solution is filtered and used as plant growth promoter.

Dosage : Spray 3.5 litre/acre. (50 ml/10 litre water)

Additional information : Meat, Beef, Pork and Cow urine are having rich nutrients. So this kunabum (powder of the said items) is used as plant growth promoter. Unsold /waste flesh can be used for the preparation of the kunabum.

Ginger + Garlic + Chillies + Cow Urine (Cotton Boll Worm)

Ingredients

Ginger-1/2 kg

Garlic-1 kg

Green chillies-1kg

Cow urine-10 litre

Preparation

Ginger (1/2 kg), Garlic (1 kg) and Green chillies (1kg) are pounded well and made into paste. This paste and Cow urine (10 litres) are put into the buried mud pot. Mouth of the mud pot is covered with waste cloth and the contents are allowed for fermentation for about 10 days. To obtain well fermented solution, every day morning it has to be stirred well by using wooden stick. After 10 days, fermented solution is filtered and used for spraying.

Dosage : 3.5 lit/ac (0.5 lit/ 10 lit water). Spraying can be done in weekly intervals.

Cost : cost incurred for the preparation of the solution is Rs.50-100/ac.

As stated by Mr. Marimuthu 40-50% of the cotton boll worms can be controlled by this solution.

Note : Fresh cow urine should be avoided.

Rice Bran + Neem Oil

Rice Brown Plant Hopper : *Nilaparvata lugens* (Delphacidae : Hemiptera)

Symptoms of Damage

Nymphs and adults congregate at the plant above the water level and suck the sap from the tiller. The affected plant dries up and gives a scorched appearance called "Hopper burn". Circular patches of drying and lodging of matured plant are the typical symptoms caused by the pest. It is vector of grassy stunt, ragged stunt and wilted stunt diseases.

Method of Control

Ingredients

Rice bran or husk-5-10 kg

Neem oil-1 lit

Procedure

Above said ingredients are mixed well and sprinkled over the rice crop at early morning i.e. 5-6 a.m to control Brown Plant Hopper damage. Cost involved for the preparation is around Rs.50/ac.

Existence of the practice : More than 50 years.

Pig Manure Smoking

Pig manure, dried neem and notchi leaves are used to make smoke. This is being followed to repel the pests and disease incidence in rice field.

Graveyard Ash Sprinkling – (Vegetables)

Method I

Ingredients

- Graveyard ash
- Cow urine – 10 litre
- Turmeric powder – ½ kg

Graveyard ash, cow urine, turmeric powder mixed solution is sprayed over the vegetable crop.

Existence : > 125 years

Method II

In early morning, ash powder is sprinkled in bhendi and brinjal field. It reduces the pest incidence.

Payir Valarchi Ookki (Plant Growth Promoters)

Method I

Ingredients I

- Cow urine-5 litres
- Neem oil- 5 litres
- Nayuruvi-1/2 kg

Ingredient II

Cow dung-10 kg

Preparation

i) Above said ingredients are taken in mud pot and the contents are allowed fermentation for about 10 days. The fermented solution is transferred into another mud pot and small hole is made on the bottom of the pot. This pot is kept in the entrance of the irrigation channel, so that crop can get the solution for its growth along with irrigation water.

ii) Fresh cow dung (10 kg) is taken in gunny bag and kept in the entrance of the irrigation channel or cow dung is mixed with irrigation water as such.

These indigenous practices can be followed during 30-40 days after sowing, when paddy is in the main field.

Method II

- 100 g Sheep manure is mixed with one litre of water and this solution is sprayed in the early morning.

General Profile of the ITK Holder

Th.Narayanasamy, Kolathancherry,Uthramerur Tk, Kanchipuram Dt

Paravai Thanki (Bird Perch)

- It is used to control insect in rice field.
- Square shaped polythene sheets are used.
- Coconut branches are kept in the field randomly.
- 'T' shaped stick is used for this purpose.
- Total numbers required : 15 Numbers/ acre.
- Expenditure : Rs.150 /ac.

Eli Viratti (Rat Control)

- Palmyra frond is tied on the stick and kept in the rice field. It will create sound due to air movement, which will scare the rat away.
- Total numbers required : 15 Numbers/ acre.
- Expenditure : Rs.150 /ac.

General Profile of the ITK Holder

Th.Rathakrishnan, Agaraperunthottam, Sirkazhi Taluk, Nagappattinam District

Eli Kattuppaduthuthal (Rat Control)

- Mud pot is kept upon 3 pegs in which some popped rice are placed. A mud lid is kept beneath with some popped rice. 4 sets of pot are placed in four corners of the field. First rat will come to take the popped rice in mud lid, after that it will go to take the popped rice in mud pot, at that time pot is fell on the lid. The rat is trapped and killed.
- Pots are filled with water and buried in the bunds of the field. When rat enter into the field, it will be trapped and killed.
- By leaving space (nearly 1 ft) in all four sides of field, rat damage will be reduced.
- Cow dung balls are immersed in kerosene and kept in 8-10 places of the field randomly. Odour comes from kerosene and cow dung will control the rat damage.

Neekal Poduthal (Rat Control in Rice Field)

During panicle initiation and milky stage, rat damage in rice field is one of the major problem. In order to avoid this problem, Neekal poduthal is one of the traditional practices followed in Kanchipuram District. Neekal poduthal is nothing but giving one feet space in 4-5 places of the field randomly when the crop is at milky stage in the rice. So that rice field will get more light penetration which prevents rat entry and movement. It also provides chance to owl to catch the rat in the rice field during night time.

Existence of the storage structure : > 100 years

General Profile of the ITK Holder

Th.Mohan, Thiruvalluvarpuram, Kaliyapettai, Uthramerur Tk, Kanchipuram Dt

Paravai Viratti (Bird Scaring)

Among all the growth stages, economic part development is the most important and crucial period. During such period, external damages especially by birds are more common and more intensive. Due to such damages farmers face loss in yield. Normally, to avoid such loss happening due to birds, various local bird scaring techniques are being followed by the farmers. Following are some of the techniques, which is being followed by farmers in Trichy district. These techniques are highly popular among small and marginal farmers, because it involves low/no cost expenses.

Tying Palmyra Fronds

Dried palmyra fronds are tied to the poles which makes sound when wind blows. In an acre nearly 5-10 such fronds are tied to a pole and kept sparsely. Thus, by producing sound, birds like sparrows and parrots are scared and the crop is saved from its damage.

Tying Unused Recordable Tapes

Unused Recordable Tapes are tied across the field in a cris-cross manner and it looks like a net spread over the field. So birds perceive that a net is bring spread to trap them and avoids to land on that field. When light falls on, it the polished sides, reflects the light scares the sparrows, parrots and other birds.

Tying Polythene Sheet

Polythene sheets/papers are tied to a pole as like that of dried palmyra fronds and kept in the field. When wind comes in contact with it, Sound is produced and this will scare the birds.

Beating Drums and Using Crackers

Sometimes to scare the birds, sound is produced by beating drums or by beating a plate with a stick. Crackers are also fired and thus sound is produced. The disadvantages of this method is, it needs some persons to perform, by firing crackers environmental pollution happens, which is not an encouraging activity.

All the above techniques, uses a simple logic i.e., by producing sound in a way or other, the birds are made to scare and thus the crop is protected from damage. Tying polythene sheet is followed both in garden land and wet land crops (Rice). Remaining other three techniques are followed in garden land only. (Ragi, Cumbu, Sorghum and Maize). The farmers of Perahambi and Pachamalai (Trichy district) are practicing these techniques.

Pasu Komiam Vithai Nerthi (Seed Treatment with Cow Urine)

1 litre of cow urine is required to treat 25 kg of seeds. This treatment prevents the moth damage during storage.

Manjal and Mor Vithai Nerthi (Seed Treatment with Turmeric and Butter Milk)

1 kg of turmeric is taken and it is soaked in 2 litres of skimmed butter or Neeragaram. It should be pounded well and the extract is used for seed treatment. This can be done in 2 days after harvesting of the rice grains. Treated seeds are dried in shade for 8-10 hours. This will control the storage pest moth and enhance the germination percentage. Treated seeds are stored in "Kottai"- storage structure made out of paddy straw (now it is not in use). Cow dung is smeared over the kottai to avoid moth damage.

Mann Pzhlu Uram (Vermi Compost)

Mr.Yallalappa, 55 years old who is residing at Bogasandiram village of Thenkanikottai taluk in Krishnagiri District. He has been practicing vermicompost for past two years.

Details

Raw materials used : FYM, Coconut sheath and Banana sheath

Vermi tank size : 2.50x2.25x0.65

Vermi wash tank size : 0.65x0.60x0.50m

Verm Name : *Perionex escavatas*

The raw materials are changed once in two months. He collected the vermi tonic from vermi wash tank at monthly once. Vermi tonic acts as a growth regulator.

The 15 litre of vermin tonic is diluted with 60 litre of water and sprayed in to the field. It reduces the pests and disease incidence. The vermi tonic applied field totally prevented from animal gracing. He sold the verms and vermi compost. Until, he sold the 30 kg of verms around the cost of Rs.5000 and he earned the money from selling the vermi compost also.

General Profile of ITK Holders

Th. Yellalappa, Bogasandiram, Irudhu kottai post, Thenkani kottai taluk,

Krishnagiri District - 635 107.

Lanther Pori (Light Trap)

Light trap is used to trap the pest in the rice field. Now a day's bulb is used as light source, previously Aricane or lanther light is used as light trap. Lanther is kept on the stool which is in one corner of the field. Below this structure, water and kerosene mixture is kept in a pan. Pests are attracted by light and fall in the pan. Once the pest fall into the pan, wings get damaged due to kerosene. So it can not fly again. Light trap can be used between 6.00-9.00 p.m.

General Profile of the ITK Holder

Th.Kumar, Vallathodu, Uthramerur Tk, Kanchipuram Dt

Poochi Pori (Pest Trap)

It is an organic method of plant protection i.e. pest trap in a low cost manner.

Purpose

This organic method of plant protection measure is used to trap the pests particularly the

Rhinocerous beetles and root grub.

Description

A mud pot is filled in with 0.5 litre castor oil and 4 litres of water. It is mixed well and buried up to pot's neck inside the field. This solution creates an odour which attracts the Rhinocerous beetles and root grubs. Thus the incidence of damage to the leaves and economic parts are diverted into this trap. If sunlight is more, the odour is increased which attracts more insects into it.

Nearly 40 pots have been buried in his field @ 1 pot per slope i.e. he has made 40 contours with very less slope in which each contours has one pot buried with the solution.

About 2000 beetles are being trapped in a day through a single pot. These beetles are in turn used as organic manure to field. Since the wings are wetted with oil, it sticks to the solution and due to weight; beetles are unable to fly out of the pot. Thus beetles are trapped and such beetles are used as organic manure by incorporating it near /around the trees.

Cost Effectiveness

Cost of pot is Rs. 10-12. Caster costs Rs.10-12/kg. So totally Rs. 20-22/ is the only expenditure to be incurred in making this pest trap. So compared to other ways and means of controlling pest, it is much cheaper as well as easier.

Advantages

This method is highly advantages as well as highly useful. By cost factor, it is much cheaper, by made of operation and maintenance; it is highly easier and comfortable to handle. Moreover, it controls the pest and makes it available as manure to crops. This method turns the negative factor, so called pest into a positive factor called organic manure to crop. Thus it is highly advantageous.

Disadvantage

Though it is highly advantageous, the only disadvantage is that, it fails to trap all pests.

Other Details

Initially, tried with curd and castor but it was damaged by dogs due to the preparation and taste. So avoid this, instead of curd water was used.

Base for Invention

Insect gets attracted to the oil applied paper tied under tube light at home. On sticking to it, unable to fly. With this idea, he has taken the castor oil which possesses more viscosity and stickiness.

Thiru.G.Sakthivel,S/oRamasamy,Gounder,Ganesa,puram,Gettavadi(Post),

Thalavadi (Via),Sathy (Taluk),Erode (District)

Mooligai Poochi Viratti (Herbal Pest Repellent)

Ingredients

a. Oomathai (*Datura metal*) leaves – 2 kg.
b. Erukku (*Calotrophis gigantea*) leaves – 2 kg.
c. Peenaarisangu (*Clerodentron inerme*) leaves – 2 kg.
d. Thumbai (*Leucas aspera*) leaves – 2 kg.
e. Water or cow urine – 10 litres.

Vermicomposting

Bird Perch

Rat Scaring

Ash Treated Seeds

Pest Repellent

Herbal Pest Repellent

Fig. 17. Crop Protection Measures

Method of Preparation

The aforesaid leaves are to be ground well by using a grinder and soaked in an earthen or plastic vessel. This is to be allowed for fermentation by keeping the vessel as such for 15 days. Then, the solution is to be filtered and added with water for spraying.

Recommendation

- The recommended ratio is 1 :10.
- It is used to control all types of pest in fruit trees and especially fruit borer.

Advantages

- The cost of preparation of pesticide is Rs.150/ per ha. But the chemical pesticide cost is Rs.500-800 per ha.
- It is eco-friendly and will not create any detrimental effect on human body while applying and ensures sustainable farming.

POST HARVEST METHODS

S.No	Particulars
1	**Vallai Nar Oorithal** (Banana sheath fibre extraction) in Marthandam of kanyakumari district. To get better quality of fiber to make toys.
2	**Pattai oorithal** (Dried cinnamon barks and leaves) in Marthandam of kanyakumari district. It is used in soap industries.
3	**Pakku pathapaduthuthal** (Arecanut processing) in Marthandam of Kanyakumari district. Separate the seed kernels from nut.
4	**Panak Karkandu** (Palm candy extraction) in Kanyakumari district. Used to cure chicken pox and through infection.
5	**Panai Olai Petti Thayarithal** (Palm leaf basket making) in Ramnathapuram district. The palm leaf mats are used in ships, drying of grains and in packing of fish ice boxes.
6	**Panai Olai Pai Pinnuthal** (Palm leaf mat weaving) in Ramnathapuram district. The jaggeries are kept in "kottans" for safe transport.
7	**Munthiri pathapaduthuthal** (Cashew processing) in Perambalure districts. Manual broken of nuts are high compared rather than the mechanical processing.
8	**Puli Pathapaduthuthal** (Tamarind processing) Method I in Trichy district. To avoid the pulp sticking the bulk quantity of pulp can be pounded simultaneously.
	Puli Pathapaduthuthal (Tamarind processing) Method II in Trichy district. Processing minimum quantity of tamarind, used for house hold consumption.

Vallai Nar Oorithal (Banana Sheath Fibre Extraction)

Well matured banana trees are selected. The pseudo stem of the trees are removed and allowed for optimum drying. After drying, with the help of scrapper-Vaazhayinaa kathi the sheath is removed. Now, the pseudo stems are converted as raw banana sheath fibre. It is repeated to get better quality fibre. The fibre is used to make toys, fancy articles, bags. Fetches better price in the market and even sometimes the products are exported. They are easy to use and highly economical.

Ms.Silvibai,Young Men Christian Association (YMCA),Rural Centre and Institute of Rural Development,Marthandam-629165.Ph : 04651-270260

Pattai Oorithal (Dried Cinnamon Barks and Leaves)

Cinnamon is a spice crop which is used in soap industries. The barks are falling off from the tree even after maturity. Similarly the leaves fall off from the tree are also collected in a gunny bag. They are weighed and sold for better price in market.

Pakku Pathapaduthuthal (Arecanut Processing)

There are two methods viz., wet and dry method

Method I

Fully matured arecanut kernels are collected from the tree and soaked in water. Change the water once in week or else they arise bad smell. Younger arecanut will decay if added to the lot and emits bad smell. The process is continued upto 3 months and storage of arecanut is extended. Whenever needs, arecanut is taken and the shell is removed used further.

Method II

It is a dry method in which water is not used. The harvested arecanut are heaped in a corner and allowed to be as such. Give reversal to the heap two days once. By this method, the arecanut can be stored upto 3 months. Now the market price of the produce is 45-50/kg-I, 70-80kg-Grade II.

Panam Karkandu (Palm Candy Extraction)

Palm candy is an alternative to palm gur- "karuppukaktti". It is used by the people of Kanyakumari district in their tea, coffee making. This is extracted from Neerah-"padhaneer".

Procedure

Fresh neerah is collected from the palm tree in early morning. No toddy and water should be added to it. The collected neerah is brought to the palm candy manufacturing centre. It should be sweeter. The PH and brix value are maintained to above 9 and 14 respectively. A filter is used to super phosphate is added to neerah and made clear. Super phosphate helped to remove the lime present in it. Again, neerah is filtered using a filter cloth. The filtered product is again given a uniform heat upto 108°C. The heated neerah is poured to the crystallizer of 30 liter capacity. Sand is kept around the crystallizer to avoid movement in it. Sometimes, paddy straw and bricks are also used. Crystalizer is kept in the same position for 41 days. After 41 days, the palm candy can be taken out from the crystallizer frames. Immediately after removing out from crystallizer the palm candy is washed with cold water and kept for sun drying. Threads on them are removed. Based on their size, the grades are separated but the quality remained the same. Finally, they are packed in polythene bags for sale. From a crystallizer 12-14 kg of palm candy is extracted in a single time.

Grades

Parumani	-	highest price with biggest size.
Salangai and Saral	-	intermediate size with moderate price
Podi saral	-	lowest price with smallest size.

Cost Effectiveness

100 liter nectar gives 14kg of palm sugar RS.20/kg and 6.5 kg of palm candy (Rs.95/kg).

Panai Olai Petti Thayarithal (Palm Leave Basket Making)

Palm leave basket and hat making process is the community based profession (of "kuravars", "Sakkiliyars" and "Paraiyars") in Uchipuli region of Ramnad district. Mat weaving is done by the rest of communities (Nadars and Thevars). They buy palm leaves from adjacent areas. Palm leaf box, hat and mat weaving are done by "Sathuvakathi" especially made for palm leaf products. The edged portion of "Sathuvakathi" is used for tearing of palm leaves and back point is used for punching of palm leaves while making palm leaf products. It is specially made curved portion on top. The palm leaf petti ranges from Rs.20, 30, 50, 60 based on size and colouring given to palm leaf petti. The plastic wire is also added to give good support to the palm leaf petti. The matured palm leaves are bought for 10 paise and "kuruthu" bought for Rs.3-5/-. The customers are allowed to take only the palm leaves. The palm leave mat is made from matured leaves and the younger palm leaf (kuruthu) is used for "petti", "Kottan" and hat making .The palm leaf mats are used in ships, drying of grains and in packing of fish ice boxes. The younger palm leaves known as "vidali" are used as brushes .The palm "Ekki, and palm"Thumbu" are exported to foreign countries from Trichy and Tuticorin areas. As rain fed farming is not profitable one, these people are interested in palm leaf product business. The bigger kuruthu palm leaves are used for "narpetti" making and smaller kuruthu palm leaves are used for "fine mat" weaving. These fine mats have the longevity of several months.

Address of the Person

Mr&Mrs.Velu, Sathya nagar, Uchipuli, Ramnad (DT).

Panai Olai Pai Pinnuthal (Palm Leaf Mat Weaving)

Palm leaf mats are produced by matured palm leaves and palm leaf "kattu" was bought from nearby places (one palm leaf kattu = 40 leaves). Some time palm leaves are alone bought leaving mattai and "eekki". There are three types of mats are made viz., ordinary mat, nice mat, and kurutholai mat. The longevity of ordinary mat is just one day (Rs.2-3) and nice mat is kept for 2 months (Rs.6/). The kurutholai mat has longevity of more than 2 months (Rs.15) and is used for drying of paddy and keeping small children inside the mat. Ordinary mats numbering 6 to 7can be made by single person in a day. Whereas only one kurutholai mat can be made in a day. Small type of "kottans" and Archani thattus are made by palm leaves. The jaggeries are kept in "kottans" for safe transport. The same sathuvakathi used for making basket is used for palm leaf mat.

Munthiri Pathapaduthuthal (Cashew Nut Processing)

Cashew Processing

Cashew is one of the important cash crop which are grown in larger area of Perambalur District of Tamilnadu. From seed selection to packing several indigenous practices are involved, they are detailed below

Seed Selection

Well matured apples are plucked from the tree by using thorati or alaku. Manually nuts are separated from apples. Nuts are dried for about 2-3 days. One kg of cow dung is mixed in three litre water and made it as solution, which is locally termed as 'Saanipaal.' Dried nuts are put into the solution, some of them floated and rest of them sunk in the solution. Floaters are discarded, because these are immature or ill filled nuts. Rest of the well filled nuts are selected for sowing.

Processing

Fresh nuts separated from apples are sun dried for about 2-3 days. (Less or more sun drying yields increased number of broken nuts) after that, nuts are ready for processing, for manual breaking of the whole nuts, plate like square shaped rock stone and small cylindrical stone are used. While breaking the whole nut is hit vertically and horizontally. So that the nuts is separated easily from the outer shell. Small iron needle is used for separation. During this operation, exudates is oozing out from the outer shell, which will create wounds on fingers, in order to avoid this wood ash is smeared over the fingers. Mostly women are employed for this cashew processing operation. One woman can process 21kg of nuts in a day. One kg whole cashew nut yields about 270 gram processed nuts. Wages for this processing operation is Rs 2.30/kg.

Only disadvantage is that, more broken nuts are obtained from manual breaking compared to machine breaking.

Peeling

After breaking, nuts with light orange colour peels are obtained. Peels are removed by using small knife, specially made for this purpose, which costs about Rs 35. One woman can peel upto 7 kg nuts/day. Wages for this is Rs. 4/kg.

Grading

After peeling nuts are graded manually, it can be done based on experience in cashew processing and grading. Whole nuts, splits, powder, granules, stained nuts are graded. Different grades and price of the processed nuts are tabulated below.

Table 5. Different Grades and Price of the Processed Nuts

S. No	Different Grades	Price (in Rs)
1	210 W	250
2	240 W	230
3	320 W	200
4	450 W	180
5	Split	220
6	Split (First quality)	250
7	Small wide pieces	165
8	Large wide pieces	150
9	Small pieces	110
10	Kurunai /Granules	65
11	Powder	25

Efficiency of one woman per day may vary from 1 kg-15 kg, which depends on experience. Wage for grading is 3 Rs / day. Graded nuts are packed in polythene bags for selling purpose.

Note

1. W-Whole
2. 210 W means 210 unbroken processed whole nuts weighing about 454 g.

3. 240 W means 240 unbroken processed whole nuts weighing about 454 g.
4. 320 W means 320 unbroken processed whole nuts weighing about 454 g.
5. 450 W means 450 unbroken processed whole nuts weighing about 454 g.

General Profile of the ITK Holder

Mrs.lakshmi, Kazhumangalam, Udayarpalayam taluk, Perambalur District.

Puli Pathapatuthuthal (Tamarind Processing)

The main concept behind tamarind processing is purification. i.e., Tamarind is deseeded and dried by which the pulp alone is separated and stored. By doing so, its shelf life increases and moreover as days passes the taste increases in such processed tamarind.

The method in purification (processing) of tamarind differs from area to area.

Method 1

Tmt. Rasathi, residing in Siruganoor, Manachanalur taluk of Trichy district processes Tamarind using stone mortar and wooden pestle. This method of tamarind processing is almost followed by most of the women in Trichy district particularly in Manachanallur taluk.

In this method, a stone mortar, both of circle as well as rectangular shape is used. Dried tamarind fruits are put into this mortar and pounded by using a wooden pestle (ulakkai). So, that the entire content is deseeded. To avoid the pulp sticking the pestle, either castor oil or gingelly oil is smeared at the edges of pestle.

Efficiency

Nearly 30 kg of tamarind fruits can be processed at a time by 4 women members. After pounding, tamarind pulp and seeds are removed separately by hand. The advantage of the method is the bulk quantity that can be pounded simultaneously.

Method 2

The dried tamarind from Arival manai (a kitchen gadget with curved iron blade fixed to a base and used for the cutting vegetable) or by using suthiyal (hammer). This method is used only at household level i.e., for processing minimum quantity of tamarind that is used for household consumption. This is most common practice followed in almost all eco systems.

GRAIN AND SEED STORAGE METHODS

S.No	Particulars
1	**Kuluppai or Kulumai** (Mud Pot) in Edayapatti village of Trichy district. It is believed that the structure has a spiritual power.
2	**Thaniyakuthir** (Grain storage Structure) Storage Method I in Kamalathipatti Village of Trichy district : The structure is spindle shaped and circular frame is used to construct it. Its cost of making is low.
	Thaniyakuthir (grain storage structure) Storage Method II in Kazhumangalam village of Nagapattinam district. This structure is cylindrical shaped and circular frame. The capacity is 250-300 kg.
	Thaniyakuthir (Storage drum) Storage Method III in Kakarunkalpatti village of Salem district. It will be purely protected from ants, rat and termite. The capacity of the drum will be 550 kg.
3	**Thombai koodu** (Grain storage structure) Vellapalam village of Nagapattinam district. The structure is spindle shaped, purpose of this structure is to retain the quantity of seeds and the whole structure is easily shifted to another location.
	Thombai (Storage Container) Method I in Kazhumangulam of Perambalur district. The seeds are used for consumption purpose. Shape of the structure is fabricated with combination of bamboo.
	Thombai (Storage Container) Method II in Pachamalai village of Trichy district. It is circular in shape.
4	**Ragi kuzhi** (Grain Storage Pit) in Mallankuzhi village of Erode district. The structure is circular or square shaped. The quality of produce is preserved longer period. Only ragi is mainly stored material, in this structure.
5	**Pakka kothi** (Seed Storage Structure) in Malasajanoor village of Erode district. The structure is barrel shaped container.
6	**Thirugai Enthiram** (Processors and Hand Hullers) in C.R. Palayam of Trichy district : This process conserves nutrients like minerals and vitamins.
7	**Seru kalanjium** (Storage Structure) in C.R Palayam of Trichy district Safety is secured and the capacity is 1200 kg, bulk quantity of grains is needed.
8	**Saal or Urai** (Storage Structure) in Kolathancherry of Kanchipuram district. To control storage pest.
9	**Ambaram** (Paddy Grain Storage Structure) in Kolathancherry of Kanchipuram district. Paddy will stored upto six month
10	**Venkaya Pattarai** (Onion Storage Structure) in Manachanallur of Trichy district. The method of construction is very simple. It is feasible as well as economic aspect.
11	**Moonkil kudai Vidhai Petti** (Seed Box) in Malasajanoor village of Erode district. Used to store the grains and sowing purpose.

Periya koodai (Bamboo container) Malasajanoor village Erode district. It is used to collect the harvested produce from the field.

Moonkil thattu (Bamboo drying plate) in Malasajanoor village of Erode district. It is used for drying the seed.

Processing and Preservation Methods

12 **Vazhaipalam Mootampoduthal** (Banana ripening method) in Berahambi of Trichy district. The ripened banana used for cooking and culinary purpose.

13 **Mampalam palukkavaithal** (Ripening Method in Mangoes) in Vellapalam village of Nagapattinam district. The keeping quality of mango is high and the taste as well as aroma is retained.

14 **Maravalikilanku pathapaduthuthal** (Tapioca Preservation) in Marthandam village of Kanyakumari district. Stored even upto 3years and cost effective when compared to other methods.

15 **Pirandai storage** (Pirandai Processing) in Vellicharanai village of Kanyakumari district. The poisonous substances in the fruits get removed.

16 **Nel vithai semithal** (Paddy seed storage) in Malavilai Village of Kanyakumari district. It emanates a pungent smell so, the grains are pest free.

17 **Pudalai vithai pathapaduthuthal** (Snake gourd seed processing) in Keevalur of Nagaipattinam district. Act as anti fungal and pest repellent.

18 **Ulunthu vithai nerthi** (Black gram + oil) in Manachanallur of Trichy district The seeds are stored in mud or plastic container.

19 **Gingilly + Vasambu** Storage method I in Manachanallur of Trichy district. It wills safe guard from pest damage, low cost technology.

Gingilly + Chillies Storage method II in Manachanallur of Trichy district. To control storage pest.

Gingilly + Ash Storage method III in Manachanallur of Trichy district. This is low cost method of pre- storage.

Neem + Pungam (Pest repellent) in Manachanallur of Trichy district. Spread over the grain that prevents pest attack.

Seed Storage Structure

Kuluppai / Kulumai (Mud Pot)

Th. Gopalakrishnan, 52 years old is residing in A.Edayapatti village in Manapparai taluk of Trichy district. He is educated up to 10th standard and has agriculture as his primary occupation. He possesses a post harvest storage structure called 'kuluppai/ kulumai'.

It is a storage structure used to store grains, which is suited to all ecosystems. This structure is invented by ancestors and maintained for more than 100 years until now.

This entire structure has a construction process similar to that of constructing a house. Stone pillars (Nearly 10 feet) are buried upto 6 feet into the ground in circular manner. The distance between each pillar is about 3-5 feet. Thus a basement is made and upon these pillars, wooden planks (Mostly of Neem – *Azardiracta indica*) are laid down. These planks are plastered by using clay. A platform is raised on the base about 1-1.5 feet height, in a similar circular fashion. Above this platform a circular bin (storage structure) is erected with 6-7 feet height, 2.5– 3 feet width. Nearly at 6 feet height of this bin, a small window is provided with a lock that serves as an entrance into the bin. The side walls are made up of bricks (Dried stalks of *Dodoneae angustifolia* – virali are also used) and both inner and outer walls were plastered with clay.

In the total 6 feet height nearly bottom 3-4 feet of the inner portion is compartmentalized into four equal compartments using wooden planks.

The top portion is closed with wooden planks, (generally neem) and plastered by using clay. Above of this, a hut like thatching is made using either sorghum or maize stalks. This thatched roof is pillared by 6-10 neem pillars (up to 5-6 feet), which are raised upon the platform margin.

Existence

Its existence is found to be more than 100 years and it is a core structure mostly erected by farmers in and around Trichy district. It is believed that the structure has a spiritual power i.e. the holders state that, members from other than their family cannot get into the structure and can't even sit/sleep during night hours. Moreover, even the female members of their own family are avoiding stepping into the platform.

Cost Effectiveness

The entire structure is more or less similar to a hut in which a four member's family can dwell. So this structure, comparatively involves some investment on pillars, wooden plank, bricks etc.

Feasibility and Relative Advantage

Though there is a spiritual belief on the structure is feasible to construct and maintain the same in other location too. The advantages can justify it.

The Advantages are,

- Size- a small hut sized structure enables to store more quantity of grains.
- Compartmental structure provides space for storing various grains simultaneously.

- At a time 100 bags of grains can be stored.
- Multipurpose- since it is constructed on pillars, where the bottom portion underneath the platform can be used to herd goats. Thus it is used both for grain storage as well as shed for small animals.

Moreover, the platform (veranda) rose around the circular structure provides enough space for the family members to use it for other purposes (like drying, staking any materials etc.)

Condition for Adoption and Limitation

This structure needs a separate place to be allotted for its construction. Since it is an out of house storage structure, cost is one of the limiting factor (due to its size). But the benefits out of this structure will definitely mask these limitations.

Though the structure is made with enough space, the entry window is a limitation i.e., the window provided is very small, that requires care. The trained person's only can enters in to this window.

Thaniya Kuthir (Grain Storage Structure)

Type I

Th.kuppamuthu, 49 years old is residing in kambalathupatty in Manaparai taluk of Trichy District. He is educated up to 10th standard and agriculture is his primary occupation. He possesses a post-harvest storage structure called **'Thaniya kuthir'.**

It is a storage structure used to store cereals in almost all ecosystems. This was invented by ancestors and carried by present generation. Nearly 5 bags of cereals can be easily stored in a safe manner within the house.

This structure is spindle shaped and circular frames are used to construct it. These frames are, made up of clay and rice bran mixture. Based upon the quantity to be stored, the size of the structure and number of frames to be prepared are determined. To attain spindle shape, each frame is reduced from the size of the centre frame. Finally, the position of the frame is fixed with dots (clay balls) i.e. bottom most frames are fixed with one dot and the second with two dots and so on. Finally top opening is based with a mud made lid. This helps to ensure balance and order of frames while positioning it. These frames are generally made by potters and bought by farmers.

These frames are positioned and plastered using clay and cow dung mixture. For outlet, a hole is made in the bottom most frames, which is plugged with paddy straw and plastered by using clay, similarly the lid is also plastered.

Existence

It exists for more than 100 years, but now faces an extinction phase due to the arrival of modern iron storage bins. Even though iron bins are introduced, due to its limitations there are lots of scopes for reintroducing such mud made storage structure in the rural houses.

It is completely made out of raw materials that are easily available. Its cost of making (including labour charge for potter) will come around Rs.200/-.

The structure is neither too small nor too big, so size is compatible for having it inside the house. The maintenance is very simple as well as damage is very meager which are the added advantages of this storage structure.

Since it is a mud made structure, it can be placed only in the rural houses which are made out of mud/concrete, but not in any other type of houses. This structure is mainly used to store grains that are to be used for seed purpose. So it doesn't fit into urban centres.

Other Details

Whatever to the number of frames, the topmost frames is encased with a symbol of growing moon (Pirai Nila) . They believe that, as like the moon grows, their grains wealth also grows, ensuring their food security.

Type II

It is a storage structure used to store cereals in almost all ecosystems. This was invented by ancestors and carried by present generation. This structure is cylindrical shaped and circular frames are used to construct it. These frames are, made up of clay and rice bran mixture. Based upon the quantity to be stored, the size of the structure and number of frames to be prepared are determined. The structure is provided with a small hole at the bottom which serves as outlet and bigger hole at the top which serves as inlet. The bottom hole is used for taking out the grain. Outlet is covered by using coconut shell and plastered with clay whereas inlet is covered with a mud made lid or wood. The capacity of the structure ranges from 250-300 kg. These frames are generally made by potters and bought by farmers. These frames are positioned and plastered using clay and cow dung mixture.

Existence

It exists for more than 200 years, but now faces an extinction phase due to the arrival of modern iron storage bins. Even though iron bins are introduced, due to its limitations there are lots of scopes for reintroducing such mud made storage structure in the rural houses.

It is completely made out of raw materials that are easily available. Its cost of making (including labour charge for potter) will come around Rs.200/-.

The structure is neither too small nor too big, so size is compatible for having it inside the house. The maintenance is very simple as well as damage is very meager which are the added advantages of this storage structure.

Type III

Purpose

This drum is traditionally used for storing paddy, cumbu, ragi etc. This method is in existence for about 20 years in kakarunkalpatti village at Salem district.

Making

This storage drum is made up of iron steel plate. The cylindrical shaped drum is having a flattened bottom and the top. The height of the drum is 7 feet and the diameter is 3 feet. The top portion of the drum have the inlet, through this grains are filled. The bottom portion of the drum is having one small opening for withdrawal of grains whenever it is needed.

Cost

The cost of the drum is Rs.1000 in 10 years ago; nowadays it is Rs.3000.

Special Feature

1. It will be purely protected from ants, rat and termites.
2. The capacity of the drum will be 550 kg.

Thombai Koodu (Grain Storage Structure)

Type I

Th. Senguthuvan, 45 years old is residing in Vellapallam village in keevalur taluk of Nagapattinam district. He is educated up to 10th standard and agriculture is his primary occupation. He has a storage structure called **Thombai koodu'** which is used to store grains. This storage structure is suited to all ecosystems but possessed mostly in coastal areas only.

The purpose of this structure is to retain the quantity of the seeds to be used either for seed or consumption purpose.

This structure is mostly spindle shaped with an average height of 4-5 feet. It is made out of bamboo palm splits, paddy straw, rock pillars and few poles. Either bamboo splits or palm frond splits are intertwined in a criss-cross fashion and a spindle shaped basket with a wide opening at the top is made. An opening of small size is made at any one side at the bottom outlet. This basket is placed upon a platform made out of poles which are raised on four small stone pillars (approximately 1 ft).then the basket is plastered by using clay and cow dung mixture. This is left to dry for one day. By plastering, it is ensured that there is no opening for pest entry.

After drying, the outlet is plugged with straw and the grains are filled into it and covered by using paddy straw. Finally the entire structure is enclosed either by paddy straw or palm fronds. Now, the entire structure is tied by using a rope made out of palm fronds. (i.e. palm fronds are split into many then and long fibre are made into rope).It is nearly stored upto 60 kgs.

Existence

The practice of having **'Thombai koodu'** has its origin beyond more than a century, as stated by him. This type of storage structure when compared to other structures followed in other location (eco system) is different i.e. it is made without completely depending either mud or wooden planks. This kind of storage structure is much prominent in coastal ecosystem (especially Nagapattinam), because of non-availability of clay that is needed for making storage structure.

The storage structures are made out of materials that are locally available like palm fronds, paddy straw etc. so it doesn't incur more expenditure for constructing this structure. The cost of making one such storage bin comes around Rs.50/- depending upon the size. Generally, the bin is bought from carpenter through orders. Thereafter, the outer plastering and covering with paddy straw/palm fronds are done by the holder.

The whole structure can be easily shifted to another location whenever needed. Moreover, it doesn't need more space to erect. Therefore space is not a constraint. Similarly, the maintenance is very less, but need to do twice every year (like plastering, cleaning etc.).Thus every handling, less maintenance, cost effectiveness is the relative advantages of this storage structure that makes it feasible to erect.Since it is an out of house structure, there is no specific condition for adopting it.

Limitations

First, when compared to other storage structure, capacity of holding is less. Secondly, when proper plastering is not done there is a chance of termite damage. Similarly when the outlet is not plugged properly, loss / damage of grain may occur in both quantity as well as quality wise.

Type II

The purpose of this structure is to retain the quantity of the seeds to be used either for seed or consumption purpose. This structure is mostly spindle shaped with an average height of 4-5 feet. It is made out of bamboo palm splits, and then the basket is plastered by using clay and cow dung mixture. This is left to dry for one day. By plastering, it is ensured that there is no opening for pest entry.

After drying, the outlet is plugged with straw and the grains are filled into it and covered by using paddy straw. Finally the entire structure is enclosed either by paddy straw or palm fronds

Existence

The practice of having 'Thombai' has its origin beyond more than a century, as stated by him. It is made without completely depending either mud or wooden planks. Cylindrical structure made by using bamboo splits and plastering with cow dung. It doesn't incur more expenditure for constructing this structure. The cost of making one such storage bin comes around Rs.75/- depending upon the size. Thereafter, the outlet and inlet are covered with paddy straw bunches and plastered with cow dung and clay mixture.

The whole structure can be easily shifted to another location whenever needed. Moreover, it doesn't need more space to erect. Therefore space is not a constraint. Similarly, the maintenance is very less, but need to do twice every year (like plastering, cleaning etc.).Thus handling, less maintenance, cost effectiveness is the relative advantages of this storage structure that makes it feasible to erect.Since it is an out of house structure, there is no specific condition for adopting it.

Limitations

When compared to other storage structure, capacity of holding is less. Similarly when the outlet is not plugged properly, loss/damage of grain may occur in both quantity as well as quality wise.

Type III

Thombai is an out door storage structure found in the villages of pachamalai. It is a circular storage structure made out of clay and bamboo splits. Wooden poles are arranged in a criss cross manner on four stone and smeared with clay. Hence it looks like a floor or thinnai. Five feet lengthy bamboo splits are erected on the floor or thinnai in a circular manner. Inner and outer side of the bamboo splits are coated with clay for about 5cm. inner side of the thombai is divided into two rooms. Roof is made on the top portion of the thombai. Cereals, pulses can be stored in thombai with out pest attack and rat damage.

Existence : More than 75 Years.

Name : Mrs.Revathi Address : Periyapakalam, Pachamalai, Trichy district.

Ragi Kuzhi-(Grain Storage Pit)

Th.M.S.Sivanna, 60 years old and 40 years experienced in agriculture is residing in Mallan kuzhi village of Sathy taluk of Erode district. His primary occupation is agriculture and the predominant ecosystem is garden land ecosystem, where all pulses are grown along with minor millets like Ragi, Cholam. As a post harvest storage practice a storage structure called Ragi pottu (Ragi crop waste) is used. This post harvest storage structure is typical serving as a multipurpose structure.

Description and Uses

A big (about 1 feet height) area is exhausted below ground level either in circular or square shaped, with a small mouth that provides as inlet. Inlet is dimensioned such that one person can enter into it. Stones are used to line the inner wall so that, soil is not eroded in the inner side. Harvested produce are stored inside the structure which ensures the quality of produce for longer period. The capacity of this structure is about 4 qtl. While storing cholam, the gunny bags are plastered by using mud and this type plastered and stored cholam is used only for consumption and not for cultivation. Banana ripening can also be done in this structure. A bunch of banana is kept inside the structure flames are raised using Sambirani. By doing this, within 3 days banana gets ripened.

Cost Effectiveness

Except excavation cost, there is no additional investment in this structure. This type of structure is entirely different because, it is not erected /constructed completely but excavated and a room like structure is made.

Advantages

No maintenance cost, is the main advantage similarly, stored produce is not affected by any pest and diseases which is the desired advantages of any storage structure.

Disadvantage

This cannot be followed in all ecosystem and all types of produce cannot be stored in it.

Existence of the ITK : more than 100years.

Pakka Kothi (Seed Storage Structure)

One such storage structure is Pakka kothi which is nothing but the modified 'kadage' –old traditional storage structure made out of bamboo splits. A bamboo made barrel shaped container or storage structure is called 'kadge'

Description

Pakka kothi is made out of brick and sand. It is a concrete structure which differs in shape but usually in square /rectangle shape. Walls are raised in four sides with an inlet on the top and an outlet at the bottom. Before fitting in the grains the inner wall is lined with plastic sheet to avoid damage to grains due to moisture. It capacity is 12 qtl.

Advantages

Prior to this structure, gunny bag storage was used (during extinction of kadaga) where rat damage was more. But it has been avoided by this structure moreover the holding capacity can be modified as per the need while construction. Grain quality deterioration also prevented due to plastic sheet lining.

Disadvantages : It occupies space with in the house.

Cost Effectiveness

It requires some expenditure to be spent on concrete work. But while calculating the benefits and advantages, the cost of Rs.360/ for making one such as structure seems to be an investment not expenditure.

General Profile of the ITK Holder

Name and address : M.Pomman, S/o Mathan, Malasajanoor,Burgur (post), Bhavani (TK), Erode district. Age : 35 years. Educational status : 5th standard.

Existence of the ITK : more than 85 years.

Thirugai/ Endhiram Processors /Hand Hullers

The stone made quern is called 'thirugu' or 'Endhiram'. The circular stones with a diameter of 30-35 cm and 15cm width are placed one upon the other. The upper stone plate is provided with a hole at the centre that serves as an inlet, through which the grain is powdered and poured in. Similarly, a wooden handle is inserted at the margin of the circumference of the upper stone. This handle is used as a grip while grinding.

The lower stone has a peg inserted at the centre of it, upon which the upper stone is placed. When the grains are poured into the centre hole and grounded. The grains get struck in between the stone plates and breaks into halves due to centrifugal movement, the split halves of grains fall out and get collected around the "Thirugai'.

Advantages of Using the Hullers/Processors

- Cereals (rice, sorghum), minor millets Cumbu, ragi, samai, kuthiraivalli, thenai, varagu), Pulses (red gram, black gram, green gram and cowpea) are processed by using these traditional processors.

- While processing through these processors nutrients like minerals and vitamins are conserved as such, whereas in machine processing some of the nutrients are removed. Because, while operating the machine lot of heat is produced, that will affect the nutrient status of the grain.

Seru / Kalanjium Storage Structure

Name and Associated Crop with Ecosystem

Seru / kalanjium are a storage structure to store cereals, mostly found in garden land ecosystem.

Description

Seru / kalanjium is a storage structure which can be erected as both in house as well as out house structure. Generally, this structure is built as a part of house, by just making a partition across the opposite walls. Such type is considered as in house structure whereas in out of house structure, entirely a new house like construction is made, fixed with doors and compartments within the structure.

Stone pillars of one feet height are laid as basement; above which wooden planks are arranged to form a platform. This is plastered with clay so that there is no gap in between the planks. Walls made up of either brick or simply mud is raised along the margins of the platform. The height and size of the structure varies depending upon the needs and requirement. Irrespective of size and height, the width of the wall remains nearly 1.5 - 2 feet. After raising the walls a room like structure is formed which is either divided into two or three compartments. Each compartment has an outlet (mostly circular with a diameter of 15cm approximately) plugged with paddy straw after filling the structure with grains. The upper portion is left uncovered which serves as inlet. But like that of thatched roof made for houses. This entire structure is roofed.

There is no provision/staircase to reach the upper opening (inlet).But by stepping on doors and windows, one can climb to the inlet and other person should help him by lifting the grain filled through gunny bags or vessels.

This entire structure is built attached to the house and a common roof is erected both for the house and the storage structure.

In the case of out house structure construction process remains the same, but both the portions are compartmentalized and a common entrance is provided in-between those two portions. Since this structure is out of house to ensure its safety, a door is fixed with a lock.

Existence

As stated by the holder, the origin of this kind of structure goes beyond a century and is mainly followed in the garden land ecosystem.

Cost Effectiveness

In the first type of Seru the cost of construction is less when compared to the second type (i.e. out of house model).The reason behind is in the first model, the storage structure forms a part of the house whereas in the second model, a separate house like structure is built. So it requires, additional investment (Rs 1000 approximately)

Feasibility and Relative Advantage

In the first model, safety is secured. Since it is within the house. In the second model, the capacity is two times more when compared to the other model i.e. each compartment has a capacity of holding 20 bags of grains (1200kg) obviously the holding capacity of second model increases.

Whenever required, based on the quantity, both outlet as well as inlet can be used for taking stored grains. I.e. if bulk quantity of grain is needed, then the inlet is used as outlet.

Constraints/Limitations

Though this structure is advantageous it fails to mitigate from rat damage, which leads to grain loss. Since, the inlet is left uncovered, pest and rat invasion is not checked and so this become a big disadvantage.

Similarly, climbing to the inlet is a risky task which can be performed only by trained persons. Also, when compared to other storage structures like (Kuthir,Thombai)this requires more space.

General Profile ITK Holder

Name & Age : Thiru.Natarajan, 62 YearsAddress : C.R.Palayam, Manachanallur Taluk,

Trichy District. Occupation : Agriculture-Farming, Farming experience : 45 years.

Saal /Urai (Traditional Storage Container)

Mud pot or mud jaadi used to store the grains. It is locally called as saal. Quantity of the saal may vary; it depends on the quantity of harvested grain availability while storing the grains in the saal. Ash, neem and notchi leaves are kept to control storage pest like moth. Inlet of the saal is covered by using cow dung.

Existence of the storage structure : > 100 years

Ambaram (Paddy Grain Storage Structure)

Five feet lengthy bamboo poles are place in the ground. With in the space available between the four poles (space will vary according to the paddy grains to be stored). Paddy straw is spreaded over the ground; above that gunny bags are spreaded. Over the gunny bags paddy grains are heaped. (After harvest, the paddy grains are dried in the sun light for about 7 days and winnowed before storage). Around the four poles paddy straw rope which is locally termed as 'piri' is used to cover the heaped grains. Over that polythene sheets / Tharpai is used to cover the entire ambaram from rain water. Upto 6 months grains can be stored.

Expenditure : Rs 200-300 / Ambaram

Existence : > 125 years

General Profile of the ITK Holder

Thiru :karuppaiah, Sukkankollai, Chengalpet tk, Kanchipuram Dt. Age :60, Experience in farming : 40years, land holding :3ac.

Onion Pattarai – (*In Situ* Onion Storage Structure)

Name with Associated Crop and Ecosystem

Onion storage structure – "Pattarai" (local name) practiced in garden land.

Origin : It seems that this practice is being followed for long past by their ancestors.

Purpose : It is used to store onion i.e. for in-situ storage.

Description

A leveled place (9'x3') at any corner of the field is selected. Four stones are placed, so that it serves as a base pillar. Over this base pillar, either bamboo or poles of any other tree is laid down. Harvested onion are uniformly spread over the poles and covered with coconut fronds (paddy straw is also used) to protect it from sunlight. This sort of storage is practiced only to store seed onions. Nearly 75 years old technology invented and followed by group. It is still existing and used by almost all onion growers.

Feasibility and Advantage

Since the material needed are locally available and the method of construction is very simple, it is much feasible as well as advantages in economic aspect i.e. *in-situ* storage and no transportation cost. Thus because of its cost effectiveness it is considered relatively advantageous.

Extent of Adoption

Due to its relative advantages, almost all of the onion growers adopt this storage practice.

General Profile of ITK Holder

Th. Raman.Address : Perahambi (P.o), Manachanallur (T.K), Trichy District.

Different Types of Bamboo Baskets

Since the farming is done in hilly area, the system and implements as well as the storage structures are somewhat different from that of plains. Among those structure various container made out of bamboo is typical and traditional to this locality.

a. Vidhai petti (Seed box)

Bamboo is splitted into many small splits of 1 cm wide and length of 40-45cm. using these splits a container like structure is made i.e. splits of 2 cm wide is kept one upon the other in a star shaped fashion. This forms as a frame/skeleton to give a shape to the box. Then the small splits are in a criss-cross fashion by inserting in between the frame splits.

Thus a basket is made to a height of 15-20 cm with a diameter of 20-25cm. It is plastered with dung and dried in sunlight. Since it is used for sowing purpose, size is made such that it is neither too big nor too small such are filled in and broadcasted in the field.

b. Periya koodai (Big size basket)

It is similar in material and making of seed box. But the only difference is its size. It varies from 1 feet height with 1-1 ½ diameter to 1 feet height with 2 feet diameter.

c. Moonkil thattu (Bamboo drying plate)

Materials and making are similar to seed box and periya koodai. But the size differs i.e. height is very less upto 10-15 cm only, but diameter is more when compared to other baskets i.e. up to 3-4 diameter. It is used to dry the seeds.

d. Moonkil muram (Bamboo winnower)

Materials and making principle are similar to seed box, but the shape differs. The outer i.e. top most split is the base frame for this winnower, it serves as a rim on which the entire structure is shaped. In this winnower, one side is raised up to 20-25 cm and it is gradually sloped on both the sides and finally finished with a flat end. It is used to collect the harvested produce from the field.

PROCESSING AND PRESERVATION METHODS

Vaalai Pazham Mootampoduthal (Banana Ripening Method)

Gereral Profile of the ITK Holder

Thiru.Vellaiyan, Perahambi,Manachanallur(TK),Trichy (DT),Age : 45 years.

Preservation

Well matured banana bunch is harvested, which is cut into several bunches. The bunches are put into water for a week. It will increase the shelf life and freshness of the banana. Mostly Monthan variety banana is preserved by this method. This technique prevents the ripening for banana, until used for cooking / culinary purpose.

Mambalam Palukkavaithal (Ripening Method in Mangoes)

For quick ripening of mangoes, normally sorakondrai flowers are used in coastal areas of Nagapattinam. As stated by Th.Senguttuvan, residing in vellapalam in keevalur taluk of Nagapattinam district. The practice of ripening mangoes using Sarakondrai (*Cassia fistula*) flowers is alone in almost all mango growers farm and it exists for more than 60-70 years.

A pit is dug in any corner of the field and the dimensions vary based on the quantity of mangoes to be ripened. The leaves of sarakondarai are put into it. Such that, it forms a bedding layer. Above this layer, the mangoes are arranged and covered with sarakondrai flowers. Finally a thin layer of soil is spread over it, so that it is completely covered. It is left undisturbed for one week after which it ripens. By doing so, the keeping quality of mango is high and the taste as well as aroma is retained.

If the number of mangoes to be ripened is less, instead of pits, mud pots can be used. The practice of doing such ripening method exists for more than 90 years.

1Maravalli Kilangu Padhapatutahal (Tapioca Preservation)

Method I

In January month, the matured tapioca is dried under sun for a day. After this boil the tapioca till the raw taste removes and gets completely cooked. Leave after sometime and again boiled. This is to maintain stableness and again sun dried. Uluppai poochi which creates holes on the tubers are kept away during this storage.

II. Tapioca Storage- Vatthal (Chips)

- The freshly harvested tapioca are cut in to pieces and boiled in water.
- After boiling they are stored in the house, if required it can also be fried and stored over years.

III. Tapioca Storage : Unakkan Kizhangu

- Freshly harvested tapioca is sun dried
- The sun dried tapioca are taken in a gunny bag and tied
- The tapioca in the gunny bag is kept for storage even upto 3 years
- Cost effective when compared to other methods.

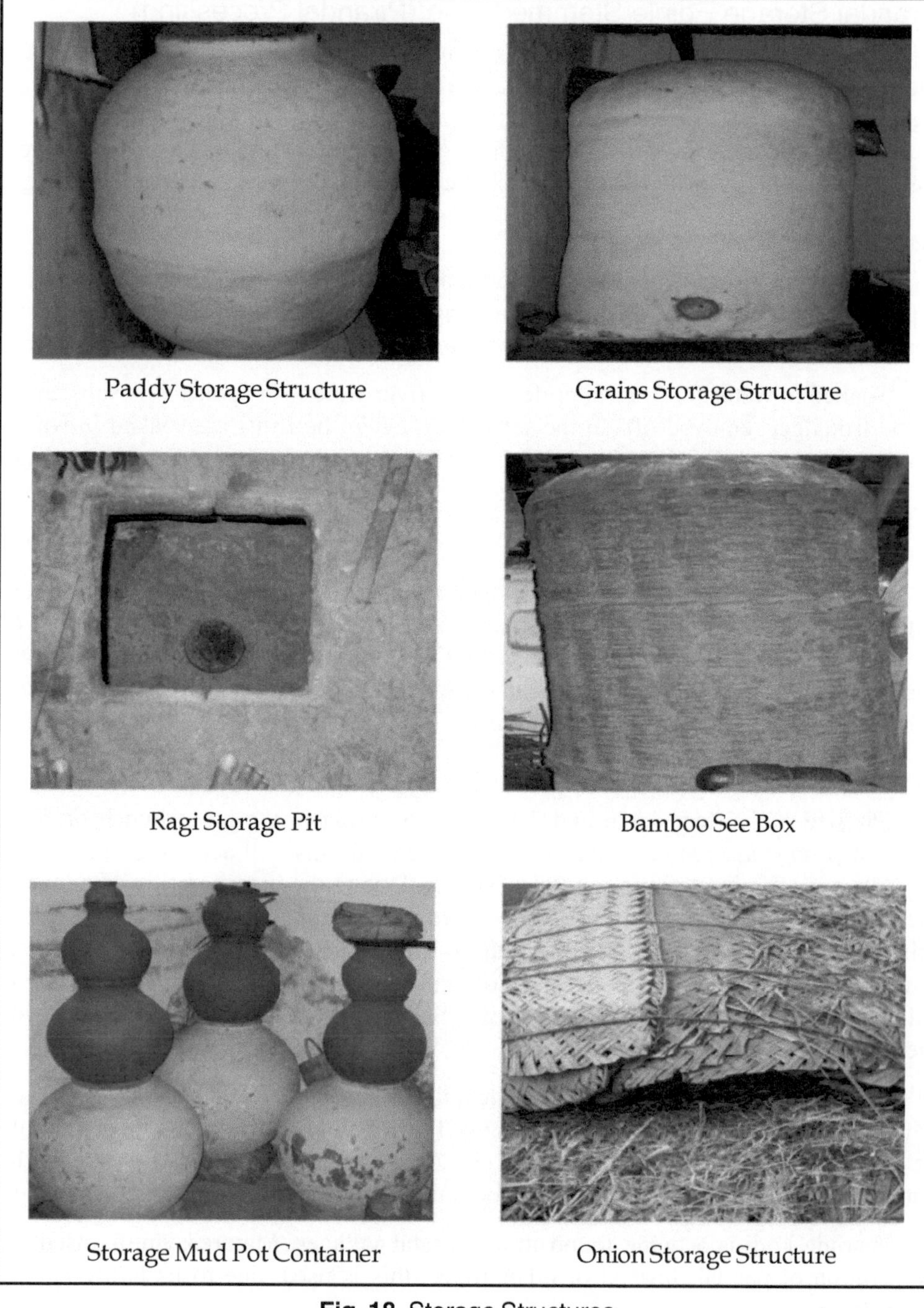

Paddy Storage Structure

Grains Storage Structure

Ragi Storage Pit

Bamboo See Box

Storage Mud Pot Container

Onion Storage Structure

Fig. 18. Storage Structures

Pirandai Storage Edible Stemmed Vine (Pirandai Processing)

The pirandai inside it is white in colour allowed to get soaked for a day, the quality or nature of the fruit is changed. Again, they are taken out and cut into small pieces and put in water for 3 days preferably in river water. The poisonous substances in the fruits get removed off. In the same way, again the fruits are soaked in water, boiled and washed. Finally added shriveled coconut pieces to it and kept for storage.

Paddy Seed Storage Method

To prevent the attack of pests in stored paddy grains, a plant 'kanjanamkorai', is kept along with the grains. It emanates a pungent smell which drives away the pests. Sometimes, neem leaves are placed over stored grains to keep the grains pest free.

Soaked in water for 3 days, preferably in river water. The poisonous substances in the fruits get removed off. In the same way, again the fruits are soaked in water, boiled and washed. Finally added shriveled coconut pieces to it and kept for storage.

Snake Gourd Seed Processing Storage

Fresh Snake gourd seeds are extracted from the well matured fruit and mixed with fresh cow dung. This cow dung treated seeds will be placed on the mud wall of the house, which absorbs excess moisture. Cow dung which acts as antifungal and pest repellent agent, prevents pest and disease attack on the seeds until the next season comes. Seeds can be stored upto one year.

Sowing

Pit size of 1.5 feet depth and 2 feet width is dug in the pit ,a bunch of fresh Pungam (*Pongamia pinnata*) leaves are incorporated ,this will acts as green manure as well as preventing termites and ant damage. Due to the incorporation of pungam, the plant will emerge quickly and grow sturdy. After that, pungam is covered with thin layer of soil, 2.5 kg well decomposed FYM is applied over it and then it is covered with 5cm of soil. The layer of soil and FYM should be mixed well and pit is rilled at centre, in which 3 seeds are sown. Sharp edge of the seed is dibbled inside the pit and then covered with soil.

Finally the entire pit is covered with leaf litters in order to avoid excess evaporation which in turn avoids damage due to rodents. Immediately after sowing, the pit should be watered. The seedlings will emerge out 7 days after sowing. After 3 weeks, the pandal is erected using Palmyra fibre/coconut coir pith rope/ Karung kodi-a twiner.

Karung kodi-is a twine found in all coastal villages. Matured vine is used for preparation of pandal. Instead of other ropes, this is used, due to its hardy nature, after drying.

From 50-60 days after sowing, it starts bearing. Naturally snake gourd grows in an irregular manner which is least preferred in the market. In order to avoid this economic loss, coconut pith is tied at the posterior end of the fruit. There are two types of local snake gourd varieties are prevailing. One is lengthy snake gourd (Neetu pudal), which grows upto a length of 6 feet and it weighs about 2kg. Second is short and stout snake gourd (Pani pudal) which weighs about ½ kg-1kg and it grows upto 1-1.5 feet length. Suitable season for sowing snake gourd is July-August. Existence of the local variety is for 75 years.

General Profile of the ITK Holder

Name&Address : Thiru V.N.ThangarasuVadaku Sali kulam,Vettaikaranirruppu Nagapatinam District Age : 67 Years, Farming experience : 47years

Ulunthu Vithai Nerthi (Black Gram Seed Storage Method)

Th.Rajalakshmi of Perahambi village in Manachanallur taluk of Trichy district follows two processing method in black gram before storage. She says that these methods are practiced for more then 75 years.

Black gram seeds to be stored are presoaked in water for one day.

- Red earth is made into slurry and to this slurry presoaked blackgram seeds are soaked for 2-3 hour.
- Then it is sun dried for 2-3 days.
- While sun drying the slurry enclosing black gram is hardened and form a thick shield like enclosure around the seeds.
- Seeds treated by this method can be kept until next season without deteriorating the quality as well as the quantity of the seeds.
- This is a no cost technology that has the advantages of improving shelf life, preventing the seeds from storage pest incidences like bruchids.

Pre storage processing practices of pulses varies widely depending upon the type of pulse. Black gram is treated with coconut oil so that all the seeds are coated with oily surface. On an average 50 ml of coconut oil is enough to treat 1kg of black gram. Thus treated black gram seeds are stored in a container made up of mud or plastic. These black grams can be stored as such or they may be split into two halves of using traditional stone splitters called 'Thirugai' splitted pulses can be sieved by using iron sieves. This type of iron sieve depends on the grain to be sieved and it varies from grain to grain. For e.g. big sized hole is used for sieving red gram, medium sized hole for black gram and small sized hole for green gram.

After splitting, when the complete mixture of splitted and unsplitted seeds are sieved by appropriate sieve. While sieving, splitted seeds falls down and unsplitted seeds remain in the sieve.

Such splitted and sieved black gram can be stored for 6-8 month. By treating black gram with coconut oil, the damage due to pulse beetle is effectively controlled. This is another low cost technology that can increase the keeping quality as well control the pest incidence in black gram.

Ellu Vithai Nerthi (Gingelly Seed Preservation)

Gingelly is mostly affected by storage beetles that drastically reduce its quality and shelf life. So to mitigate from such damage, Tmt. Rajalakhsmi follows many methods that involves very low or no cost. They are found to be eco friendly, and the process is alone with out deteriorating the quality of gingelly seeds.

a. **Using vasambu** : Just by placing Vasambu (*Achorous calamus*) on stored gingelly will safeguard it from pest damage. I.e. vasambu has a property of repelling the pest through which the seeds are safeguarded. This is a low cost technology that costs only about Rs.10/-.but the disadvantage of this method is that, only a little quantity of seeds can be processed and stored.

b. **Using dried chillies** : In this method, a handful of dried chillies are spread over the surface of gingelly. The concept of repellation is used here to control storage pest. This is also a no cost technology.

c. **Using ash** : The concept of mixing is similar but the content is different. Here ¼ kg of ash is used to treat 1 kg of seeds. Seeds are mixed well with ash and stored separately. This is a low and no cost technology that has various benefits in such pre-storage.

d. **Without treatment :** Neem and pungam leaves are known to have pest repellent property. So invariably neem and pungam leaves are just spread over the grains that prevents it from pest attack. This technique is very much eco friendly and so it doesn't deteriorate the quality of seeds. Similarly even nochi leaves are used. It may be spreaded into the grain or the twigs of notchi may be inserted between the gunny bags filled with grains.

AGRICULTURAL TOOLS

S.No	Particulars
	PLOUGHING AND LEVELING TOOLS
1	**Katti udaikum karuvi (Clod breaking tool)** in Mallakuli village of Sathy taluk in Erode district : Bullock drawn the tool, the iron rods are break the clods

2 **Kaluvai (Payaru ootum karuvi)** Mallakuli village of Sathy taluk in Erode district : Facilitate uniform spread of seeds and easy sprout of seeds

3 **Parampu palagai (Wooden leveling board)** Type I V. Mettuppalayam village of Salem district : Used for spreading and sweeping of Paddy and Groundnut during sun drying and also used for leveling the paddy field before transplanting the paddy seedling

Rack (Leveling board) Type II in Solamadevi post, Udayar palayam taluk of Perambalur district. Used to level the land.

4 **Manvetti (Spade)** Palmyra workers Development society (PWDS), Crystal Street, Marthandam in Kanyakumari district. Lengthier handle and broader main plate portion which lifts the soil from the field

Palmanvetti in Palmyra workers Development society (PWDS), Crystal Street, Marthandam in Kanyakumari district. Dig the highly compacted soil and break the clods, remove the bushes and shrubs

Mammaty (Spade) in Keevalur taluk of Nagapattinam district. Using this Mammatty for digging the field.

5 **Koru Kalappai (Koru + Seed box)** in Koolathanchrrry, Uthiramerure taluk of Kanchipuram district. It is used for sowing and weeding operation.

6 **Nattu Kalappai (Country Plough) I** in V. Mettuppalayam village of Salem district. It is comfortable for straight and easy ploughing.

7 **Melore / Bose Kalappai** in Ilankudi village of Sivagangai district. It cuts a new furrow slice.

WEEDING AND THINNING TOOLS

8 **Kalai Kothi (Small hand hoe)** in Mallakuli village of Sathy taluk in Erode district. Used to uproot the weed completely

9 **Koorgai (Thinning tool)** in Mallakuli village of Sathy taluk in Erode district. Used for thinning ragi crop.

10 **Bagunda (Weeding tool)** in Mallakuli village of Sathy taluk in Erode district. Used for weeding in ground nut field as well as for earthing up activity.

11 **Araganagunda (Weeding tool)** in Mallakuli village of Sathy taluk in Erode district. It is used for thinning operation, in crops like gingelly and ground nut.

12 **Payarukunduvay** in Sambantha kottai, Santhana palli post in Thenkanikottai taluk of Krishnagiri district. It is used for Thinning and weeding.

13 **Marapolluku** in Sambantha kottai, Santhana palli post in Thenkanikottai taluk of Krishnagiri district. Used for thinning and weeding operation. But it is used at the later stage of the crop.

14 **Uzhavaram (Weed remover)** in Solamadevi post, Udayar palayam taluk of Perambalur district. Used to remove the weeds like grasses and broad leaved weeds.

15 **Burme Saturn (Dry land weeder cum mulcher)** in Kappalur, Kannangudi post, Devakottai taluk of Sivagangai district. Used for weeding and mulching operation.

16 **Mammaty kalaikothi (Spade Weeder)** in Kolathancherry, Uthiramerur taluk of Kanchipuram district. Used for weeding in widely spaced crops like ground nut.

17 **Oosi Kalaikothi (Needle type weeder)** in Kolathancherry, Uthiramerur taluk of Kanchipuram district. Used for weeding in the closed spaced crops like ragi and vegetables etc.

18 **Kothu Kalai Kattu (pick Axe)** in Kolathancherry, Uthiramerur taluk of Kanchipuram district. Used to dig the land left while ploughing i,e corner of the field.

19 **Paluvai Maram (Ragi weeder)** in Kolathancherry, Uthiramerur taluk of Kanchipuram district. Used for weeding in ragi field.

20 **Kondi ('C' Shaped Hook)** in Pallikonda village of Vellore district. Used for pulling the thorn push.

21 **Kala karuvi (Dry land cycle weeder)** in Kothali village of Thoothukudi District. Used to remove the dry land weeds.

MEASURING TOOLS

22 **Vallam (Measuring Device)** in K.N.pudhur village of salem district. Used for measuring the cereals (Paddy, cumbu, sorghum, ragi, groundnuts and millets) and materials.

23 **Tharasu (Indigenous Weighing balance)** in Kolathancherry, Uthiramerur taluk of Kanchipuram district. Used weighing grains and vegetables.

24 **Padi (Brass measuring device)** in pallikonda of Vellore district. Used to measure the grains.

HARVESTING TOOLS

25 **Alaku/Thorati (Hook)** Sambantha kottai, Santhana palli post in Thenkanikottai taluk of Krishnagiri district. Used to pluck the leaves and fruits.

Thoratti (Hook) in Pallikonda of Vellore district. It is used to plucking the tree branches (younger branches) and harvesting tamarind fruits and drumstick etc.

26 **Arivalmanai (Cutting knife)** Sambantha kottai, Santhana palli post in Thenkanikottai taluk of Krishnagiri district. Used for weeding and harvesting.

27 **Aakathi (Harvesting knife)** in Palmyra workers Development society (PWDS), Crystal Street, Marthandam in Kanyakumari district. Used for harvesting operation in coconut and Pine Apple.

28 **Manjal Vettum Karuvi (Turmeric Harvester) type I** in K.N.pudhur village of Salem district. It is used for harvesting the turmeric rhizome.

Manja kothi (Turmeric Harvester) type II in Karunkalpatti village of Salem district. Used for harvesting the turmeric rhizome.

29 **Korai Aruva (Korai harvester)** in Kunnipalayam village of Namakkal district. Used to harvest the korai grass.

30 **Iruppu Nigam (Bete Vine Leaf Harvester)** in Kunnipalayam village of Namakkal district. Used to harvest the betel vine leaves.

31 **Kodikkal Kathi (Sesbania pruner)** in Kunnipalayam village of Namakkal district. Used to harvest the Agathi branches.

32 **Pachakal Eani (Bamboo lader)** in Kunnipalayam village of Namakkal district. Used to remove the betel vine leaves and to remove the side branches of Agathi.

33 **Vanngu Arival (Sickle)** in Keevalur taluk of Nagapattinam district. Used to cut thorny bushes and used to clean coconut trees.

34 **Arival (Cutting Tool)** in Sambantha kottai, Santhana palli post in Thenkanikottai taluk of Krishnagiri district. Used weeding and harvesting.

35 **Vaashayinaa Kathi (Scrapper)** in Palmyra workers Development society (PWDS), Crystal Street, Marthandam of Kanyakumari district. Used to remove the banana sheath and fiber extraction.

36 **Sathuvakathi (Palm leaf making knife)** in Ramanadhapuram district. Used for palm leaf mat weaving and basket making.

37 **Korai cutter (*Cyprus sp*) (Koduval)** in Kunnipalayam village of Namakkal district. To cut the edges of Korai.

IRRIGATION DEVICE

38 **Kaakattai** in Palmyra workers Development society (PWDS), Crystal Street, Marthandam of Kanyakumari district. It is used for water carried over the required place.

39 **Thooku (Mudpot)** in Keevalur taluk of Nagapattinam district. Technique of irrigating nursery plot.

40 **Pattai (Indigenous Irrigation Water carrier)** in Kunnipalayam village of Namakkal district. Used to irrigate (Wetting the field) the betel vine field.

SIEVES

41 **Kadalai Salladai (Groundnut grading sieves)** in V. Mettuppalayam village at Salem district. Grading the good quality ground nut seeds.

42 **Ariya Salladai (Ragi Grading Sieves)** in V. Mettuppalayam village of Salem district. Cleaning the ragi Grains (To remove the dust and other unwanted materials)

43 **Palathania Salladai (Grains and Pulses Grading Sieves)** in V.Mettuppalayam village at Salem district. Cleaning and grading the good quality seeds from immature seeds.

44 **Mallattai Salladai (Ground nut sieve)** in Vettuvanam village of Vellore district. Used for grading the ground nut seeds.

45 **Thuvarai Salladai (Red gram sieve)** in Vettuvanam village of Vellore district. It is used to clean the broken red gram Dhall.

46 **Arisi Salladai (Rice sieve)** in Vettuvanam village of Vellore district. It is used to clean the rice from stone dust and other impurities.

TRAPS

47 **Anil kanni (Squirrel Traps)** in V. Mettuppalayam village at Salem district. For trapping squirrels in coconut trees as they damage tender coconuts and the top portion of coconut trees.

48 **Peruchali Kanni (Rodent traps)** in V. Mettuppalayam village of Salem district. Used to trap the rodent in the field and shape guard the plants from damage.

49 **Eli koondu (Rat traps)** in V. Mettuppalayam village of Salem district. Used to trap the rats in home and field.

OTHERS

50 **Vilangu Veratti (Animal scaring device)** in Malasjanoor, Burgur (post) of Bhavani (TK) at Erode district. Used to scaring the animals.

51 **Thengai Mattaiuri (Dehusker)** in Palmyra workers Development society (PWDS), Crystal Street, Marthandam of Kanyakumari district. To remove the outer husk of the coconut.

52 **Padhaneer kuduvai (Neerah jar)** in Crystal street of Marthandam of Kanyakumari district. To maintain the chillness of the Neerah.

53 **Panai Vellam Kopparai (Palm Candy Kopparai)** in Aandipalayam village of Salem district. Used to prepare palm candy.

Man panai Aachu (Mud Pot for Palm Candy preparation) in Aandipalayam village of Salem district. Used to prepare palm candy in a different shape.

Mara Aachu (Wooden Box for Palm Candy preparation) in Aandipalayam village of Salem district. Used to prepare palm candy in a different shape.

54 **Kopparai (Cane Jaggery)** in Devoor village of Salem district It is used to prepare cane jaggery.

Karumpu Vettum Karuvi (Sugarcane Harvesting Knife) in Devoor village of Salem district. Used for cutting the sugarcane.

Matti Pattai in Devoor village at Salem district. It is used to remove the dust from the sugarcane juice during the time of jaggery preparation.

Kuruppi in Devoor village of Salem district. It is used at the time of cane treacle cooling, to avoid sticking.

Muttukattai in Devoor village of Salem district. Used for broke the clot and also used to get fine powdered jaggery during the time of jaggery preparation.

Pagu Aathi in Devoor village of Salem district. It is used to stir the cane juice and help to avoid sticking of treacle.

55 **Mathu (Curd Mathu)** in V. Mettuppalayam village of Salem district. Used for prepare butter milk preparation.

56 **Aria kal (Ragi Stone Grinder)** in V. Mettuppalayam village of Salem district. Used for grinding grains and to break the pulse grains.

57 **Mara ural and ualkkai (Wooden Pestle and Mortar)** in Malayalappatti village in Salem district. Used to pounding the grains and pulses.

58 **Thali (Animal watering pot)** in Kalathiyur village of Vellore district. Used to watering the animals.

59 **Koonai (Traditional Water Lifting device)** in Vellapaneri village of Thoothukudi district. It is used to carry water from well.

Kavalai (Urulai) in Kalathiyur village at Vellore district. It is used to lift the water from the well.

60 **Theevanapetti (Feed box for smallruminants)** in Kothali village of Ottapidaram Block of Thoothukudi district. Used to given feeding to the small ruminants.

61 **Manja kopparai (Indigenous turmeric boiling stove)** in Gettavadi village of Erode district. Used for boiling turmeric rhizome.

Ploughing and Levelling Tools

Katti udaikum karuvi (Clod breaking tool)

Th.Sakthivel, 45 years old residing in Ganesapuram village in Sathy taluk of Erode district is practicing agriculture for the past 6 years. He has been following organic farming method and shows himself as a sample to his neighboring farmers.

He has invented a tool for breaking clods named clod breaking tool.

Description

The tool is made up of wood and iron rods. Following are the components of the tool.

a) **Clod breaking roller** : A big cylindrical shaped roller with a diameter of 20 cm and length of 1.5 m Velamaram (*Acacia spp*) used for making it.

   Iron rods of 10 cm length is sharpened and made flat like a chisel on both ends. These rods are inserted into the roller such that one end projects out. Rods are inserted in an alternative row arrangement. At the centre of the cylindrical roller and iron rod/ long wooden pole is inserted in the centre.

b) **Connecting pole :** The roller is connected to two poles of both the ends. The iron rod / long wooden pole inserted in roller are fitted into a hole provided in the bush, which is fitted with the poles.

   The connecting pole is slightly bended at the anterior end which facilitates bullock while drawing i.e. prior it was used as a straight pole, but while drawing the tool, bullock legs striked to it and got wounded. The poles are slightly bended by which it was avoided.

c) **Shaft/yoke** : the connecting poles are connected to the shaft/ yoke which is tied to the bullocks.

d) **Seat** : A wooden plank is fitted on the connecting pole for sitting while operating.

Working Principle

While bullock draws the tool, while rolling is rolled and the iron rods break the clods which are smashed by the roller. Since the iron rods are inserted alternatively there is no chance of leaving the clods without breaking.

Moreover, the operator when sits seat provided near the roller, weight is increased which ensured clod breaking as well as leveling.

Cost Effectiveness

It costs about Rs.3000/- Which includes wood and iron. Compared to the implements that does clod breaking activity and leveling, this is relatively cheaper method for clod breaking.

Advantages

1. Its efficiency is 1 ac per 2 hrs, which is sufficient to operate with minimal investment.
2. Moreover it needs only one person to operate or handle the entire activity.
3. Since it is bullock drawn, and also with reduced weight (approx. 100 kg; including operator).
4. In the implement, no iron bushes are used, instead wooden bushes are used which needs only rare grease application. So maintenance is less.

Disadvantages

1. Compaction is prevented.
2. The seating height of the operator will be round 1-1/2 feet from ground level while clod breaking. So dust rose from the ground while operating will be a disturbing factor for the operator.
3. With the present set up, there is a risk for the operator regarding safety. Since there is no safety provision at the back, care is needed such that roller doesn't cause any harm while operating.

Other Details

While moving to the field, coconut fronds are placed under the roller such that, roller is rested upon the broad base of the frond. Similar set up is made at both the ends. It has also been planned that this tool can be used for pulling nut grasses, by modifying the rod's ends like a hook like structure. It has been estimated that Rs.1000/ - will be additional cost for inserting it. Similarly an seating arrangement to be made using 3 planks (Wood) to ensure safety.

Reason for Invention

The soils of Nilgris extension of Erode district are lateritic. Lateritic soils are hard enough, that they cannot even get completely broken by the rotavator. As well, the weight of the tractor is more which creates additional soil compaction. The amount of microorganisms also gets decreased. For this, he designed a clod breaking device with less cost, easily operated and less weight.

Kaluvai (Payaru Ootum Karuvi) - Leveller

Kaluvai is used in Mallan kuzhi village of Sathy taluk at Erode district. It is a wooden tool of about 5-6 feet length; it is used to level the ragi field. A long wooden pole is fitted with a handle in the middle, facing upwards and nearly 10-15 pegs are fitted facing downwards. It is made out of any wood but kaluvai is preferable.

Uses-After sowing Ragi it is used to level the field. This facilitates uniform spread of seeds and easy sprouting of seeds.

Cost- It costs about Rs.200 for making a single kaluvai

The Wooden Leveling Board

a. Parambu palagai

This tool is used in V. Mettuppalayam village at Salem district; it is existence about 50 years

Purpose

1. It is made up of neem wood and it is used for leveling the paddy field after puddling
2. And also used for spreading and sweeping paddy, groundnut, sorghum grains while sun drying.

Manufacturing

It has a wooden blank which measures 6 feet long and 1.5 feet wide with a handle. Chains are providing at the end of the blank for hitching. Leveling board is prepared by local carpenters. It takes one day for completing this task.

Operation

When the operator stand over the board, it stinks soil lightly in to the loose mud, excess soil is collected and released to the undulated place. Since it is field operation it is attached with cattle but in drying floor it will operate manually.

Cost

The cost of the board is Rs.200/-

b. Rack (Leveling board)

This rack is used in Thenkanikottai village at Krishnagiri district. Wooden leveler is used to level the dry lands. It is a bullock drawn tool, made out of both iron as well as wood, with the size of 3.5 feet length and 1 feet width. In wooden rack two poles are attached which facilitates the bullock to pull the rack easily. Two iron rings are provided in which the yoke and neck rope of bullock is tied. At the centre of the plank wooden handle is attached with the help of iron bolts. While operating the rack, the handle should be hold tightly to get well leveled field. (One member can dig the soil and another one for operating the rack). Two members can level an acre of land in 4 days. Iron rack cost about Rs.1000 whereas wooden rack costs about Rs.800.

Spade

a. Manvetti

In kanyakumari district the spade is of different type with a lengthier handle and broader main plate the portion which lifts the soil from the field. The handle is of 3-3 ¼' and the bottom plate is 1'x ¾' the wood used for making spade are usually from Vaahai (*Albezia lebbeck*). The lengthier handle allows the farmers to work longer time without getting tired.

b. Palmanvetti

It is used in garden land eco system of Kanyakumari district with the handle size of 3-3 ¼' and the bottom plate is 1'x ¾' and the tines are 15 cm in length and is four in number. This is to dig the highly compacted soil and break the clods, remove off the bushes and shrubs.

c. Mammaty

Square shaped iron blade is inserted in the 3-4 feet lengthy handle. Handle is made out of Pinnai tree's wood. Because, it will not affected by frequent exposure to water and sun light due to it‘s lengthy handle less strain to back bone. Normally farmers in keevalur taluk are using this Mammaty for digging the field which is coastal sandy soil.

Total weight of the tool is 2-2.5 kg, Cost of the tool is Rs.150/- it is Existence about 100 years in Keevalur taluk at Nagapattinam district.

Koru Kalappai (Indigenous Seed Drill – 'Koru')

It is a bullock drawn tool used for sowing ground nut and black gram and also used for weeding operation.

Parts

Mozhi – Handle used to hold the plough / koru

Thindu – wooden plant (4 feet length) in which mozhi and iron pegs are inserted

Munsattam – connecting pole

Erukaal – wooden pole, attached to the "Nukathadi"

Kuzhai – wooden tube made out of bamboo shoot through which seeds are sown.

Sera/kar – iron pegs, used for ploughing.

Existence : > 100 years

Coverage : 2 ac/ day

Wages : Rs.100/day

Seed box : It is made out of bamboo splits are used to keep the seeds while sowing. This seed box is tied on the waist of the man; he can also walk along with the koru for sowing. While sowing two hands are used simultaneously.

General Profile of the ITK Holder

Name and address : Th.Narayanasamy, Kolathancherry,Utheramerur Tk, Kanchipuram Dt

Nattu Kalappai (Country Plough)

a. Country Plough I

Purpose

This plough is traditionally used for ploughing the agricultural lands.

Making

It is made by local carpenter, once in a year it will be modified. Newly made plough is used for practiced animals; old plough (small sized plough) is used only trained animals.

Manufacturing

The plough consists of a wooden body to which a handle and a shaft pole are attached. The body is made with a bent piece of hard wood, like Neem and Acacia sp. With arms making a angle about 120 ^{0} to 135 ^{0}. A small piece of bar share made of mild steel is fixed at the plough bottom. It actually penetrates the soil and protects the wooden wedge against wear and tear. The length of the shaft pole varies from 2.5 to 3.5 m. One end of the shaft pole is fixed to the body and the other end is free to be secured with yoke during working. The handle is fixed with the back of the body.

Special Feature

Wooden plough is comfortable for straight and easy plough; soils are pulverized up to 20- 25 cm depth. It covers the area of 2 ac / day.

Cost

The cost of the plough is Rs.300/-

b. Country plough II

The indigenous plough consists of wooden body, to which shaft pole and handle are attached, the body is made with a bent piece are hard wood, like **Acacia spp**, with arms making an obtuse angle (135 degree). It is given wedge shape. A small piece of flat iron (share) serves as the piercing point of the plough and it is fixed over the plough body with one or two clamps (staples).the one end of the shaft pole is fixed to the body and other end is free, to be secured with the yoke during working. the nature of the ploughing or breaking up the soil performed by this type of plough consists in the opening v-shaped furrow. The depth of furrow and its width may vary according to the size and set up the working part or plough bottom.

Uses

1. For opening soil and preparation of seedbed.
2. For covering the manure, spread on the land.
3. For inter cultivating wide spaced crop like cotton, redgram etc.
4. For removing weeds.
5. For sowing seeds.

Bose Plough or Melur Plough

This is a simple plough skillfully made to have the advandages of wooden and iron plough. It is a wooden plough which takes a mould board cum share instead of the usual small iron share. The mould board share is slightly tilted. It cuts a new furrow slice. The slice is twisted slightly and pushed to s side. There is a complicated fitting or adjustments. The draught is loe and less than that of mould board plough and country plough.

Weeding and Thinning Tools

Kalai Kothi (Small Hand Hoe)

Suitable for Eco system : garden land and dry land eco system

Fig. 19. Ploughing and Leveling Tools

Technological Details

Thiru. Selvaraj is Rtd teacher cum organic farmer who have designed a small size handy hand hoe. It is mainly meant for weeding the garden and dry land. It is made up of iron and weighs about ¼ kg.

Advantages of Hand Hoe

- It is used to uproot the weed completely.
- Less weight compared to ordinary hand hoe.
- No occupational health hazards to farm women.
- One woman can cover 1 cent/ hour and 6 cents /day.
- There is no damage to root of the crop.

Reason for Intervention

Previously used hand hoe weighs more than the modified new one. So, the farm women get tired up quickly. Previously used hand hoe removes the grass and other monocot weeds at top level only. But modified one removes the shrubs completely. This is the reason behind the intervention.

Koorgai (Thinning Tool)

It is a wooden tool of about 5-6 feet length; it is used to thinning the ragi field. A long wooden pole is fitted with a handle in the middle, facing upwards and nearly 4-5 pegs are fitted facing downwards.

Uses

- It is mainly used for thinning ragi crop.
- It is also used in sowing when the soil is wet condition.

Bagunda (Weeding Tool)

The entire structure looks like a plough. More or less this is also similar to the koorkai but the difference is, the wooden pole in which the pegs are fitted is slightly bigger in size i.e. the diameter is up to 15 cm. Two pegs are fitted at the bottom which is fitted with a flat iron strip. It is fitted such that, two pegs are connected at the both end of the strip/blade and the mid portion of the strip/blade alone gets buried into the soil while operation.

Use : It is used for weeding in groundnut field as well as for earthing up activity. It is also used in other crops like Avarai, Motchai.

Arganakunda (Weeding Tool)

This is similar set up of bagunda, but instead of one such wooden pole, a pair of set up i.e. wooden pole fitted with two pegs and an iron strip/blade is fitted and operation handled with two hands. A rope is tied to connect the pair, at the handle portion.

Uses : it is used for thinning operation, in many crops like gingelly, groundnut, especially in ragi crop. While bullock draws the implement, person who operates it should press well so that the thinning is performed effectively.

Advantages

- Implements can be made out of locally available forest trees.
- Less preparation cost.
- Easy for operation.
- Eco friendly and involves less cost i.e. in mechanized farming diesel and petrol usage leads to air pollution and needs additional cost.
- Less maintenance cost.

General Profile of the ITK Holder

Th.M.S.Sivanna, 60years old and 40 years experienced in agriculture is residing in Mallan kuzhi village of Sathy taluk of Erode district. His primary occupation is agriculture.

Existence of the implements : More than 80 years.

Payaru Kunduvay (Weeder)

It is a tool used for thinning and weeding operation in the field at the earliest stage of the crop i.e. 11 days after sowing. Six number of one feet iron pegs are inserted in the wooden pole of about 2 feet length facing downwards. Number of pegs will vary according to the size of the wooden pole, capacity of the bullock and stage of the crop. The iron pegs are gradually broadened towards the bottom. The bottom of the peg is sharpened for efficient and effective weeding and thinning operation. 'T' shaped wooden handle of about 3 feet length is attached at the centre of the pole which faces upwards. Two bamboo poles are attached as shaft. It covers 1.5- 2 acres / day.

Note : Shape of the peg can vary. Sharp edged pegs can be used.

It cost is about Rs.5000/-.

Mara Polluku

It is also used for thinning and weeding operation. But it is used at the later stage of the crop i.e. 45 days after sowing. It is similar to that of Payaru kunduvay. But dimension will vary .Wooden pole of about 3.5 feet, pegs are made up of wood about 1 feet length. Number of pegs attached are 12. Architectural works are done in the wooden pole. It costs about Rs.200 (approximately).

After finishing the one stripe of the entire length of the field the thinned and weeded plants are collected and disposed.

Uzhavaram (Weed Removing Tool)

It is an indigenous tool used to remove the weeds (grasses, broad leaved weeds). It has two parts one is wooden hand second is iron blade. A handle is made out of wood, it‘s’ shape is cylindrical and the length is 1.5-2.0 feet. One end of the handle is sharpened. ‘V’ shaped iron blade which has the 30 cm lengthy cylindrical provision is attached to the sharpened end of the handle. Total weight of the uzhavaram is 1.5 kg (approximately). The length of the handle, size of the iron blade and total weight of the uzhavaram may vary it depends upon the efficiency of the member/person.

If any damage may occur we can change the handle or iron blade. Continuous use of the tool may reduce efficiency of the iron blade. We can also sharpen the iron blade.

Advantages

- It is a less weight tool.
- Easy for handling and operation.
- No /less repairing cost.
- Durability is more.

Burme Saturn (Dry Land Weeder Cum Mulcher)

Introduction

Generally, weed growth is a menace in dry lands. It affects the crop drastically and reduces the yield. To overcome this weeding at critical stage is important. Thus, Mr. Chinnathambi has developed a design to uproot the weeds and mulching them simultaneously in the field.

Design of the Machine

The new design has been developed similar to that of cultivator tynes. The wedge shaped angle designed at 60 degree and it has designed with a hole and knob. The plough end is similar to that of chisel plough to remove deep rooted weeds. It is made up of galvanized iron. The entire length of plough is 30 - 40 cm and the curved portion length is 15-18 cm. Similary, 10 tynes have been designed and fitted into the cultivator. If it is operated by bullock, 3-5 tynes are sufficient. The depth of the plough should be 5-10 cm only.

Operation Mechanism

The 10 tynes should be fitted instead of cultivator tynes and operated for burial of weeds. By using tractor lever mechanism the depth should be maintained at 5 cm since the plough is to be operated in the paddy field one week after sowing. Thus, the initial stage weeds are uprooted and mulched as manure to the emerging paddy seedlings. It can not be operated in water stagnated field.

Salient Features

1. The cost of plough (10 tynes) is Rs.1500/-
2. It is issued to the villagers on rental basis at the rate of Rs.150/- per hour (including the labour cost)
3. Coverage is 2 hrs / acre. Thus, the cost of coverage per acre is Rs.300. But the manual weeding costs Rs.1000/- (20 w @ Rs.50/-).
4. One time weeding is sufficient for the crop. But, the manual method needs 3-4 times.
5. Bullock drawn method needs 3-5 tynes and the coverage is 4 hr/acre. Here also the depth can be adjusted accordingly.
6. As this instrument is compact in size it can be operated in the corners areas also.
7. It reduces the manure cost as it mulches the weed

Name and Address of the Innovator

Mr. Chinnathambi,S/o. Seyya,Kappalur,Kannangudi post,Devakottai taulk, Sivagangai dt, Ph : 04561 – 274 212.

Mamaty Kalaikothi (Spade Weeder)

It is used in Kolathancherry village of Kanchipuram district. It is in existence for more than 10 decades. One feet wooden handle is inserted in iron blade whose length and width is 10 cm and 10 cm respectively. Iron blade looks like a square shaped one. It is a type of hand hoe used for weeding in the widely spaced crops like ground nut.

Oosi Kalaikothi (Needle Type Weeder)

It is used in Kolathancherry village of Kanchipuram district. In Oosi kalaikothi one feet wooden handle is inserted in iron blade whose length and width is 15 cm and 5-7 cm respectively. It is a type of hand hoe used for weeding in the closely spaced crops like ragi, vegetables etc.

Kothu Kalai Kattu (Pick Axe)

It is used in Kolathancherry village of Kanchipuram district. Pick axe is half feet lengthy iron blade with hole at top in which the wooden handle is inserted. Wooden pole serves as handle. It is a tool used to dig the land left while ploughing i.e. corner of the fields. It is also used to remove the interference of the root to the field and used to loosen the soil in hardy surface.

Cost :Rs. 200/-

Palluvai Maram (Ragi Weeder)

Used for weeding in ragi field.

Parts

- Mozhi – Wooden handle
- Thindu – wooden plank – 3-4 feet length, in which mozhi is inserted on the upper portion, iron pegs are inserted on the lower portion.
- Pal : 10-12 iron pegs. Length – 1- 1 ½ feet
- Coverage : 2 ac/day
- Wages – Rs.200/ac
- Existence > 150 years

General Profile of the ITK Holder

Name and address : Th.Raman, Kolathancherry,Utheramerur Tk,Kanchipuram Dt

Kondi ('C' Shaped Hook)

Purpose

It is used for pulling the thorn push after cutting the plant. It will help the farmers from thorn injury.

Making

This kondi is intervention by the farmer himself. It is made up of bamboo stick and iron hook. A question mark shaped hook is attached with bamboo stick.

Cost

It cost is Rs.15/-.

Kala Karuvi (Indigenous Dry Land Cycle Weeder)

In Kothali village of Thoothukudi District a farmer Shri. Arumugam has indigenously developed dry land weeder by modified bicycle wheel. It is easy to operate with the help of long handle and the bottom side is fixed with iron plate which is used to remove the dry land weeds.

Working Principle

The operation of the implement is just by walking and application of gentle pressure by hands. Both men and women can easily operate this implement in line sowing crops. This implement is the farmer's own innovation and is being used for the past 10 years.

Uses

This weeder is much useful in crops which are planted in ridges and furrows.

Cost

It costs about Rs. 750 /-.

Measuring Tools

Vallam (Iron Measuring Device)

Purpose

It is specially used for measuring the grains (cereals, Pulses and oil seeds etc.). The capacity of this equipment is 4kg.

Making

This equipment is made up of iron and it is round in shape. The height of the equipment is 23 cm, the diameter is 20 cm.

Cost

The cost of the measuring device is Rs. 200/-.

Tharasu (Indigenous Weighing Balance)

Parts

Kazhi-one feet cylindrical wooden stick on which measurements (1/4, 1/2, 1 kg) are marked. It is made out of Neem or Nuna or Puvarasu tree.

Pidi kayaru- Small jute rope is tied on the Kazhi to hold the balance.

Thattu (Plate) - Made out of bamboo which is used to hold weighing materials.

Tharasu kayaru- One feet lengthy three ropes tied on the three sides of the plate

Padi (Brass Measuring Device)

Purpose

It is an measuring equipments used for measuring grains like Cereals, Millets, Pulses and oil seeds

Fig. 20. Weeding and Thinning Tools

Making

It is made up of brass. It has role shaped flattened bottom and round shaped top portion. The height of the equipment is 25cm and the diameter is 8cm. It is golden in color. The capacity of the equipment is 1kg.

Cost

Cost of this equipment is Rs. 150 /-.

Harvesting Tools

Thorati

a. Alaku/Thorati (Hook)

It is used in Thenkanikottai taluk of Krishnagiri district. 'C' shaped iron knife / 'hook' like iron is tied at one end of the long bamboo pole which is used to pluck the leaf, fruit and plucking.

b. Thoratti

Purpose

It is used to plucking the tree branches (younger branches) and harvesting tamarind, drumstick, etc.

Making

It is made up of bamboo stick and iron hook. The iron hook having 2cm thickness with the height of 1ft. the top of the hook is blended in down ward position and the bottom is attached with bamboo stick.

Cost

This cost of the hook is Rs.15/-.

Arivalmanai (Cutting Knife)

It is used in Thenkanikottai taluk of Krishnagiri district. It is a cutting tool / knife used to the cut the fodder for cattle feeds. Earlier it was used for cut the mulberry leaves for silkworm rearing. One feet iron knife is attached at one end of the wooden plank. Length of the wooden plank is 3 feet and width is 0.5 feet.

Aakathi (Harvesting Knife)

It is used in Marthandam of Kanyakumari district. The knife is an iron made with curve at it top and bottom to get expanded 'U' shape. Knife is sharpened at one end and other side blunt. The knife is mostly used for harvesting operations, in coconut, pineapple.

Manjal Vettum Karuvi-1 (Turmeric Harvester)

Purpose

It is used for harvesting the turmeric rhizome.

Making

It looks like a hand hoe. The handle is made up of tree pole, the length of the handle is 60 cm and the metal blade are attached with the end of the pole. The length of the blade is 30 cm. The longevity of the tool is 4-5 years.

Cost

The worth (Cost) of this harvester is Rs.80/-.

Manja kothi-2 (Turmeric Harvester)

Purpose

It is mainly used for harvesting the turmeric rhizome

Making

It looks like a wood cutter and it is made up of iron. The handle is made up of tree pole. The bottom of the harvester is 8 cm thick and the top of the portion is sharp ended. Area under operation can cover according to man power.

Cost

It costs of the tool is Rs.150/-

Korai Aruva (Korai Harvester)

Purpose

It is specially used for harvesting korai.

Making

It is made up of very thin iron blade and wood. The iron blade has a breath of about 6cms and length is 35cms one side of the serrated edges. The handle is made up of neem or teak wood. The wooden handle is machine made.

Cost

It is available in local market and its cost is Rs.150/-.

Iruppu Nagam (Betel Vine Leaf Harvester)

It is used in Kunnipalayam village of Namakkal district.

Iron Water Carrier

Irrigation Device

Measuring Equipments

Rat Trap

Weighing Balance

Palm Candy Preparator

Fig. 21. Measuring and irrigation tools

Purpose

It is used to harvest the betel vine leaf. In a single cutting some fifty vine leaves can be cut depending upon the age and capacity of the users.

Making

It is made up of iron. A ring shaped tool is put to the thumb finger. The edge of the tool look like an inverted blade the ring is of 2cms width. The length between the iron ring and shaped of the inverted blade is 5cms.

Cost

The cost of the tool is Rs.25/-

Kodikkal Kathi (Sesbania Pruner)

Purpose

It is used to remove the side branches of the Agathi plants.

Making

It is a small sized knife with a 'C' shaped iron curve attached to a wooden handle. It is made up of wood and iron. The length of the curved iron piece is 8cm and the wooden part is 7cms. The wooden handle has several ring shaped curves giving grip to the users' hand.

Cost

The cost of the Sesbania pruner is Rs.100/-.

Pachakkal Eani (Bamboo Ladder)

Purpose

The ladder is used to harvest the betel vine leaves and the side branches of the agathi. It is also used to tie the branches of agathi plants and betel vine plants with the help of banana fibre.

Making

It is made up of bamboo poles. Two 10ft length bamboo poles are attached with vertical serving as steps in between them. This is called a ladder. Another two poles of same size are tied at the top of the ladder in the shape of an inverted "V" these poles help to fix the ladder in the betel vine field.

Cost

The cost of this bamboo ladder is Rs.250/-.

Vanku Arrival (Sickle)

It is used in keevalur taluk of Nagapattinam district. The Length of the tool is 1-1.5 ft. in the top 'C' shaped portion is sharpened for cutting rest of the portions are having blended end to safeguard hands during cutting operation. It is used to cut thorny bushes and used to clean coconut trees.

Arrival (Cutting Tool)

It is used in Thenkanikottai taluk of Krishnagiri district. It is a 'C' shaped traditional cutting knife made out of iron. The length of the iron knife is 15 cm. Handle is made up of wood and special provision is made in the handle at the lower end to give grip to the hand. Knife is inserted in the upper end of the handle and covered with bras ring. This knife is used for weeding and harvesting.

Vaashayinaa Kathi (Scrapper)

It is used in Marthandam of Kanyakumari district. It is a rectangular shaped one, made of iron and has a wooden handle. The length and breadth of the scrapper is 10 cm and 6 cm respectively. The handle is to hold on the knife and remove the sheath. Wet banana sheath preferably un-diseased one is taken and kept for extraction of fibre. The scrapper is kept on it and scrolled towards the bottom, the sheath is removed and the fibre is extracted. Extracted fibre is used for making toys and other fancy articles.

Sathuvakathi (Palm Leaf Making Knife)

It is used in Thuthivalasai village of Ramanadhapuram district. It is a knife used only for Palm leave mat weaving and palm leaf basket making. It is specially designed by black smith as per the instruction of palm leaf mat weaver and the cost of the knife is Rs.20/- only. There are two portions in the knife. The curved frontier portion is used to tear the palm leaf mat and the back long portion is used to punch and weave the mat/basket. As it is tiny and long it is very easy to handle while weaving. The weight of the knife is 200-250 g.

Koduval -Korai Cutter (*Cyprus sp*)

Purpose

It is specially used for korai cutter. After harvesting the korai grass the edges of korai grass is cut by using this cutter (koduval).

Sugar Cane Harvester
Pine Apple Harvester
Cutting Knife
Korai Grass Harvester
Betelvine Harvester
Sesbania Pruner

Fig. 22. Harvesting Tools

Making

This tool is made up of iron blade and wood. The iron blade is very thick and has a breadth of 8cms the total length of the implements is 35cms the wooden handle has a length of 12cms. It has shaped and sharpened edges.

Cost

The cost of the tool is Rs.150/-.

Irrigation Device

Kaakattai

It is used in Marthandam of Kanyakumari district. It is an irrigation device designed for irrigating vegetables. A bamboo stick of 5'-6' is taken and to its both sides palmyrah leaf made – kaakattai are tied. The kaakattai is opened fully at the top. Water is taken in the kaakattai and poured to the vegetables. This is used in areas where water scarcity arises. Easily makes in the farm itself. Water is directly poured to the root zone and plants get their required water. An alternate to this now a days iron tin are tied to both ends and water is carried over to the required place.

Thooku (Mudpot)

Thooku is nothing but a mud pot with a dimension of 45 cm width and 50cm height with holding capacity of 10 litres. Generally Thooku is made by the potters and it is bought by farmers at the rate of Rs.10/Thooku.

While bringing up the water filled pots (Thooku) due to its weight and sloppy nature of the pond, balance, two (Thooku) are brought simultaneously together.

This technique of irrigating nursery plot with "Thooku" has been followed even by the ancestors more than 75 years, but the technique of pouring on straw bunch has been recently used by him.

Pattai (Indigenous Irrigation Water Carrier)

Purpose

It is boat shaped wooden tool used to irrigate the betel vine field.

Making

It is made up of Vadhanarayanan tree. The length of the Pattai is 40 cm and the width is 20 cm, an inner curve of the tool is 10cms depth.

Cost

The cost of this device is Rs50/-.

Sieves

Different types of Sieves

S. No	Name of the sieves	Made up of	Size of the sieves	Diameter of the mesh hole	Purpose of the sieve	Cost of the sieve
1	Groundnut Grading Sieves	Iron plate	2.5ft length and 1.5ft width.	2.5cm x 8cm (Rectangular)	Grading the good quality ground nut seeds	150/-
2	Ragi Grading Sieves	Iron plate with wooden stick	3 x 3 ft	0.4cm (Round)	Cleaning the ragi Grains (To remove the dust and other unwanted materials)	300/-
3	Grains and Pulses Grading Sieves	Iron plate	1.5 ft	0.4cm (Round)	Cleaning and grading the good quality seeds from immature seeds	150/-

Mallattai Salladai (Ground Nut Sieve)

Purpose

It is used for grading the ground nut seeds. The good quality seeds are stag over the sieving equipment. The immature and damaged seeds are fall down from the equipment.

Making

This sieve is made up of iron. The round shaped sieve having the diameter of 45cm. Bottom of the sieve have vertical thin holes up to 60nos. the length of the hole is 5cm and the width is 1cm.

Cost

The cost of the sieve Rs50/-. This equipment is being used by the farmer for the past 80 years in Vettuvanam village of Vellore district.

Thuvarai Salladai (Red Gram Sieve)

Purpose

It is used to clean the broken red gram Dhall.

Making

This sieve is made up of iron. The shape of the sieve is round. The diameter of the sieve is 45cm the sieve has vertical thick holes up to 33 nos. the length of the bole 7cm and the width is 1.5cm.

Cost

The cost of the sieve is Rs.50/-. This equipment is being used by the farmer for the past 80 years in Vettuvanam village of Vellore district.

Arisi Salladai (Rice Sieve)

Purpose

It is used to clean the rice from stone dust and other impurity seeds.

Making

This sieve is made up of iron. The shape of the sieve is round and the diameter of the hole is 0.4cm.

Cost

The cost of the sieve is Rs50/-. This equipment is being used by the farmer for the past 80 years in Vettuvanam village of Vellore district.

Traps

Anil Kanni (Squirrel Traps)

Purpose

These traps are used for trapping squirrels in coconut trees as they damage tender coconuts and the top portion of coconut trees.

Manufactured

It is made up of iron wire which is used for power looms. Waste iron wire are collected and used for making traps (Kanni).

Mechanism

If the squirrels runs on the coconut trees it gets trapped in the traps are tied around the coconut trees.

Special Feature

It is no cost-effective and easy to prepare.

Peruchali Kanni (Rodent Traps)

Purpose

It is used to trap the rodent in the field and shape guard the plants from damage

Manufactured

It is made up of iron wire which is used for power looms. Two iron wires are tied on a 15 cm stick which acts as a trap. It is prepared by farmer itself.

Mechanism

This is placed near a rat burrow or on the rat path. If rat comes in /goes out of the tunnel it gets caught in the trap.

Cost

No needs to spend money for prepare the rodent traps.

Eli Koondu (Rat Traps)

This trap is existence for 25 years from now in K.N.puram village at Salem district.

Purpose

Used to trap the rats in home and field.

Manufacturing

It is made up of iron wire. The trap is semi circle in shape. This trap is having an inlet and outlet opposite to each other. It is 45 cm long and 20 cm wide and the weight is 1.5 kg.

Mechanism

The trap can be placed where the rats problem is predominantly achieve both home and field. In order to trap the rat food materials like Groundnut, Tomato, Vada are kept inside the trap. In one acre of land four traps are required. The trap should not be kept in one place for a long time and it should be changed once in a day.

Special Feature

In this trap is 5-10 rats will be caught at a time.

Killing

After trapping the rats, are collected in gunny bags in the outlet of the trap. Then it will be killed by beating it the ground or dipped in water.

Cost

The cost of the trap is Rs.150/-

Others

Vilangu Veratti (Animal Scaring Device)

Th. Pomman, 45years old is residing in Malasajanoor village of Anthiyur taluk of Erode district and practices agriculture for about 25 years. The hilly track cultivation

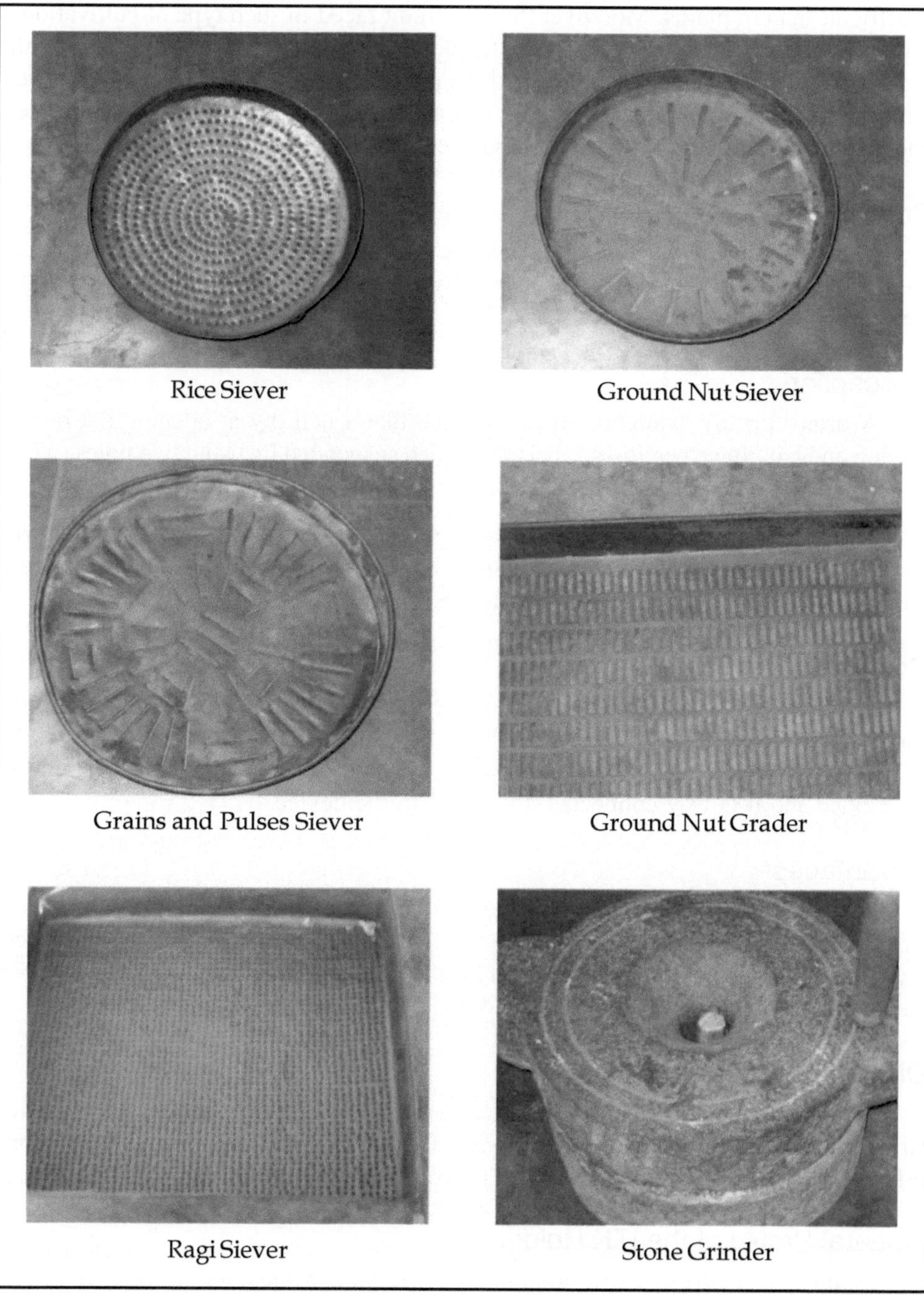

Rice Siever

Ground Nut Siever

Grains and Pulses Siever

Ground Nut Grader

Ragi Siever

Stone Grinder

Fig. 23. Sieveing and grading tools

is difficult than in plains. Moreover the constraint faced in such type of cultivation is difficult to solve. One such problem is the damage due to grazing animals especially pigs and birds. Since Tapioca and garden land rice crops are suited for hilly track, almost all the farmers go for cultivating it. But pigs and birds become a great problem and as a solution a kind of low cost device called 'veratti' are used.

Principle

The principle behind is avoiding damage due to pigs and birds by producing sound i.e. while sound is produced, pigs and birds are scared and so they run away from the field.

Description

A drum/ locally available oil tin is made like a bell, by at opening the bottom portion and hanging a peg inside. This tin is then suspended by using two poles (where in between the poles tin with peg is tied).

The peg is tied with a long string and held by the operator, who is provided with a shed like structure made out of bamboo splits and coconut fronds. This structure is erected upon four wooden pillars to a height of 3-4 feet.

Cost Effectiveness

It is low cost technology, where all the needed materials are available locally, for which no expense is to be spared. Regarding the device cost, it is almost nil. But the shed in which the operation costs up to Rs.150-200. But it is only a part in operating the device and it is way connected directly with the device.

Advantages

Crop is protected from the damage without more risk, usually either cracker is fired or sound is produced from one end of the field using plates. But such practices are either harmful to environment or not so effective which doesn't serve the purpose. But this 'Veratti' is placed in centre and operated from any end of the field

Disadvantages

It can be implemented in the fields up to one acre and not more than that. Because string tied to the tin, when goes beyond certain distance (approx 20mt) its sound efficiency decreases and thus effective utilization of the device can't be obtained.

General Profile of the ITK Holder

Name and address : M.Pomman, S/o Mathan, Malasjanoor, Burgur (post), Bhavani (TK), Erode district.

Thenkkai Mattai Uri (Coconut Dehusker)

It is an iron made having a long rod as its main portion. To it a sharp edged "Uri" is attached with a screw connecting them. A handle is present at the side to remove the outer husk i.e. to get coconut. At bottom is a straight platform like structure is present which is to place the 'Uri' stable in a place. Cost of the implement is Rs.75/ efficiency is 200-250 coconuts / hr can be dehusked / removed from its shell. Easily operated by an single person.

Padhaneer Kuduvai (Neerah Jar)

It is purely made out of palm and coconut leaf and fibre. Inside, the jar is coated with red earth to maintain chillness. Neerah collected from the Palmyra tree is directly poured in it and kept for use. Its capacity is 2.5 litres, has a handle at the top to carry from one place to other. No cost is incurred in making the jar. It is naturally made one, no chemical is added. Neerah kept in this jar remains its quality and taste.

Panai Vellam Kopparai (Palm Candy Kopparai)

This tool is used in Aandipalayam village at Salem district.

Purpose

It is used for palm candy preparation

Making

The vessel is made up of iron. It is round in shape and the diameter of the vessel is one meter. The top portion of the vessel is wide and it narrows down to the bottom. The capacity of the equipment is 54 liters.

Cost

The cost of the equipment is Rs.350/-

Man panai Aachu (Mud Pot for Palm Candy preparation)

Purpose

A small sized round shaped pot is traditionally used for moulding the palm candy.

Mechanism

Palm treacle is poured into the pot to get the pit condition. The weight of the candy is around 2 kg.

Cost

It is a low cost method because the cost of the pot is low.

Mara Aachu (Wooden Box for Palm Candy preparation)

Purpose

It is used to prepare palm candy

Making

It is made up of wooden piece. The length of the wooden piece is 1 meter, in this box 5 hallow pits are made, and the diameter of the pit has15 cm and the depth is 10cm. In this pit only treacle is poured. Longevity of the tool is 3-4 years.

Special Feature

It is used for longer time and un- broken.

Traditional Cane Jaggery Making Tools

a. Kopparai

Purpose

It is used for cane jaggery preparation.

Making

The kopparai is made up of iron. It is round in shape. The diameter of the kopparai is 8 feet. The top portion of the kopparai is wide which narrows down to the bottom in these kopparai 800 liters of cane juice can be poured.

b. Karumpu Vettum Karuvi (Sugarcane Harvesting Knife)

Purpose

It is traditionally used for harvesting the sugarcane.

Making

It looks like axe. The flat shaped iron blade is attached with wooden pole. The length of the blade is 15 cm, width is 12 cm.

c. Matti Pattai

Purpose

It is used to remove the dust from the sugarcane juice during the time of jaggery preparation.

Making

The dust remover is made up of metal plate. The shape of the dust remover is inverted round in position. Holes are present in this plate and the diameter of the hole is 30cm. The handle is made up of wooden pole and the length of the handle is up to 2 meter.

d. Kuruppi

Purpose

It is used in jaggery making process. It is used at the time of cane treacle cooling, to avoid sticking.

Making

It is made up of metal plate. The length of the plate is 23 cm and the width is 14 cm. It is attached with 2 meter wooden stick.

e. Muttukattai

Purpose

It is used for broke the clot. It is used to get fine powdered jaggery during the time of jaggery preparation.

Making

It is made up of tree wood. The wooden roller is attached with the wooden stick. The size of the roller is 30cm length and 12cm width. The length of the stick is up to 2 meter.

f. Pagu Aathi

Purpose

It is used to stir the cane juice and help to avoid sticking of treacle.

Making

It is made up of wood. The wooden board is attached with wooden stick. The size of the board is up to 35cm length, 18cm width and the length of the stick is up to 2 meter.

Mathu (Curd Mathu) - Churner

Purpose

Mathu is used for prepare butter milk preparation.

Manufacturing

It is made up of tree wood and bamboo sticks. A wooden stick with a hemispherical bottom having a grooves collect the butter. In that bamboo sticks streaks are made to fix the rope because the rope is used for rolling the curd mathu. The length of the bamboo sticks is 5 feet and the width of the groove is 15 cm.

Process

For butter milk preparation, mud pot or vessels are used; suitable time for preparing butter is early morning. So that easy preparation of butter from the curd. When the process is delayed, the butter formation is reduced because of heat intention. Nowadays mixie is used for butter separation, but the quality of the butter will differs. Collected butter is put it in to the container having the water, which helps in thickening of butter. Collected butter is kept for one week after that ghee is prepared. During the ghee preparation, moringa leaf is added to give good flavor and taste.

Cost

The cost of the curd mathu is Rs.50/-

Aria Kal (Ragi Stone Grinder)

Purpose

It is used in V. Mettuppalayam village at Salem district. This equipment is existence about two generation. Aria kal is traditionally used for grinding grains and to break the pulse grains. Mainly it is used for grinding ragi, red gram, black gram, green gram, cow pea and lab lab etc. The pulses are soaked in water and then sun dried. Red gram and lab lab (Avarai) are soaked in the whole night and then in the next day it will be sun dried. So that it will be in easy for breaking. Some times pulse seeds are fried and then broken, and it gives good flavor and helps easy broken.

Special Feature

Dhal obtained through this type of grinding process possessed good taste.

Mara Ural and Ulakkai (Wooden Pestle and Mortar)

Purpose

Used to pounding the grains and pulses. This tool is existence of two generation.

Manufacturing

It is made up of teak wood. Shape of the tool is cylindrical with slight curve in both the side. The height of the pestle is 3 feet. Centre of the top portion have one feet shallow depth with the diameter is 15 cm. The capacity of the pestle is 2 kg. It is used to pounding the grains like paddy, combu, wheat, thenai and samai etc. For pounding mortar is used, it has the 6 feet long and 7 feet wide. Fixed with an iron ring both side of the motor. It helps to remove the outer skin of the grains.

Special Feature

Hand pounded grains are compared to milling hand pounded grains are minimizing the grains broken. It has good taste compare to mill pounding.

Cost

It is prepared by the local carpenter. The cost of the tool is Rs.150/-

Thali (Animal Watering Pot)

This structure is build with red soil and bricks. The height of the structure is 4 ft. The diameter of the pot can be 60cm. It cost about Rs.80-100/-. It is traditionally used in Vellore district of Tamil Nadu.

Traditional Water Lifting Device

a. Koonai

It is used to carry water from well. It is made up of iron plates and the mouth is tied with leather tube for uniform flow of water. By using long ropes and with the help of Bullocks it carries around 60 litres of water at a time. Because of the invention of diesel and electric motor pump sets this koonai is just the thing in museum. Now a day in rural areas the so called koonai is used for Poultry shed or kozhi koodu.

b. Kavalai (Urulai)

Kavalai is made up of wood (Neem, Teak,) shape of the roller is round. The diameter of the roller can be 30cm the centre of the roller holes are provided it's moulded with iron blade. It is used to lift the water from the well. The cost of the roller is Rs150/-.

Theevanapetti (Indigenous Feed Box for Small Ruminants)

Iron trunk boxes which are previously used for keeping clothes and valuables now a day used as the feed box for small ruminants. Goats and Sheep will not eat the grasses which are mixed with the soil, dung and urine. So this iron plate box is very much useful for the safe feeding of small ruminants. This practice has been followed in Kothali village of Ottapidaram Block of Thooththukudi district.

Manja Kopparai (Indigenous Turmeric Boiling Stove)

The size of this stove is 5feet length and 3 feet breath .The depth is 6feet from the top of the earth. The harvest turmeric leaves are used as the fuel material. The fuel material are inserted in one way (3 Feet breath area) and out let smoke in other side of the breath area. Indigenous boiling drums having the capacity of 90 Kg each, can be kept on the Indigenous turmeric boiling stove for a period of 45minutes.this is the cost effective stove locally made in the field itself. They are using the harvested dried turmeric leaves as the fuel which is having no cost. To set up an Indigenous turmeric boiling stove 2 laborers are needed and the total cost of this set up is Rs 500 only.

MEDICINAL PLANTS

S.No	Particulars
1.	**Karuvel Tree** *(Prosopis julifera)* in Athireddiyur village of Erode district. Local medicine for scorpion bite.
2.	**Vishnu kiranthi** *(Evolvulus alsonoides)* in Athireddiyur village of Erode District. Cures the tuberculosis, cold and fever.
3.	**Purusalachanam** in Bhavani taluk of Erode District. It reduces the body heat and gives cooling to the body.
4.	**Ariyanangai** in Athireddiyur village of Erode District. Local medicine of Snake bite.
5.	**Unni mul thazhai** – Wild sage *(Lantana camera)* in Kolli hills of Namakkal district. Leaf extract is used to heal the cut wounds.
6.	**Karuumathai** - Indian Thorn Apple (*Datura fastuosa)* in Kolli hills of Namakkal district. Used for curing the asthma.
7.	**Notchi** - Five leave chaste tree *(Vitex negunda)* in Perahambi village of Trichy district. Used to cure fever and cold.
8.	**Mudakatran-**Balloon Vine (*Cardiospermum halicacabum)* in Perahambi village of Trichy district. Reducing joint pains and constipation.
9.	**Perumarunthu** - Indian Birth Wood (*Aristolochia indica)* in Perahambi village of Trichy District. To cure the poison bites.
10.	**Aanai nerunji-** Bathala Mooligai (*Pedalium murex)* in Perahambi village of Trichy district. Fed to the animal affected by diarrhoea.
11.	**Sangu** (Azima *tetracantha)* in Perahambi village of Trichy district. Added to the eye of dull and drowsy animal.
12.	**Adathoda** -Malabar nut (*Adathoda vasica)* in Perahambi village of Trichy district. Best relief for asthma.
13.	**Veli paruthi** (*Fergularia daemia)* in Perahambi village of Trichy district. Relief for indigestion.
14	**Thulasi** (*Ocimum sanctum*) in Perahambi village of Trichy district. Local medicine for cold.
15.	**Nayi vellai** -Wild mustard *(Cleome viscosa)* in Perahambi village of Trichy district. To use the dull and drowsy animal.
16.	**Nanchu murichan** -Emetic swallow wort *(Tylophora indica)* in Perahambi village of Trichy district. Animal affected by poisonous bite.
17.	**Papaya** (*Carica papaya)* and **Guava** (P*sidium gujava)* in Perahambi village of Trichy district. Cure the stomach ulcer.
18.	**Vella poondu** - Garlic (*Allium sativum)* in Perahambi village of Trichy district. For easy indigestion.

19. **Aamanakku** - Castor (*Ricinus communis)* in Perahambi village of Trichy district. Pain killer / reliever.

20. **Banana pseudo stem** in Perahambi village of Trichy district. Cure kidney stone.

21. **Arugampul** - Bermuda grass *(Cynodan dactylon)* in Perahambi village of Trichy district. Against cold, fever and poisonous bite.

22. **Kuppai meni**-Indian Acalypa (*Acalypa indica)* in Perahambi village of Trichy district. Curing scabies and internal wounds after delivery period in human beings.

23. **Aanaikathaalai** - Century Plant *(Agave americana)* in Myrada kvk, Acade Palli, Hodur village of Krishnagiri district. Purifies the blood, prevents the formation of pus and promotes the flow of urine.

24. **Pilapalsm** - Jack tree (*Atrocarpus heterophyllus)* in Myrada kvk, Acade Palli, Hodur village of Krishnagiri district. Arrest diarrhoea, fever and asthma.

25. **Neer mulli** - Long leaved barleria (*Hygrophila auriculata)* in Myrada kvk, Acade Palli, Hodur village of Krishnagiri district. Cure the jaundice, rheumatism and urine - genital diseases.

26. **Sudukadu mallikai** - Periwinkle *(Catharanthus roseus)* in Myrada kvk, Acade Palli, Hodur village of Krishnagiri district. Anti carcinogenic vermifuge.

27. **Mahuli keerai** -Wool flower (*Celosia argentea)* in Myrada kvk, Acade Palli, Hodur village of Krishnagiri district. Treatment of tuberculosis, acute abdominal pains and inflammed stomach.

28. **Christmas star** -Poinsettia *(Euphorbia pulcherrima)* in Myrada kvk, Acade Palli, Hodur village of Krishnagiri district. Increase of milk secretion of lactating mothers.

29. a. **Tottalvadi** - Sensitive plant (*Mimosa pudica)* in Myrada kvk, Acade Palli, Hodur village of Krishnagiri district. Used for constipation, liver disease and also mouth and throat inflammations.

    b. **Thotta sinigi** –Touch me not (*Mimosa pudica)* in Vettuvenni village of Kanayakumari district. To cure piles and diabetes.

30. **Andhimalligai** - Four O' clock plant (*Mirabilis jalapa)* in Myrada kvk, Acade Palli, Hodur village of Krishnagiri district. Promotes flow of urine.

31. a. **Marul** - Bowstring hemp *(Sansevieria zeylanica*) in Myrada kvk, Acade Palli, Hodur village of Krishnagiri district. Used as a purgative.

    b. **Marul** -Bowstring hemp (*Sansevieria zeylanica*) in Tandarai village of Kanchipuram district. Controls ear problem.

32. **Pachaiyalari** -Yellow oleander (*Thevetia peruviana)* in Myrada kvk, Acade Palli, Hodur village of Krishnagiri district. Used to reduce / eliminate fever.

33. **Thali Keerai** *(Ipomea sepiaria)* in Myrada kvk, Acade Palli, Hodur village of Krishnagiri district. Cure diarrhoea in animals.

34. **Thaivelai** (*Cleome gynandra)* in Chinnamuthulur village of Perambalur district. To cure scorpion bite.

35. **Omavalli** (*Coleus aromaticus*) in Chinnamuthulur village of Perambalur district. To against cold and fever.

36. **Vallarai** - Indian peeny wort (*Centella asiatica*) in Chinnamuthulur village of Perambalur district. Increase the memory power.

37. **Perukurinchan** in Chinnamuthulur village of Perambalur district. For control diabetes.

38. **Sathisaranai** (*Trianthema decandra*) in Chinnamuthulur village of Perambalur district. Relief from gas trouble.

39. **Manathakgali** (*Solanum nigrum*) in Chinnamuthulur village of Perambalur district. Cure Uterus problem and ulcer.

40. **Thavasi murungai** (*Jusitica tranquebariensis*) in Chinnamuthulur village of Perambalur district. For control dog and cat bite.

41. **Karunthulasi** (*Ocimum basilicum*) in Karadikulam village of Perambalur district. To control diabetes and blood pressure.

42. **Tree barks of Nagal** (*Syzygium* cumini), **Arasu** (*Ficus religiosa*), **Aathi** (*Bauhinia acemosa*), **Uthiyan** (*Lannea coromandelica*) in Chinnamuthulur village of Perambalur district. To cure uterus problem.

43. **Thaneervittan kizhangu** - Wild asparagus (*Asparagus racemosus*) in Chinnamuthulur village of Perambalur district. Increase the milk secretion for lactating women.

44. **Thiruneetrupachalai** - Sweet basil (*Ocimum basilicum*) in Karadikulam village of Perambalur district. Curing earache and dullness of hearing.

45. **Thaluthalai** - Wild killer glory tree (*Clerodendrum phlomides*) in Karadikulam village of Perambalur district. Demulcent in gonorrhea.

46. **Eluthani poondu** (*Launaea* sarmentosa) in Karadikulam village of Perambalur district. Diuretic and aperients.

47. **Palperuki** (*Euphobia heterophylla*) in Karadikulam village of Perambalur district. Protect new born babies from germs.

48. **Keelanelli** (*Phyllanthus amarus*) in Karadikulam village of Perambalur district. Used as treatment of jaundice.

49. **Seenthil** in Karadikulam village of Perambalur district. The fried mixture is given to the person affected viraivatham.

50. **Periyaanangai** (*Andrographis alata*) in Ilankudi village Sivagangi district. Treatment for snake bite.

51. **Karunkurinji** in Kulusekaram village of Kanyakumari district. Used as stroke, wounds created by hitting on stones and thorns and also used in preparation of "health oil"

52. **a. Sivappu nayuruvi** - Prickly chaff flower (*Achyranthes aspera*) in Kulusekaram village of Kanyakumari district. Used for tooth ache in human and retention of placenta.

    b. Nayuruvi -Prickly chaff flower (*Achyranthus aspera*) in Thakandarayapuram village of Tiruvannamalai district. Remedy for scorpion bite.

53. **Naaga malli** *(Rhinacanthus nasutus)* in Kulasekaram village of Kanyakumari district. Anti cancerous and root purifies the blood.

54. **Pirandai** - Edible stemmed vine *(Cissus quadrangularis)* in Kulasekaram village of Kanyakumari district. Anorexia in animals.

55. **Sadhavari kizhangu** *(Asparagus racemosus)* in Kulasekaram village of Kanyakumari district. Increases milking, cures stroke.

56. **Visha Naarayini** in Kulasekaram village of Kanyakumari district. Poison bite treatment.

57. **Oridhazh Thamarai** (*Hybanthus enneaspermus)* in Vettuvenni village of Kanyakumari district. Remedy in tuberculosis.

58. **Gundumani** -Wild liquorice *(Abrus precatorius)* in Vettuvenni village of Kanyakumari district. Remedy for tuberculosis, diuretic, anoestrus and suboestrus in animals.

59. **Musumusukkai** (*Mukia maderaspatana)* in Vettuvenni village of Kanyakumari district. Remedy for intestinal worms in animals.

60. **Neela amiri** in Vettuvenni village of Kanyakumari district. It enhances the hair growth and acts as a hair conditioner.

61. **Manjal Karisaali** in Vettuvenni village of Kanyakumari district. Remedy for tuberculosis and curing the cut wounds.

62. **Otrai Moolihai** in Vettuvenni village of Kanyakumari district .Remedy for tuberculosis, wounds and abrasions. It cures intestinal worms in human.

63. **Anali** in Vettuvenni village of Kanyakumari district .Remedy for tuberculosis poisonous bite and snake bite.

64. **Cinchona** (*Cinchona succirubra)* in Vettuvenni village of Kanyakumari district. Treatment for tuberculosis and malarial fever.

65. **Vasambu** -Sweet flag *(Acorus calamus)* in Vettuvenni village of Kanyakumari district. Antipyretic, nerving tonic.

66. **Paruppu keerai / koluppu keerai** - Purslane (*Portulaca oleracea*) in Vettuvenni village of Kanyakumari district. To control excess fat in human body.

67. **Kumari/ Sotru kathalai** - Indian aloe *(Aloe vera)* in Vettuvenni village of Kanyakumari district. To cure piles, jaundice and menstrual problem.

68. **Arokya pachilai** - Kuravan kanda mooligai in Vettuvenni village of Kanyakumari district. Controls appetite and improves immunity power.

69. **Thippili** – Long pepper *(Piper longum)* in Vettuvenni village of Kanyakumari district. Used to cure bronchial infection of throat and respiratory disease.

70. **Uzhinchai** - balloon vine in Vettuvenni village of Kanyakumari district. To control rheumatism, nervous diseases and chronic bronchitis.

71. **Athi** – Gular fig/cluster fig *(Ficus racemosa*) in Vettuvenni village of Kanyakumari district. To cure local inflammations, diabetes, nausea and mouth ulcer.

72. **Thuthi** – Country mallow (*Abutilon indicum)* in Vettuvenni of Kanyakumari district. Nerving tonic, Aphrodisiac and to cure dysentry.

73. **Vilvam** - Bengal Quince, Bael Fruit (*Aegle marmelos)* in Vettuvenni village of Kanyakumari district. To cure high blood pressure and joint pains.

74. **Sathakuppai** – Dill (*Anethum graveolens)* in Vettuvenni village of Kanyakumari district. Remedy for tuberculosis, colic pain, vomiting and hiccoughs.

75. **Biramadand-** Thuththi (*Argemone mexicana*) in Vettuvenni village of Kanyakumari district. Act as a rat bite.

76. **Aduthinnapalai** - Worm-killer (*Aristolochia bracteolate)* in Vettuvenni village of Kanyakumari district. Anti-inflammatory.

77. **Karpura Tulasi** - Camphor Basil (*Ocimum kilimandscharicum)* in Vettuvenni village of Kanyakumari district. To control cough and wounds.

78. **Pudhina** -Mint *(Mentha arvensis)* in Vettuvenni village of Kanyakumari district. To control cough, headache and fever.

79. **Maruthonri** –Henna/Cypress shrub (*Lawsonia inermis*) in Vettuvenni village of Kanyakumari district. Used as a hair tonic.

80. **Kattamamakku-**Purging Nut/Physic Nut (*Jatropha curcas)* in Vettuvenni village of Kanyakumari district. To increase secretion of milk.

81. **Vetchie-**Jungle Flame (*Ixora coccinea)* in Vettuvenni village of Kanyakumari district. Used as a sedative in hiccough.

82. **Kalyana Murunkai** -Indian coral tree (*Erythrina variegata)* in Vettuvenni village of Kanyakumari district. To increase secretion of milk and regulates the menstrual flow and to cure sterility in women.

83. **Karisalankanni** -Trailing Eclipta (*Eclipta prostrate)* in Vettuvenni village of Kanyakumari district. To cure viral hepatitis.

84. **Korai Kizhangu** -Nut Grass *(Cyprus rotundus)* in Vettuvenni village of Kanyakumari district. Remedy for diarrhoea and indigestion.

85. **Sirukurinjan-**Sarkarai kolli *(Gymnema sylvestre)* in Vettuvenni village of Kanyakumari district. Remedy for snake bite.

86. **Perunkurinja** -Sneezing Silk Cotton *(Wattakaka volubilis)* in Vettuvenni village of Kanyakumari district. To cure headache after child birth.

87. **Vellai Kundarikum** -Indian Copal Tree (*Vateria indica)* in Vettuvenni village of Kanyakumari district. To cure impure blood.

88. **Venthayam** -Fenugreek (*Trigonella foenum)* in Vettuvenni village of kanyakumari district. Used as a carminative.

89. **a. Siriyanangai** - Creat (*Andrographis paniculata)* in Vettuvenni village of Kanyakumari district. Used for curing jaundice

    b. Siriyanangai (*Andrographis paniculata*) in vandavasi village of Tiruvannamalai district. Poison treatment plant.

90. **Senthalai** -Screw pine (*Pandanus odoratissimus*) in Tandarai village of Kanchipuram district. Used to cure leprosy, scabies and diseases of heart and brain

91. **Aavuri** -Wild Indigo (*Tephrosia purpurea)* in Tandarai village of Kanchipuram district. Used in bronchitis, boils, pimples and bleeding piles.

92. **Manjal karisalankani** (*Wedelia chinensis)* in Thakandarayapuram village, Vandavasi taluk of Tiruvannamalai district. Leaves used to cure jaundice.

93. **Navakonchi** (*Rauvolfia tetraphylla*) in Tandarai village of Kanchipuram district. Use roots as snake-bite medicines.

94. **Poonai mesai leaves** (*Orthosiphon spiralis*) in Tandarai village of Kanchipuram district. To control kidney problem.

95. **Elumpotti thalai** (*Ormocarpum sennoides*) in Tandarai village of Kanchipuram district. Leaves used to regulate the blood circulation.

96. **Sennithulasi** (*Stevia spp*) in Tandarai village of Kanchipuram district. To control diabetes.

97. **Malai vembu** (*Melia azedarach*) in Tandarai village of Kanchipuram district. To cure stomach problem.

98. **Thandri** *(Terminalia bellirica)* in Tandarai village of Kanchipuram district. General health care medicine.

99. **Paruthi** -Shoe flower plant (*Gossypium spp)* in Vandavasi taluk of Tiruvannamalai district. Cure menstrual problems.

100. **Thumbai** -Thumbe (*Leucas aspera*) in Vandavasi taluk of Tiruvannamalai district. Remedy for scorpion bite.

101. **Ranakalli** in Vandavasi taluk of Tiruvannamalai district .Used to reduce the body heat.

102. **Kallimulli** in Kattuchivaghri village of Villupuram district .Used for the treatment of chest pain and body heat.

103. **Karunaikilangu** in Kattuchivaghri village of Villupuram district. Used as medicine for piles as well as constipation.

104. **Murukan** in Kattuchivaghri village of Villupuram district. Dewormer in human beings.

105. **Perunerinjal** in Kattuchivaghri village of Villupuram district. Used as endometritis.

106. **Masipathri** in Kattuchivaghri village of Villupuram district. To cure chronic respiratory diseases.

107. **Nalvelae** in Kattuchivaghri village of Villupuram district. Remedy for one sided headache.

108. **Nanjumurunjan** in Kattuchivaghri village of Villupuram district. To catch the fish from the ponds.

109. **Uthamani** in Kattuchivaghri village of Villupuram district .Used to scrape the scar tissue formed in wounds.

Karuvel Tree-*Prosopis julifera*

Half matured side branch of *Prosopis julifera* stem is taken and leaves are removed. One end of the stem is kept in the Chulah, after sometime due to increase in temperature liquid will ooze out from another end of the stem. That liquid is applied over the portion which is affected by poisonous bite. (Eg. Scorpion bite).This should be continued 3-4 times.

Vishnu Kiranthi -*Evolvulus alsnoides*

A fistful of Vishnu kiranthi leaves, 25 gram pepper and 25 gram seeragam are boiled well. It should be boiled until one litre of water comes to half litre. It can be taken twice a day. It can be mixed with milk also. This medicine cures the Tuberculosis, cold and fever.

Purusalachanam

A fistful of Purusalachanam root is taken and cleaned. To this little quantity of seeragam is added and ground well. Paste is made out of it and it is mixed with 100 ml of water. This medicine should be given twice a day for 2-3 days to women who is having uterus problem. It controls menstrual problems.

Precaution while administering the medicine : It reduces the body heat and gives cooling to the body. Continuous consumption may lead to cold problem.

Ariyanangai

A fistful of Nilakunni root, a fistful of Ariyanangai leaves and little quantity of pepper are ground well and paste is made out of it. This paste is given twice a day to person who is affected by snake bite.Care to be taken while administering the medicine.Food items to be avoided upto three days (sweet, salt and sour items). Up to three months the person should not take fish, dried fish, chicken, lab lab. These treatments were learnt from their ancestors.

Unni Mul Thazhai- Wild Sage *(Lantana camera)*

It is a thorny bushy plant whose leaf extract is used to heal the cut wounds and other wounds.

Karu Umathai -Indian Thorn Apple *(Datura fastuosa)*

Flower of the karu umathai is used to cure suffocation due to asthma. Inhaling of smoke comes out from dried flower will give relief from asthma.

Notchi - Five Leave Chaste Tree *(Vitex negunda)*

Handful of Notchi leaves are taken and ½ lit of water is added to it. Notchi decoction has to be prepared using hot coal in the absence of burning. This should be done in previous day night .Next day morning the filtered decoction will be given to the person suffering from fever and cold. It should be taken, before intake of food and to be continued for 3-5 days. It should be given twice a day.

Mudakatran- Balloon Vine – *(Cardiospermum halicacabum)*

a) Mudakatran decoction is used to cure joint pains and constipation. Handful of Mudakatran leaves is taken and ½ lit of water is added. The contents are boiled and filtrate is given to persons affected by joint pain and constipation. This can be taken in twice a week.

b) Handful of Mudakaran, 4 bulbils of garlic, 4 dried chillies, required quantity of dried ginger (*Gingiber officinale*), Asafoetida, Tamarind and salt are taken and paste is prepared out of these. This paste is also used to cure joint pains. This can be taken with normal food, twice a day.

Perumarunthu -Indian Birth Wood *(Aristolochia indica)*

Handful of Peru marunthu leaves are taken and it has to be fed to the off feed animal with dub grass (*Cynodon dactylon*).This should be done twice a day for 3 consecutive days.

Aanai Nerunji- Bathala Mooligai *(Pedalium murex)*

Handful of Aanai nerunji leaves are taken and soaked into ½ lit of butter milk for 10-15 minutes. The contents are smashed well and the filtrate is fed to the animal affected by diarrhoea. This should be given in twice a day.

Sangu *(Azima tetracantha)*

A fistful of sangu leaves, 2 numbers of betel vine leaves and required amount of salt are ground well. The extract is added to the eye of dull and drowsy animal. It should be administered till the animal come to normal condition.

Adathoda (*Adathoda vasica*) Malabar Nut

A fistful of Adathoda leaves are soaked in half litre of water. This should be boiled well and the decoction is to be given twice a day, for one month. This medicinal extract will give best relief for Asthma.

Veli Paruthi *(Fergularia daemia)*

Handful of fresh veli paruthi leaves are fed to the animal thrice a day. This will give relief for indigestion.

Thulasi (*Ocimum sanctum*)

10-15 fresh Thulasi leaves are given to person suffering from cold. This is recommended twice a day for a week.

Nayi Vellai - Wild Mustard *(Cleome viscosa)*

A fistful of Nalla vellai leaves, 4 bulbils of garlic, 2 nos. of dried chillies, required amount of Asafoetida and salt are macerated well and extract is taken. Two drops of extract is added to the eye and ear of the dull and drowsy animal. This should be administered once a day upto three days.

Nanchu Murichan -Emetic Swallow Wort *(Tylophora indica)*

A fistful of Nanchu murichan leaves, 5 nos of pepper and 3 nos of betel vine are taken and made into paste which is then fed to the animal affected by poisonous bite. This should be continued for a week, thrice a day. The same paste recommended to human beings.

Papaya -*Carica Papaya* and Guava *(Psidium gujava)*

Continuous consumption of papaya and Guava will cure the stomach ulcer.

Garlic *(Allium sativum)*

Five to six bulbils of garlic should be consumed and should drink enough water. After one hour this will cure indigestion.

Aamanakku / Castor *(Ricinus communus)*

Mustard, chopped onion, kuppai meni leaves are fried in castor oil. This medicine is given usually to the women after delivery for a period of 3-5 days. It prevents puss formation. Applying castor oil over stomach and massaging will relieve the pain.

Banana Pseudo Stem

A fistful of sprouted horse gram and cooked banana pseudo stem consumption will cure kidney stone.

Arugampul-Bermuda Grass *(Cynodan dactylon)*

100 ml of Arugampul (*Cynodan dactylon*) consumption will serve as first aid against cold, fever and poisonous bite.

Kuppai Meni -Indian Acalypa *(Acalypa indica)*

It is used to cure scabies and internal wounds after delivery (in human beings).Handful leaves are taken and water is added to make it into paste. This paste is smeared over the affected portions. This should be continued until scabies heels.

Feasibility and Relative Advantages

Since all the ingredients are locally available. It is very easy to prepare the medicine. Similarly, the method of preparation is also very simple which doesn't involve cumbersome process. Herbal cure is relatively advantages over other types of medicines due to its nil side effects. Moreover, while taking this treatment, the patients need not adhere to food restrictions.

Aanaikathaalai- Century Plant *(Agave americana)*

Brief Description

Agave is a fibre-bearing plant which grows in tropical countries. It produces leaves upward from the ground, forming a massive rosette. Both sides of the leaves are smooth, having prickly edges, and at the upper extremity, a thorny end. The plant produces a long and erect stem which bears bulbils that fall to the ground and grow into new plants. The whole plant is used for medicinal purpose.

Medicinal Use

- Purifies the blood.
- Prevents the formation of pus.
- Promotes the flow of urine.
- Stimulates the bowel-mild laxative.
- Checks excessive falling of hair.
- Checks internal wounds and swellings.
- Soothes irritated eyes and helps in general eye diseases.
- Used against tuberculosis of the lungs.
- Used for jaundices and infective livers.
- Checks anemia, kidney diseases and liver problems.

Preparation Methods

1. For cleansing the blood an infusion or decoction of the finely chopped leaves can be used. Use two grams of leaves for a cup of water. One cup taken a day is enough. The preparation should be taken slowly in sips.
2. The water in which the leaves are soaked for a day can be used externally as a disinfectant and tonic, in case of falling hair.
3. The sap of the leaves is used internally in case of wounds and inflammations.
4. The infusion of this plant with the addition of a teaspoon of honey can be used to soothe irritated eyes.
5. The decoction of the leaves can be used in the form of lotions for general eye diseases.
6. The sap of the plant can be used at times of pulmonary tuberculosis, diseased liver and jaundice.

7. The powder of the dry leaves of the plant, in water, is recommended for anaemia, kidney diseases, and liver problems. The recommended dose is one teaspoon of powder a day.

Pilapalsm - Jack Tree *(Atrocarpus heterophyllus)*

Brief Description

The jack fruit tree is well known for its very large fruit which sometimes weighs over 30 kg. It grows well in the tropics and is loved by most of the local inhabitants. The tree grows to a height of 10 meters or more with many heavy branches and foliage which provide a beautiful shady picnic or resting spot. The edible pulp inside the fruits is fleshy and yellow when ripe. The unripe fruit is also eaten as a vegetable. The whole plant is used for medicinal purpose.

Medicinal Use

1. Used for diarrhoea, fever and asthma.
2. Used for ulcers, wounds, glandular swellings and insect bites.
3. Also used as a laxative, for giving tone and vitality to the body and for acute abdominal pain.

Preparation Methods

1. The decoction of the root is a good remedy for diarrhea, fever and asthma.
2. The milky latex of the tree is applied externally for the cure of ulcers, wounds, glandular swellings and insect bites.
3. Excessive consumption of the ripe fruit works on the bowels as a laxative.
4. The consumption of roasted or boiled seeds gives tone vitality to the body and helps in checking abdominal pain.

Neer Mulli -Long Leaved Barleria *(Hygrophila auriculata)*

Brief Description

It is a hardy, upright, herbaceous plant which is hydrophytic in nature, or found growing in wet and marshy areas. The hairy stem is square and thickened at the nodes. Six stalkless, lanceolate leaves forming a whorl are produced from each node. Four pairs of purplish- blue bilabiate, or two-lipped, flowers are put forth from the nodes. The capsule which is produced contains 4 to 8 seeds with white hairs. Mostly leaves are used.

Medicinal Use

- Useful as a tonic to give tone and vitality to the body.
- Useful for diarrhoea and dysentry.
- Useful for inflammations and biliousness or congestion of the liver.

- To cure diseases of the eye.
- For pains and other abdominal problems.
- To take care of constipation and anuria, or the absence of urine excretion from the body.
- Useful to cure the jaundice, rheumatism and urine - genital diseases.

Preparation Method

A decoction or infusion of the leaves will take care of all the disorders.

Sudukadu Mallikai- Periwinkle *-Catharanthus roseus*

Brief Description

It is often linked with the graveyard as it thrives there without much care. The flowers and leaves of this herbaceous plant would never find its place in a garland, bouquet or in a maids hair as the odour emitted by the plant is so bad that it always brings out a negative response in any individual. Later, this plant has gained great popularity of its medicinal use. The plant may have a few branches and grows to about 50 cm high. The glossy leaves are oblong in shape. The flowers are normally white or a purplish- pink in colour. The fruit is somewhat cylindrical with a length of about 2-3 cm. It is seen growing in some gardens, seashores and sandy areas. The whole plant is used for medicinal purpose.

Medicinal Use

- Said to be anticarcinogenic or an agent which works against cancer.
- Used for diabetes.
- Used for diarrhoea.
- Used as a vermifuge or a substance this expels worms from the body.
- Good remedy for toothache and works well as a purgative.
- The decoction of the flower is given for treating tuberculosis.
- The decoction of the plant is used as a wash for mouth ulcers and tired eyes to clear the vision.

Caution : women should not use this plant during menstruation period.

Mahuli Keerai -Wool Flower *-Celosia argentea*

The wild cock's comb grows in waste land as an annual, smooth, erect, branched herb, reaching upto 1 metre high. The leaves are linear to ovate lanceolate. The flowers are silver-white, with a tinge of pink in erect or drooping spikes. It is rarely cultivated plant. The tender flowers and leaves are eaten in tropical countries as a vegetable.

Medicinal Use

1. Used in cases of dysentry, diarrhoea, acute abdominal pains and inflamed stomach.
2. Heals wounds, sores, ulcers, skin eruptions and relieves itching.
3. Used in the treatment of tuberculosis.
4. Used to cure the ulcer of the mouth and for tired eyes to clear vision.

How to Use

1. As an infusion, the whole plant is used for cases of dysentry, diarrhoea, acute abdominal pain and inflamed stomach.
2. For external use, the pounded leaves, flowers and seeds are applied over wounds, sores and skin eruptions.

Dose : 12 to 20 grams of plant material for 1 litre of water is taken in spoonful doses.

Christmas Star – Poinsettia *(Euphorbia pulcherrima)*

Euphorbia pulcherrima is an erect branched shrub, 2 to 4 meters height. The plant has a milky juice or latex. The leaves appear somewhat violin or fiddle shaped. The modified leaves are called bracts sub end the terminal flowers in clusters. They are bright red in colour and are most showy in the month of December. The leaves and flowers are used for medicinal purpose.

Medicinal Use

- Used in skin diseases.
- Increases the secreation of milk in nursing mothers.
- Used as a mild laxative.

How to Use

1. The normal crushed leaves are applied in skin diseases as poultice.
2. The flowers are said to increase secretion of milk and also act as a mild laxative.

Tottalvadi- Sensitive Plant -*Mimosa pudica*

Sensitive plant grows in the low lying land and land tropics. It is a common diffused spreading prickly, an herb which grows everywhere in southern India. The stems are branched, with bristly hairs. The leaves are small leaflets on a stalk, and when touched they fold together. They are sensitive to the touch. The flowers are numberous, long-stalked, pink, solitary globose heads. The pods are small, curved and numerous. The whole plant is used for medicinal purpose.

Aanai Nerunji (*Pedalium murex*)

Adathoda (*Adathoda vasica*)

Nalla Vellai (*Cleome gynandra*)

Anaikathaalai (*Agave americana*)

Neer Mulli (*Hygrophila auriculata*)

Sudukadu Mallikai (*Catharanthus roseus*)

Fig. 24. Medicinal Plants

Medicinal Use

- Used for liver disease, constipation, dysentry and kidney problems.
- Used for mouth and throat inflammations.
- Used for cancerous ulcers, skin diseases and glandular swellings.
- Used in scorpion sting.
- Used for urinary complaints and as a purgative.
- Used for treatment of mumps.

How to Use

1. The infusion of the leaves is used for liver disease, constipation, dysentry and kidney problems.
2. The decoction of the root takes care of the urinary complaints. The infusion of the whole plant is used for mouth and throat inflammations.
3. The seeds are used for sore throat and hoarseness.
4. Externally used, the juice of the plant makes a good lotion to be applied over cancerous ulcers and skin diseases.
5. The crushed leaves made into paste are applied topically on glandular swellings.
6. Use cold compress on swollen glands for 30 minutes at least 3 times a day. Apply crushed past of leaves after each cold compress session.
7. The crushed leaf and stem is applied over the scorpion sting to reduce the pain and swelling.

Dose

10 to 15 grams, for 1 litre of water. Use 3 to 4 cups a day.

Caution

In large quantity the root is toxic.

Thotta Sinigi- Touch-me-not- *Mimosa pudica*

Habit : Herb

Parts used : Root, Leaves and Seeds

Uses : Minor skin wounds, Diarrhoea, Piles, Diabetes

Andhimalligai - Four O' Clock Plant -*Mirabilis jalapa*

The four O' clock plant is a perennial, herbaceous, bushy plant which grows about one meter height. The leaves are narrow, pointed, and ovate. The flowers are variously coloured, showy and sometimes striped. The black seed is twice the size of a pepper corn, and the tubers, when fully grown, are more than kg in weight. It is grown in gardens for its shown flowers.

Medicinal Use

- Promotes flow of urine.
- Used in any inflammatory viral disease of the skin.
- For the treatment of abnormal accumulation of liquid in the cellular tissue.
- Treatment of boils, inflamed and enlarged lymph-nodes, and for the accumulation of pus in a cavity.
- Heals wounds, bruises, and allergy of the skin caused by indigestion.
- Used for earaches.
- Used to remove freckles on the skin.

How to Use

The root tuber is used in any inflammatory viral disease of the skin, treatment of abnormal collection of liquid in cellular tissue, boils, and inflammed lymph-nodes. Use 20 grams of root tuber in 1 litre of water, by decoction and 4 to 5 cups are to be taken per day.

1. A hot poultice of the leaves is used to mature and resolve boils, inflamed lymph – nodes, and accumulation of pus in a cavity.
2. The juice of the fresh leaves is a healing gent for wounds, bruises, skin allergy, caused due to indigestion also used for earaches in children.
3. The flowers are applied topically over the skin in the case of any inflammatory virus disease and for freckles. The powdered seeds are also used for the same problem.

Parts Used

The root tubers, leaves, flowers and seeds.

Dose

10 to 15 grams, for 1 litre of water. Use 3 to 4 cups a day.

Marul - Bowstring Hemp- *Sansevieria zeylanica*

The bow string hemp is an herbaceous perennial plant, used in the garden as border plants. The leaves are sword like, stiff and smooth. Some of the leaves have marble patterns, while others are gold edged and sometimes concave on the upper surface. The flowers are tubular, white to yellowish, in groups of 4 to 6, scattered along the stalk. Strong fibers are used from the leaves for making cloth, twine, and paper.

Medicinal Use

- Gives tone and vitality to the body.
- Promotes the removal of mucous secretions from the bronchial tubes.
- Used as a purgative.

How to Use

The decoction of the plant root is used as a tonic, purgative and as an expectorant.

Marul-Bow String Hemp (*Sansevieria zeylanica*)

Preparation

The succulent portion of marul is fried; the liquid that comes from it should be allowed to cool. And the liquid is dropped into the ear.

Dosage : One time /day

Pachaiyalari -Yellow Oleander *(Thevetia peruviana)*

The yellow oleander is an erect smooth branched shrub, with milky latex. The leaves are linear and shiny. The calyx is green and pointed. The yellow corolla is bell shaped. The fruit is green, smooth and rounded. It is an ornamental plant which is cultivated in gardens.

Medicinal Use

- Used as an insecticide.
- Used to reduce eliminate fevers.
- Mild laxative.

How to Use

1. The decoction of the seeds can be used as an insecticide.
2. The decoction of the bark of the plant, in moderate doses, is used to reduce or eliminate fever.
3. The leaves, act as a mild laxative.

Parts used : Bark, seeds, and leaves.

Dose : For internal use only with medical supervision. Externally as needed.

Caution

Unless in extreme need, this plant should be used under medical supervision for internal use. The plant is poisonous.

Thali Keerai *(Ipomea sepiaria)*

Leaves are triangular-cordiform, 2-4 x 2-3 cm.Flowers in sub umbellate cymes; peduncle to 10 cm. calyx-lobes sub equal. Corolla cream, with a purple throat, 4 cm across, salver-form. Flower opening by 9 a.m., very evanescent. Fruit Capsule is glabose, 0.8 cm across; seeds are white-velvety. Propagation through Seeds.

Medicinal Use

Used to cure diarrhoea in animals.

Thaivelai

A handful of leaves are taken, cleaned, ground and juice is extracted for about 20 ml. 10 ml of gingelly oil added to it. This mixture is warmed and given to the person who is affected by cold and fever. A fistful of thaivelai leaves are taken, cleaned and juice is extracted. 4-5 drops of juice is poured into ears of affected person. With in 10 minutes pain due to scorpion bite can be cured.

Dosage : Three times /day.

Omavalli – Karpuravalli *(Coleus aromaticus)*

A fistful of Omavalli leaves are boiled with ½ litre of water. It has to be boiled until ½ litre comes to ¼ litre of water. This treatment can be continued for about 5-7 days.

Dosage : ¼ litre/day.

Uses : Juice mixed with sugar is given to children in asthma, chronic cough and fever.

Vallarai- Indian Peeny Wort *(Centella asiatica)*
Preparation

A fistful of green vallarai leaves are collected, cleaned and taken as such internally. The regular intake will increase the memory capacity.

Perukurinchan – Sneezing Silk Cotton *(Wattakaka volubilis)*

A handful of perukurinchan leaves are boiled in ½ litre of water. It has to be boiled until ½ litre comes to ¼ litre of water. Filter the contents and regular in take of the extract/decoction will control diabetes. This can be taken up to 6 months until sugar level comes to normal.

Dosage : ¼ litre/day

Mahuli Keerai (Celosia argentea)

Christmas Star (Euphorbia pulcherrima)

Andhimalligai (Mirabilis jalapa)

Marul (Sansevieria Zeylanica)

Omavalli (Coleus aromaticus)

Vallarai (Centella asiatica)

Fig. 25. Medicinal Plants

Sathisaranai *(Trianthema decandra)*

A fistful of sathisaranai root, dried chillies 2 nos, a little amount of salt are taken and ground well. Finally paste is made out of these contents. It gives relief from gas trouble.

Dosage : Twice a day.

Manathakgali *(Solanum nigrum)*

A fistful of manathakali leaves, a fistful of murungai leaves, 2-3 numbers of onion, 5 garlic bulbils, 1 tomato, a little quantity of seeragam and goat's intestine are fried. These substances are ground well and paste is made out of it. This should be continued upto three days for curing uterus problem.

Dosage : Thrice a day.

Thavasi Murungai *(Jusitica tranquebariensis)*

A fistful of leaves are taken as such internally and same leaf is squeezed well and kept on the affected portion and tied with cotton cloth. After 4-6 hours, person should take bath in tank. The colour of the water changes into blue or discolouration of normal colour indicates the reduction in poison level. For controlling dog and cat bite

Dosage : A fistful of green leaves are taken internally as such for about three times a day.

Karunthulasi- *Ocimum basilicum*

Regular consumption of this leaves (a fistful) controls diabetes and blood pressure.

Dosage : Twice a day.

Tree Barks

Nagal – *Syzygium* cumini

Arasu – *Ficus religiosa*

Aathi – *Bauhinia racemosa*

Uthiyan – *Lannea coromandelica*

Medicinal Preparation

Each 1 kg of above said tree's barks are shade dried for about 1 week. It has to be pounded well and made it as powder. A fistful of powder is boiled in ½ litre of water. The filtered decoction is given to women who is having uterus problem.

Dosage : Twice a day.

Thaneervittan Kizhangu- Wild Asparagus *(Asparagus recemosus)*

A fistful of tuber is cleaned, add required amount of water and ground well. This has been filtered by using cotton cloth. The filtrate is given to lactating women for about four days. In order to increase the milk secretion for lactating women. Flowering Season is May to December

Dosage : once a day.

Additional Information

Fresh root juice is mixed with honey and given for dyspespsia; roots form a constituent of medicinal oil used for nervous and rheumatic complaints. It cures menstrual problems and for freeing urinary tract. Root and leaves are used for curing urinary disorders.

Thiruneetrupachalai -Sweet Basil *(Ocimum basilicum)*

An annual herb with a square branched stem and numerous opposite, stalked, ovate, slightly toothed and glabrous leaves, often reddish in colour. The small white, yellowish or pinkish flowers are arranged in whorls in the upper leaf axils. All parts of the plant are hairy and aromatic.

Parts used : Herbs and seeds

Uses

- Leaves are used for flavouring purpose.
- Juice of the leaves is dropped in the ear in earache and dullness of hearing. Mixed with a little ginger and black pepper the leaf juice is given during the cold stages.

Thaluthalai -Wild Killer Glory Tree *(Clerodendrum phlomides)*

Roots are aromatic and astringent, decoction used as a demulcent in gonorrhea.

Eluthani Poondu *(Launaea sarmentosa)*

Uses : Tonic soporific, diuretic and aperients.

Palperuki *(Euphobia heterophylla)*

Uses : The filtrate of the leaves is used to wean infants from their mothers and it's protecting new born babies from germs.

Keelanelli *(Phyllanthus amarus)*

Erect, annual herb, leaves closely arranged, flowers yellowish green in colour, pendent from the axils of leaves.

Uses

- The drug is highly reputed as a single drug remedy in the treatment of jaundice in traditional medicine. The whole plant is medicinal.
- It is used in chronic dysentery, cough, indigestion, diabetes, urinary tract diseases and skin eruptions.
- Whole plant pounded, is used as a poultice for ulcers, sores, swellings and skin affections. In practice, it is mainly used for the treatment of jaundice.

Ingredients

Keela nelli-A handful

Cow milk- 100 ml

Medicinal Preparation

A handful of keelanelli leaves are taken, cleaned and ground well. Required amount of water is added to it and made it colloid. This has to be filtered and the extract/juice is obtained out of it. This juice is mixed with 100 ml cow milk. This treatment can be continued up to 3 days for about thrice a day.

Dosage : 150ml/ time

Additional Care

While administering the medicine, the person affected by jaundice should not take salt, sour and spicy items in food.

Seenthil *(Tinospora cordifolia)*

Ingredients

- Egg- 1
- Seenthil leaves

Preparation

A fistful of seenthil leaves are cleaned and fried well. With this egg (without egg yoke) is added and again fried well. This fried mixture is given to the person affected by viraivatham. This can be administered for about 3 days, thrice a day.

Additional Care

While administering the medicine, the person affected by jaundice should not take salt, sour and spicy items in food.

Periyaannangai -*Andrographis alata*

Name of the disease : "Snake bite in human"

Symptoms

The snake bitten person is often looking fainted.

Conditions

Within 30 minutes period the medicine is to be administered into the body of person. Before this, the type of snake bite is to be identified by seeing the intensity of wound.

Ingredients

1.	"Periyaanangai" (*Andrographis alata*)	- a whole plant with root.
2.	"Syrianangai" (*Andrographis paniculata*)	- a whole plant with root.
3.	"Muluvanangai"	- a whole plant with root.
4.	"Vellarugu" root (*Enicostema auxillare*)	- a single plant root
5.	"Arugampul" root (*Cynodon dactylon*)	- a single plant root.
6.	"Kurundhampulluruvi" (*Dendrophtoe falcata*)	- a single plant root.
7.	"Neykaaya" root (*Cucumis trigonus*)	- a single plant root.
8.	"Ponmustai" root (*Sida acuta*)	- a single plant root.
9.	"Vellaipoodachedi" leaves (*Sterculia urens*)	- 20-50 g.
10.	"Semmanampullurivi" leaves (Indian mistletoe)	- 20-50 g.
11.	"Karunkattukodi" leaves (*Coculus hirsutus*)	- 20-50 g.
12.	"Keeripachilai" leaves (*Ophiorhiza mongoose*)	- 20-50 g.
13.	"Paapaaranangai" leaves	- 20-50 g.
14.	"Kalarchikaai" (*Cesalpinia bonduc*)	- a whole plant
15.	'Kilichi" leaves	- 20-50 g.
16.	"Muthirikkai" root (*Ampelocissus latifolia)*	- 20-50 g.
17.	"Kothupaalai" leaves (*Wrightia tinctoria*)	- 20-50 g.
18.	"Aduthinnaapaalai"leaves (*Aristolachia bracteolata*)	- 20 – 50g.
19.	"Vekkanotchi" leaves (*Vitex sp.*)	- 20-50 g.
20.	"Usilai" flower (*Albizia amara*)	- 1
21.	"Madukaarai" root (*Randia dumetorum*)	- 1 (2-3") length.

Sathissaranai (Trianthema decandra)

Karunthulasi (Ocimum basilicum)

Thaluthalai (Cerodendrum phlomides)

Eluthani Poondu (Launaea sarmentosa)

Palperuki (Euphorbia helerophylla)

Keelanelli (Phyllanthus amarus)

Fig. 26. Medicinal Plants

The aforesaid ingredients are to be taken, dried under sunlight for 2-3 days and ensure complete drying of ingredients. This is to be ground well either by grinder or "ammi" and sieved to prepare the fine powder. The resultant content is dull green colored.

Administration

The aforesaid prepared ingredient is to be administered orally. One teaspoonful of medicine is recommended immediately after snake bite and from the next day onwards during morning and evening hours for 7 days continuously. The time is between 6-7 a.m. in the morning and 3-4 p.m. in evening hours.

Conditions to be Followed

1. Chemical medicine should be strictly avoided. If it is taken a tumbler of coconut milk would be sufficient to weaken the medicinal effect of the content.
2. Salt is strictly avoided in the food during the course of treatment.
3. Non-vegetarian food is strictly avoided and rice 'kanji' (gruel) is recommended.
4. The intake of medicine in an empty stomach has better healing capacity.
5. If the symptom is severe the treatment period goes upto 15-18 days.

Relative Advantages

- It slowly cures the bite and doesn't create any side effects.
- The treatment fetches free of cost as Mr. Muthadaikan renders free service.
- The patients' remarks are in favourable manner and it explains the vigour of medicine.

Karunkurinji

Flowers are greenish in colour and appear similar to crossandra. Propagation through seeds.

Uses : stroke, wounds created by hitting on stones, thorns. Also used in preparation of "health oil"

Parts used : leaf, flower and seeds

Sivappu Nayuruvi-Prickly Chaff Flower *(Achyranthes aspera)*

Habit	:	shrub
Propagation	:	seeds
Parts used	:	whole plant
Flower	:	reddish in colour

Uses : demulcent, asthma, bronchitis, leprosy, skin diseases, renal and vesicle calculi, cardiac disorders, anaemia, general debility, conjunctivitis, cuts and wounds, dog bite, tooth ache in human.Retention of placenta, wounds or abrasions, eye ailments in animals.

Nayuruvi *(Achyranthus aspera)* Prickly Chaff Flower

Nayuruvi plant is uprooted from which root portion is cut and outer skin is peeled off. Peeled outer skin is shade dried for about 7-10 days. Ten gram of Neeli root's outer skin, a little quantity of Seeregam and 800ml of water are taken in new mud pot. The contents are boiled well until 800 ml comes to 200ml.Extract is filtered and the decoction is given for the poisonous bite.

Dosage : 10ml / time / day

Naaga Malli *(Rhinacanthus nasutus)*

Habit : shrub

Parts used : whole plant

Uses : Anti cancerous, root purifies blood. The roots are removed from the plant and cut into pieces. The cut pieces are ground and mixed with lime juice. It cures skin diseases. The oil can also be extracted from the plants and used for skin diseases.

Pirandai-Edible Stemmed Vine *(Cissus quadrangularis)*

Propagation : stem

Parts used : stem

Uses : Dyspepsia, general weakness and bone fracture in human. Swelling, wounds or abrasions, intestinal worms, anorexia in animals.

Sadhavari Kizhangu *(Asparagus racemosus)*

Parts used : root and tubers

Uses : increases milking, cures stroke

Visha Naarayini

Uses : veppalai, coconut oil mixed together with visha naarayini and kept in sunlight for 3days till the appearance of violet colour. Tubers are the economic part and used in poison bite treatment.

Oridhazh Thamarai- *Hybanthus enneaspermus*

Habit : herb

Propagation : whole plant

Uses : affections of urinary organs, to improve memory and vitality, a Remedy in tuberculosis, asthma, fever, leprosy, eye diseases.

Gundumani-Wild Liquorice - *Abrus precatorius*

Parts used : roots, leaves and seeds

Uses : To cure Diuretic, inflammations, wounds, asthma, leucoderma, skin diseases, stomatitis, hyperdipsia. Bloat (tympany), hemorrhagic (gastroenteritis or dysentery), anoestrus and suboestrus in animals

Special Feature

Kept inside sheep bones and covered with clay, allowed to cook for an hour. If the seeds turn white and come out easily from the bones, it is taken out and used as a medicine for diabetes. If seeds get broken inside, a stick is used to take out the seeds. Then, they are ground and the same is continued for a week.

Musumusukkai *(Mukia maderaspatana)*

Use : It is an important ingredient in kayathirumeni thailam. Intestinal worms in animals. Asthma, to stop continuous bleeding of blood from wounds its leaf mixed with pepper, "aayilya pattai" and paste is made.

Neela Amiri

Uses : Seeds are dried under shade and added with hair oil. It enhances hair growth and acts as a hair conditioner. The roots are used to remove the poison from the body.

Manjal Karisaali

Parts used : leaf

Uses : Cut wounds, abrasions. It is also an important ingredient of "kayathirumeni thailam".

Special Feature

It is believed to have gold in its leaf. Oil extraction by "neetrumurai" i.e., the leaf extract is taken directly and allowed to boil. On attaining 'sand like' texture (by feel) the boiling is stopped. The concentration of the oil depends on the extract taken for boiling. If higher the level of required extract, the concentration will also be higher. Add 3litre coconut oil, 1ltr gingelly oil to 1ltr of extracted oil.

Otrai Moolihai

Used to cure wounds and abrasions, 20 leaf taken and add 1ltr of water mixed to get leaf extract. It cures intestinal worms in human. It has a bitter taste.

Anali

Parts used : leaf

Uses : Poisonous bite, snake bite

Cinchona : (*Cinchona succirubra)*

Propagation : seeds, grafting

Parts used : quinine isolated from tree bark

Uses : malarial fever

Vasambu- *Sweet flag (Acorus calamus)*

Family : Araceae

Habit : Rhizomatous herb

Propagation : Rhizomes

Parts used : Rhizomes

Uses : Antipyretic, nervine tonic, sedative, epilepsy, mental disorders, skin disorders, skin diseases, general debility. Thikkuvaai, to avoid snakes entering in to house-solution sprayed.

Paruppu Keerai / Koluppu Keerai (*Portulaca oleracea*) Purslane

Propagation : stem cuttings

Uses : Cooking with dal used as a medicine, to control excess fat in human body

Kumari/ Sotru Kathalai *(Aloe vera)* Indian aloe

Habit : herb

Propagation : suckers

Parts used : leaf juice (sorus), roots

Uses : leprosy, piles, mental disorders, skin diseases, liver ailments, jaundice, menstrual problem

Arokya Pachilai/ Kuravan Kanda Mooligai

Use : found by khani people of Kanyakumari district, khanis look young
Even they are aged, controls appetite, sterility control, immediate
Energy source, improves immunity power,

Thippili (*piper longum)* – Long Pepper

Propagation : nodal cuttings

Parts used : fruit, root, stem

Uses :bronchial infection of throat, respiratory diseases, carminative, Sedative, insomnia, epilepsy and abortifacient.

Uzhinchai Balloon Vine

Propagation : seeds

Parts used : roots, leaves, seeds

Uses : Rheumatism, nervous diseases, chronic bronchitis, sore eyes and Herinia.

Athi *(ficus racemosa)* : Gular Fig/Cluster Fig/Fig

Parts used : bark, fruit, latex, root

Use : local inflammations, polyurea, diabetes, darkening of face and around eyes, burns, nausea, mouth ulcer.

Thuththi-Country-Mallow (*Abutilon indicum)*

Parts used : Leaves, Roots, Seeds

Uses : Nervine tonic, Aphrodisiac, Dysentery, Piles, Psoriasis, Diuretic

Vilvam-Bengal Quince, Bael Fruit *(Aegle marmelos)*

Parts used : Leaves, Roots, Seeds, Bark

Uses : Diarrhoea, Constipation, Worms, Skin diseases, High blood pressure, Joint pain

Sathakuppai-Dill *(Anethum graveolens)*

Parts used : Leaves, Seeds

Uses : Colic pain Vomiting, Hiccoughs

Seenthil kodi (Tinospora cordifolia)

Periyaanangai (Andrographis alata)

Gundumani (Abrus precatorius)

Vasambu (Acorus calamus)

Paruppu Keerai (Portulaca oleracea)

Sirukurinjan (Gymnema sylvestre)

Fig. 27. Medicinal Plants

BiramadanduThuththi- Prickly Poppyllow *(Argemone mexicana)*

Habit : Herb
Parts used : Seeds

Uses : Leprosy, Psoriasis, Rat bite

Adu-Tinna-Palai- Worm-Killer *(Aristolochia bracteolate)*

Parts used : Roots, Leaves
Uses : Anti-inflammatory, intermittent fevers

Karpura-Tulasi-Camphor Basil *(Ocimum kilimandscharicum)*

Habit : Shrub
Parts used : Leaves

Uses : Cough, Wounds, Bronchitis, foul ulcers

Pudhina-Mint *(Mentha arvensi)*

Parts used : Leaves, Flowering tops, stem extracts

Uses : Cough, Head ache, Fever, General weakness

Maruthonri-Henna Cypress Shrub *(Lawsonia inermis)*

Habit : Shrub
Parts used : Bark, Leaves, Flowers and Seed

Uses : Hair tonic, burning of soles and Palms, Local Inflammation and Jaundice

Kattamamakku-Purging Nut, Physic Nut (*Jatropha curcas*)

Parts used : Leaves, Seeds, Oil, Stem

Uses : Laxative, to increase secretion of milk, inflammations, Ear-ache

Vetchie-Ixora, Sarced Ixora Jungle-Flame *(Ixora coccinea)*

Parts used : Roots, Leaves, Flowers
Uses : Head ache, Boils, Nausea, and Sedative in hiccough

Kalyana Murunkai-Indian Coral Tree *(Erythrina variegate)*

Parts used : Barks, leaves

Uses : Increase secretion of milk and Menstrual flow, Cure sterility in Women, Diabetes, Eye infections, worms, joint pains

Karisalankanni-Trailing Eclipta *(Eclipta prostrate)*

Parts used : Seeds, Juice of leaves, Herb, Oil

Uses : Viral hepatitis, Hair hygiene, memory disorders, minor cuts and burns

Korai Kizhangu-Nut Grass *(Cyprus rotundus)*

Parts used : Bulbous Root

Uses : Diarrhoea, Indigestion, Deodorant, Dysentery, Fever

Sirukurinjan-Gurmar *(Gymnema sylvestre)*

Parts used : Leaves

Uses : Snake-bite, malarial fever, Eye diseases, Diuretic

Perunkurinja-Sneezing Silk Cotton *(Wattakaka volubilis)*

Parts used : Roots and Leaves

Uses : Snake-bite, to cure head-ache after child birth

Vellai Kundarikum, Vellai Kungiliyam, Indian Copal Tree *(Vateria indica)*

Parts used : Bark, Resin, oil

Uses : cure impure blood, Burns and Ulcers

Venthayam Fenugreek *(Trigonella foenum)*

Habit : Herb

Parts used : Seeds

Uses : Skin inflammation, Carminative, Galactagogue, Ulcers

Siriyanangai-Creat *(Andrographis paniculata)*

- Roots commonly used as a tonic and febrifuge.
- Plant is astringent, anodyne, tonic and alexipharmic, used in dysentery, cholera, diabetes, consumption, influenza, bronchitis, itches and piles.
- Decoction used for sluggishness of liver and in jaundice.
- Irulas use the parts of the plant for snake-bite and for curing jaundice.
- Flowering Season : September to March

  Flowering Season : August to October

Siriyanangai *(Andrographis paniculata)*

5-10 nos are given as such to persons who are affected by poisonous biter. Whereas in the case of animals, juice is extracted from handful of leaves and given as internal feed. Five numbers of Siriyanangai leaves are given internally as such for about 2-3 days.

Dosage : 5 nos. / time / day.

Care should be taken while administering the medicine : while taking the above said medicines intake of non vegetarian items should be avoided.

Senthalai-Screw Pine *(Pandanus odoratissimus)*

Tender leaves eaten raw or cooked. Seeds also edible. Leaves employed for covering huts, mat-making, cordage, hats, baskets, and other fancy articles. Also used for making umbrellas. Leaf fibre used for making sacks for coffee, sugar, grain. Leaves are a source of paper-making material, used in leprosy, scabies, diseases of heart and brain. Iruas offer the flowers to their goddess and irula mid-wives use it for abortion. Roots used as brush for painting.

Flowering Season : February to June.

Aavuri-Wild Indigo *(Tephrosia purpurea)*

Leaf extract used in bronchitis, boils, pimples and bleeding piles. Roots and seeds insecticidal. Decoction of roots given in dyspepsia, diarrhoea, rheumatism, asthma, and urinary disorders. Ointment prepared from the roots is used in elephantiasis. Decoction of pods used as a vermifuge and to stop vomiting. Seed yield an oil said to be a specific against scabies, itch, eczema and other skin eruptions. Irulas use the leaves for curing skin diseases and roots for treating snake-bite.

Manjal Karisalankani (*Wedelia chinensis)*

Leaves are used for dyeing hair and for promoting hair growth. Roots yield a black dye. Leaves are a tonic and also used in cough, cephalagia, and alopecia. Decoction of herb used in menorrhagea. Ethanolic extract of herb inhibits the growth of carcinoma. Irulas use leaves for curing jaundice.

Flowering season : year round

Method I

Preparation

Karisalankanni leaves are taken, cleaned and dried well. Then make it into a fine powder by using Grinder. One teaspoonful powder is mixed with milk.

Dosage : One time/day. It is continues up to 10 days.

Method II

Ingredients

- Karisilankanni leaves - A handful
- Small onion *(Allium cepa)* – 5
- Tomato - One
- Chilly(*Capsicum frutescens*) – 3
- Oil – Required amount

Preparation

Fry the leaves of karisilankanni along with onion, tomato and chilly. And made it into a fine paste.

Dosage : In the morning, the paste is taken in empty stomach for 10 days.

Additional Information About Karisilankanni

Leaves are used for dyeing hair and for promoting hair growth. Roots yield a black dye. Leaves are a tonic and also used in cough, cephalagia, and alopecia. Decoction of herb used in menorrhagea. Ethanolic extract of herb inhibits the growth of carcinoma. Irulas use leaves for curing jaundice.

Flowering season : year round.

Navakonchi *(Rauvolfia tetraphylla)*

Roots often used as a substitute and adulterant of those of *Rauvolfia serpentine* for nervous disorders; their reserpine content, however, is low. Plant may cause fatal poisoning. Extract of the herb mixed with castor oil is applied to skin aliments. Fruit yields a black dye. Irulas use roots as one of the snake-bite medicines.

Flowering Season : May to September

Poonai Mesai Leaves *(Orthosiphon spiralis)*

- Anil poo leaves – A handful
- Poonai mesai leaves – A handful

Preparation

Grind the leaves and prepare a decoction. And this decoction can be taken in the morning or evening in empty stomach.

Dosage : one time / day.

Elumpotti Thalai *(Ormocarpum sennoides)* Leaves – A handful

- Egg (Country hen) – One
- Milk – 100 ml

Preparation

Grind the leaves of elumpotti thalai along with egg and make it into a paste. And this paste is mixed with milk.

Dosage : This is to be taken for one time / day until it cures.

Additional Information About Elumpotti

Roots used as a tonic and stimulant. Used for lumbago, rubbed in oil and applied in paralysis. Leaves used for blood circulation. It is also used as fodder for cattle. Irulas use the bark and leaves for treating bone-fractures.

Flowering Season : December to May

Sennithulasi (*Stevia spp*) Leaves

Leaves a handful are taken, dried and made it into a powder.

Malai Vembu *(Melia azedarach)* Leaves – A Handful

Malai vembu leaves are ground well by using water. It is taken for one time for 3 days.

Thandri-Belliric Murobalan *(Terminalia bellirica)*

Wood used for rough shafts, carts, and in coal mines as pit-props; also used for boards, packing-cases, boats, small crafts and catamarans and ploughs. Suitable for second quality slate frames. Fruits used for tanning. Also applied to rheumatic swellings and for eye-care. Bark diuretic and the gum demulcent and purgative. Irulas use the fruit for stomach ulcers, preventing white discharge and as a general health care medicine.

Flowering Season : March to January

Ingredients

- Thandri (*Terminalia bellirica*) leaves – A handful
- Kadukkai (*Terminalia chebula*) – A handful
- Amla powder (*Emblica officinalis*) – A teaspoonful
- Sukku (Dried ginger)

Preparation

Grind the leaves of Thandri and kadukkai and it is mixed with amla powder and sukku.

Dosage : Take this medicine one time/day until it cures.

Paruthi (Gossypium spp)

Hundred gram flower is taken and 400 ml of water is added to it .then the content is boiled well until it comes to 100ml. it has to be taken internally for about 15 days. It will cure menstrual problems.

Dosage : one time/day

Thumbai *(Leucas aspera)* Thumbe

Leaf extract Hand ful of Thumbai leaves are taken, cleaned, and 10 ml juice is extracted which has to be given to person the who is affected from scorpion bite.

Dosage : 10 ml / time/day.

Ranakalli

A leaf is fried in gingelly oil and kept on the wounds and tied with cotton cloth will cure the wounds quickly.

Kallimulli

Top portion of plant can be used to make the pickle after proper cleaning and processing. This medicinal plant makes the body cool during hot climatic conditions. It is used for the treatment of chest pain and body heat. This plant abundantly seen in villupuram area.

Karunaikilangu

It is a tuber crop it is used for the treatment of piles as well as constipation. It is given in the form of labium to the patients suffering from body temperature.

Murukan

Seed of the medicinal plant is used to stimulate diarrhoea. It is used as the dewormer in human and animals. Plant secretion is mixed with cow milk can be given to weak person for over all health improvement.

Perunerinjal

Plant leaves and upper stem portion are kept in water for three to 4 hours and filter. The filtered water can be given to the women suffering from endometritis. It is having the capacity to cure endometritis, body heat and renal diseases.

Siriyanangai *(Andrographis paniculata)*

Senthalai *(Pandanus odoratissimus)*

Manjal Karisalankani *(Wedelia chinessis)*

Navakonchi *(Rauvolfia tetraphylla)*

Poonai Mesai Leaves *(Orthosiphon spiralis)*

Elumpotti Thalai *(Ormocarpum sennoides)*

Fig. 28. Medicinal Plants

Masipathri

The leaves of this medicinal plant are partially roasted and extracted. The extracted fluid is mixed with honey and given to patients who are ffering from chronic respiratory diseases. It is also used for garland making purpose.

Nalvelae

The whole plant including root is grinded well and fluid extracted. The grinded plant with out fluid is kept in forehead of the person who is suffering from one sided headache, is tied with help of gauze bandage cloth for a period of 30minutes. Then the tied cloth opened and the plant material is full of fluid which is absorbed from the head of the diseased person. The diseased person will get relief from the conditions like one sided head ache, nostril block and upper respiratory tract infections.

Nanjumurunjan

It is used to catch the fish from the ponds. The root of this plant is made to float top of the ponds. The fish present in particular pond is get anaesthetized, and they floated on the top of the water. Then we can easily harvest the fish. This plant has no deleterious effect .So; we can safely consume the fish.

Uthamani

It is used to scrape the scar tissue formed in wounds. For this purpose the plant milk is administered on the scar tissue formation. After healing of the wound the surface area of the skin will be smooth and uniform because of the plant milk applied on it.

ETHNOVETERINARY PRACTICES

S.No	Particulars
1	**Unni** (Lice control) in Odungampatti village of Pudukottai district : Effective curative technique with locally available ingredients.
2	**Silanthi** (Mite Control) in Odungampatti village of Pudukottai district. Mite control with in one week.
3	**Kalichal** (Fever) in Odungampatti village of Pudukottai district. Ingredients : Kundumani, Required quantity of kitchen ash, cures fever within 3 days.
4	**Kunnal Noi** (Poison disease) in Perahambi village of Trichy district. Ingredients : Garlic, Asafoetida, Thippili, Herbal mixture paste is very effective; and helps in relieving body pain.
5	**Elumpu Murivu** (Bone fracture) Treatment Method I in Vellapallam village, Kanyakumari district. Ingredients : Impara root, Gingelly oil, Quick setting of bones.

Elumpu Murivu (Bone fracture) Treatment Method II in Bogasandram village of Krishnagiri district. Ingredients : Thuthi, egg, Quick setting of bones.

Elumpu Murivu (Bone fracture) Treatment Method III, Ingredients : Karu velam, bamboo stick, aathi. It will not create any side effect.

6 **Vairu Uppusam** (Bloat) Treatment Method I Perahambi of village Trichy district. Ingredients : ginger soda, baking soda and soda water, palm gur it cures the bloat problems of animal.

Vairu Uppusam (Bloat) Treatment Method II. Krishnagiri district. Ingredients : kolingi garlic bulb, pepper, salt. Herbal cures relatively advantage, no side effect.

7 **Treatment for poisoned animals** (insects or snakes) in Kampalathupatti village of Trichy district.Ingredients Thumbai, Poolaikeerai, To cure the affected animal.

8 **Vishakkadi** (Poisonous Beetle bite) Treatment Method I in Manakudi village of Sivagangai district. Ingredients : Aduthinnaapaalai, Pirandai,

Vishakkadi (Poisonous Beetle bite) Treatment Method II in Manakudi village of Sivagangai district. Ingredients : Nanchilaiparichaan root, Reduce the venomous effect of the animal.

Vishakkadi (Poisonous Dog bite) Treatment Method III in Manakudi village of Sivagangai district. Ingredients : Sirugurinchaan, Thalaisuruli, Thumbai, To cure the dog bite.

9 **Kunthu** (Anthrax) Method I in Pasavapuram village Krishnagiri district. Ingredients : Ferula asafoetida, Raw materials are grocery items no side effect.

Kunthu (Anthrax) Method II Krishnagiri district. Ingredients : Pepper, garlic, radish, coriander, Raw materials is grocery items, best cure the anthrax.

Kunthu (Anthrax) Method III Perampalur district Ingredients : Veliparuthi, kundumali, A raw material is available in local. Easily will be cure.

Kunthu (Anthrax) Method VI ingredients : Acalyapa indica, salt, Perampalur district. Raw materials are available in local. Easily will be cure.

Kunthu (Anthrax) Method V in Perampalur district. Ingredients : Aaveli, cow urine, Raw materials are available in local. Easy curing technique.

10 **Vaitrupokku** (Diarrhoea) Method I in perahambi of Trichy district. Ingredients : gur and water Effectively control diarrhoea.

Vaitrupokku (Diarrhoea) Method II in Perahambi of Trichy district. Ingredients : castor oil and cotton Herbal cure treatment no side effect.

Vaitrupokku (Diarrhoea) Method III in Perahambi of Trichy district. Ingredients : Eggs Trichy district. Effectively control in diarrhoea.

Vaitrupokku (Diarrhoea) Method IV in Muniayathirayanpatti of Perampalur district. Ingredients : Agathi leaves Effectively control in diarrhoea.

Vaitrupokku (Diarrhoea) Method V in Muniayathirayanpatti of Perampalur district. Ingredients : tobacco, pepper, chillies, Effectively control in diarrhoea.

11 **Patchai Narampu** (Green nerves) in Kotarakundru of Perampalur district. Ingredients : Tamarind, salt, Treatments are sufficient and cure.

12 **Cold-Symptom** (Respiratory disease in poultry birds) in Sivagangai district. Ingredients : small onion, butter milk. The best method of prevent the disease.

13 **Mooku Adaippan** (Sneezing in Poultry birds) in Sivagangai district. Ingredients :small onion, water. The occurrence of disease has best effect

14 **Serumal kaichal** (Cold and Fever) Krishnagiri district. Ingredients : Chitra moolam root, Garlic bulb, Pepper. It is safer to use and the result of the mixture is good. The method is more economically.

15 **Kannai** (Foot and Mouth Disease) Method I in Sampanthakottai of Krishnagiri district. Ingredients : Nalabal asthi, Herbal cure is relatively advantage.

Kannai (Foot and Mouth Disease) Method II Trichy district. Ingredients :

Neem oil, raw rice gruel, poovan banana, Easy to prepare the medicine.

Kannai (Foot and Mouth Disease) Method III Trichy district. Ingredients : Turmeric powder, neem oil, camphor. Raw materials locally available, herbal cure treatment.

Kannai (Foot and Mouth Disease) Method IV Trichy district Ingredients : Neem Oil camphor hot fine sand. (Foot lesions) banana, sesame oil, kasakasa, (mouth lesions) It will not create any side effects

Kannai (Foot and Mouth Disease) Method V, in S.V Mangalam of Sivagangai district. Ingredients : Konrai tree, Vishamalli, Thirugukalli, Herbal cure is relatively advantage.

Kannai (Foot and Mouth Disease) Method VI, S.V Mangalam of Sivagangai district. Ingredients : Fish waste, dried fish, It will not create any side effects

Kannai (Foot and Mouth Disease) Method VII, S.V Mangalam of Sivagangai district Ingredients : Cumbu grain, salt, water. Herbal cure is relatively advantage.

Kannai (Foot and Mouth Disease) Method VIII, S.V Mangalam of Sivagangai district. Ingredients : Mushroom, Neem oil. (Foot lesions) Hill banana, pig milk. (Mouth lesions) The cost of medicine is very low.

16 **Rattha kalichal** (Dysentery or Hemorrhagic Gastero Enteritis) Treatment Method I in Bogasandiram village of krishnagiri district. Ingredients : Gur, ipomea, To cure the dysentery.

Rattha kalichal (Dysentery or Hemorrhagic Gastero Enteritis) Treatment Method II in Sivagangai district. Ingredients : Kiluvai tree bark,cumin,water, To cure the dysentery.

Rattha kalichal (Dysentery or Hemorrhagic Gastero Enteritis) Treatment Method III in Sivagangai district. Ingredients : Cumin, coconut spadix, water, in The required fibre content to cure dysentery.

17 **Kulambu noi** (Hoof lesion) in Sivagangai district. Ingredients : Kadukkaai, Kathirimanjal, Small amla. It will not create any side effect

18 **Pulippu noi or Thullumari noi** (Struck) in Sivagangai district. Ingredients : Sambiraani, aduthinapalai, Raw materials locally available, herbal cure treatment.

19 **Moochupidipu** (Tramatic pericarditis) in Sivagangai district. Ingredients : milagaranai, onion, water. Easy preparation it will cure in effectively.

20 **Vatha noi** (Anorexia) Treatment method I in Sivagangai district. Ingredients : coconut, fenugreek, cumin, dates. Simple method easy to apply.

Vatha noi (Anorexia) Treatment method II in Sivagangai district. Ingredients : veeli leaves, small onion, butter milk, dates. Simple method easy to apply.

21 **Pullarithal** (Ephermal Fever) Method I in Sivagangai district. Ingredients : Omam, Betel leaves, Pillaikilangu. The same medicine is also used to cure FMD in cattle.

Pullarithal (Ephermal Fever) Method II in Sivagangai district. Ingredients : Karuncheeragam, Kathrimanjal, White Kungilium. The same medicine is also used to cure FMD in cattle. Material available in local petty shop.

Pullarithal (Ephermal Fever) Method III in Sivagangai district. Ingredients : Vasambu, sukku, Vaalmilagu, Karuncheeragam, It will not create any side effects. Easy preparation.

Pullarithal (Ephermal Fever) Method IV in Sivagangai district. Ingredients : Small onion, Dried ginger, Sidharathai. Easy preparation.

22 **Kaal odithal** (Leg fracture) Method I, Sivagangai district. Ingredients : tamarind seed, paddy grains, cumbu grains, thesavaarai. Quick setting of bone, no side effect.

Kaal odithal (Leg fracture) Method II, Sivagangai district. Ingredients : Sotrukatraalai, Egg, black gram flour. Quick setting of bone.

Kaal odithal (Leg fracture) Method III Sivagangai district. Ingredients : Coconut mesocarp, Palm jaggery, Sunnaambu. Quick setting of bone.

23 **Veekkam** (Boils / Swelling or Abrasions) in Kirungakottai village of Sivagangai district. Ingredients : Avuri, water. This treatment is sufficient to cure the disorder.

Veekkam (Boils / Swelling or Abrasions) in Kirungakottai village of Sivagangai district. Ingredients : Pirandai, neem oil. This treatment is sufficient to cure the disorder.

24 **Thalai pungal/ Kayangal** (Wounds /Abrasions in Head) in Sivagangai district. Ingredients : Thaivelai leaf, Kuppaimeni, Dry ginger. Continuous treatment is must.

25 **Mutti Veekkam** (Swelling of Joints) in Sivagangai district. Ingredients : Pirandai, Camphor, Garlic bulblets, Lime fruit. This treatment is medicinal effect. Reduce the swelling.

26 **Madi Veekkam** (Mastitis) in Sivagangai district. Ingredients : Parboiled rice, Sunnaambu, Paakku. To cure the swelling.

27 **Kaluthu Veekam** (Swelling of Neck) in Sivagangai district.Ingredients : Ridge gourd, salt, water. To reduce the yoke gall.

28 **Kaluthupun** (Neck wounds) in Sivagangai district. Ingredients : Tharaipugailai, Veeram, Small onion. To cure the wounds.

29 **Kaal Pungal** (Leg Wounds) in Sivagangai district. Ingredients : Neem oil, Hen's feather, Naphthalene balls. As a good result so it's widely followed.

30 **Kaambu Adaithal** (Teat Blockage) in Sivagangai district. Ingredients : One time treatment is sufficient to cure the disorder.

31 **Kombu odithal** (Horn Fracture) Methods I in Sivagangai district. Ingredients : Palm sugar, Sunnaambu, Three days treatment is sufficient to cure.

Kombu odithal (Horn Fracture) Methods II in Sivagangai district. Ingredients : Ottadai, Hair, Veeram, Three days treatment is sufficient to cure.

Kombu odithal (Horn Fracture) Methods III in Sivagangai district. Ingredients : Neem oil, Human hair, Palm sugar. Three days treatment is sufficient to cure.

32 **Reduction in milk yield** in Sivagangai district. Ingredients : Coconut kernel, Parboiled rice.To induce milk yield as they stimulate milk secretion.

33 **Manjal Kaamaalai** (Jaundice) in Sivagangai district. Ingredients : Cow urine, Butter milk, Keelanelli. Cost of medicine is low, easy preparation no side effect.

34 **Sandhu Noi** (Black Quarters) in Sivagangai district. Ingredients : Marudhamtree, Seemaikaruvai, Karuvai tree bark. To arrest the disease,

35 **Komaari disease** (Entero toxemia) in Sivagangai district. Ingredients : Small onion, Omam, Jaggery. To arrest the disease effectively.

36 **Elangodi Veliyearaamai** (Retention of Placenta) in Sivagangai district. Ingredient : Sesamum, pirandai, palm sugar. Act as fluxing agent as it has high viscosity.

37 **Sinaipidikkaamai** (Oestrous cycle failure) in Sivagangai district. Ingredient : Yaanai nerinjii, water. To induce reproduction organ.

38 **Kannil Poochi Viludhal** (Eye Infection Due to Insects) in Sivagangai district. Ingredient : Kanchira leaves, neem oil, To cure the eye swelling and eliminate eye insect.

39 **Serumal / Jaladhosham** (Cold) Kanyakumari district. Ingredients : Onion. The ingredients are rich in capsacin content, active is high. This treatment is very effectively.

40 **Ajeeranam** (Indigestion) in Kanyakumari district. Ingredients : Betel vine, garlic, moringa bark, It is relatively cheaper when compare to allopathic medicine.

41 **Madi Noi** (Teat Obstruction) in Kanyakumari district. Ingredients : Tuemeric, nannari kizhangu, Kollankovai kizhangu, The disease get quickly cured.

42 **Koppulam** (Vesicle) in Sivagangai district. Ingredients : Ottadai, turmeric, neem oil. It is sufficient to cure the infection. If necessary we can go for three more days continuously.

43 **Avoidance of Breeding in 'goats** in Kalichettipatti village of Namakkal district. Male goat from any disturbance during the gestation period.

44 **Arecanut Juice (De wormer)** in Kalichettipatti village of Namakkal district. Controlling worms in poultry.

45 **Kangayam local** (Indigenous bullock and cow) in Kangayam block of Erode district. The shape of the body is straight, legs are height and straight and four hoofs are equally flat in nature

46 **Ramnad Sheep (Ram Aadu)** in Ramnad district but spread thorough out southern Tamil Nadu state. The Ramnad male sheep is crossed with Kilakarisal female sheep for better body weight gain, because Kilakarisal sheep is comparatively lesser weight than the Ramnad white sheep.

47 **Salem Black or Pallai Aadu** in Salem District. The disease resistance capacity of the animal is high. Twinning is the common characteristic feature of this animal. These animals are mainly meant for meat purpose.

48 **Sandai Seval (Fighter Cock)** These birds are reared in the areas of Cumbam, Theni, Tirunelveli, Madurai, Usilampatti, Sivagangai, Paramakudi and Ramanathapuram areas. The cost of the cock is ranging from Rs. 2000 to Rs. 3000. It fetches good income to the owners.

49 **Vempur Sheep** It is the native of the Tirunelveli District and now in Thoothukudi District. Vempur breed is famous for its thigh muscle. This breed has good adoptability in tropical areas.

50 **Kilakarisal Sheep** in Tirunelveli District, now Kilakarisal village is in Thoothukudi District. The animals have very good adoptability in tropical areas. The meat derived from this breed is tastier than other sheep breeds.

51 **Burgur Local** (Burgur Sheep) in Malasjanoor Village of Erode district. White and light brown in colour.

52 **Manaparai local** (Local breed) in Manaparai village of Trichy district. Predominant for several local breeds of cattle, Mayilai, Sevalai, Karukka mayilai / kari.

53 **Krishnagiri local** (Local breed) in Pasavapuram village of Krishnagiri district. It is used for dual purpose less maintenance compared with cross breed cattle

54 **Cheng Kanni Adu** (Local breed) in Tuticorin district. This breed is highly suitable for rain fed region. This variety is the tastiest meat producer in Tamil Nadu.

Unni (Lice Control)

Curative Method

Ingredients

Camphor – 5 nos

Neem oil - ½ litre

Preparation and Administration

Camphors (5 nos) are powdered and ½ lit of Neem oil is added to it and make a paste out these. This paste is smeared all over the body of the goat. It will effectively control lice. It is to an applied for 7 days.

Silanthi (Mite)

Curative Method

Ingredients

- Salt - ½ kg
- Water – required quantity

Preparation and Administration

½kg salt is dissolved in required amount of water and this salt solution is smeared over the body of goat which is infested by mite. This is to be applied twice a day, for one week.

General Profile of the ITK Holder

Miss. Marikannu, LEISA Kolinji Farm, Odugampatti, Pudukottai (Dt).

Kalichal (Fever)

Symptoms

- Shivering of body and limbs.
- Errect hairs.
- Watery eyes.
- Recumbent and unable to get up.
- Off feed and reduction in the milk yield.

Treatment

Ingredients

1. Kundumani (*Abrus precatorius*) leaves with wine.
2. Required quantity of kitchen ash.

Method of Administration

Kundumani leaves powder with wine are tied over the neck portion of the affected animal and a handful of kitchen ash is sprinkled over the body of the animal. This will reduce the temperature of the affected animal. This should be done twice a day till the fever cures.

Relative Advantage

This is a no cost technology which cures fever within 3 days. No side effect while administering the treatment and materials used for treatment are locally available.

Kunnal Noi (Poison Disease)

Symptoms

If cattles are diseased with 'Kunnal' symptoms of head pressing, cynosis and salivary mucilage secretion from mouth will be found. It may be lead to death, if left uncared.

Medicine Preparation

To cure this disease, a medicine with a combination of herbal mixture called 'kunnal nooi marundhu'is prepared and administered. The ingredients are as follows;

Ingredient I (Herbal Products)

Each 5g to be taken.

1. Thippili (*Piper longum*)
2. Omam (*Trachyspermum ammi*)
3. Garlic (*Allium sativam*)
4. Asafoetida (*Ferula asafetida*)
5. Karunseeragam (*Cuminum cyminum*)
6. Vellaimilaguu (*Piper sp*)
7. Valmilagu (*Piper sp*) Idampuri, Valampuri, Kadalnurai, thesavaram,

Ingredients II (Herbal Plants)

Each one handful of leaves.

1. Thumbai (*Leucas aspera*)
2. Thulasi (*Ocimum sanctum*)
3. Notchi (*Vitex negundo*)
4. Uppilangodi (*Pentatrophis capensis*)
5. Vellarugu (*Eniccostema axillare*)
6. Velli paruthi (*Pergularia daemia*)
7. Nalla vellai (*Clome gynandra*)
8. Sangu (*Azima tetracanth*)
9. Peru marundhu (*Aristolochia India*)
10. Mudakatan (*Cardiospermum halicabam*)

Method of Preparation

For any medicine preparation, purification of the ingredients is must, without which its effectiveness cannot be obtained. So, the ingredients are purified. In the case of Ingredient I, the contents are bought from market. So all impurities like foreign materials and other crop seeds should be removed. Incase of Ingredients II, care should be taken while collection i.e. disease free and clean plants are collected and washed thoroughly with water. Thus the ingredients are purified.

Next to purification, the process of grinding starts. In this process, all the ingredients are added in a stone made motor and pestle (kuzhavikal) and ground well.

While grinding little quantity of water is added to the ingredients for preparing it into a paste. Thus, finally herbal paste is prepared. This paste is diluted with water to make it about ¼ litres. To prepare this medicine it costs about Rs 65/-.

Administration

1. Mode : It is administered internally as oral medicine.
2. Dosage : 1/4 litre /time (for large animals) and half of the dosage for small animals.
3. Number of time : It is administered in morning and evening of first day and the next day morning.

Care to be Taken While Preparing

Before preparing the medicine, care should be taken during purification process. All the ingredients should be purified i.e. ensured that no impurities are adhered or mixed with the ingredients.

Conditions and Precautions

As stated by Thiru.Oosaladiyan, the vaidhiyar (local practitioner) who administers this medicine, the herbal leaves plucked for preparing medicine should only be from northern side facing. In a plant, leaves on the northern side alone should be plucked. He substantiates that such north direction facing leaves have more efficiency in curing the diseases.

Medicines prepared are fed to the cattle using "kottan" (A bamboo made feeder, with one end closed and the other end cut opened in a slanting manner) while feeding, the tongue is held with hands and the medicines in fed into the mouth. If it is not done so, there is chance for the cattle to push out the kottan and the medicines.

Relative Advantage

This particular herbal mixture paste is very effective against this particular disease as well as it helps in relieving body pain, which is much more needed for draught animals.

ITK Holder's Profile

Thiru.Oosaladiyan, 60 years old is residing in Perahambi village in Manachanallur taluk of Trichy district. He administers medicines for both human and cattle. He treats the animal based on the knowledge gathered from his father, who was a Vaidhyar and from the vaidhyar books.

Elumpu Murivu (Bone Fracture)

Method I

Symptoms

Fracture of bone and sprain on legs of animals occurs due to hitting and falling down. As a result swelling of the affected portion makes the animal unable to walk.

Ingredients

1. Impara root – (*Oldenlandia umbellate)* fistful
2. Gingelly oil – 100ml

Preparation

Fistful of Impara root (*Oldenlandia umbellate*) is taken, it should be pounded well and finally 100 ml of gingelly oil is added to it. This paste is smeared over the affected portion and covered with cotton cloth. This should be continued up to one week.

Conditions and Precautions

While administering the medicine three to four bamboo splits are used to secure the point of fracture and to prevent mobility. So that quick setting of bones will be ensured.

Relative Advantage

In this method cost is the main advantage ie, it costs about Rs. 10 (approximately)

General Profile of the ITK Holder

Name and age : Thiru Sengutuvan, Age : 45 years, Vellapallam, Kanyakumari Dt.

Method II

Ingredients

1. Thuthi (*Abutilon indicum)*- A handful
2. Egg (two numbers)
3. Indigenous local bird) - 1 number.

Treatment

A handful of thuthi leaves are collected, cleaned and ground well. Paste is made out of it and egg is added and mixed well with the paste. This mixture is applied / smeared over the affected portion and covers the portion with clean cotton cloth. This should be kept as such for a week. If it is not cured this treatment should be continued for another week.

General Profile of the ITK Holder

Name & address : Th.Bellalappa, Bogasandram, Irudhu kottai (Post), Thenkanikottai taluk, Krishnagiri district, Educational status : Nil, Experience : 57 years.

Method III

Ingredients

- Karu velam – *Acacia nilotica*
- Bamboo stick
- Aathi – *Bauhinia racemosa*

Administration

Round shaped Karu Velam bark is taken and grounded well. It is applied over the affected portion and covers the portion with small length of bamboo sticks. This is to be tied with Aathi leaf sheath. This should be kept as such for a week. This is done once in a week till the bone setting is ensured.

Relative Advantage

- The ingredients used for treatment are locally available.
- It will not create any side effects.

Vairu Uppusam (Bloat / Tympany)

Method I

Symptoms

- Stomach will be distended and bulged.
- Watery discharge from eyes and nose, cattle will find difficult in breathing and becomes restless.
- It often kicks at the sides of the abdomen.

Cause

It is caused by either sudden change in feed materials, excess intake of leguminous fodder or due to intake of feeds like kitchen and dining waste.

Treatment

Medicine and its Ingredients

A medicine called "Vayiru uppusa marundhu" is administered for curing bloat.

Ingredients

1. Ginger soda – 1 bottle
2. Baking soda – 25g
3. Soda water – 1 bottle
4. Palm gur – 250 g

Preparation

Palm gur (250 g), baking soda (25g), soda water (1 bottle) and ginger soda (1 bottle) are mixed well. This solution is fed orally using 'kottan' or 'coconut shell'.

It is continued twice or thrice a day, @ 250ml per time.

Care to be Taken While Preparing

Purification of palm gur is the only care to be taken while preparing this medicine, while adding soda with palm gur impurities in gur will float, which can be easily removed.

Conditions and Precautions

While administering the medicine, the cattle's stomach should be empty. Similarly after administering the medicine, it should be allowed to walk for ½ - 1 km and should not be fed for 2-3 hours.

Relative Advantage

Bloat is mainly caused due to improper feed or contaminated feed. This creates indigestion problem in cattle, for which ingredients like ginger, soda will induce digestion. So in order to set it free from restlessness, this medicine is the best.

Additional Information

As stated by this Vaidhiyar, after doing this treatment, the medicine container is rotated thrice around the head and a hair from tail is pulled off.

General Profile of the ITK Holder

Thru .Osaladiyan, 60 years, Perahambi village, Manachanallur taluk, Trichy district.

Method II

Ingredients

- Kolingi- *Teprosia purpurea* - A fistful
- Garlic bulb (*Allium sativum*) - ½ fistful

- Pepper (Pe*pper nigrum*) - ½ fistful
- Salt - required amount

Preparation

A fistful of kolingi bark, garlic bulb ½ fistful, pepper- ½ fist ful and required amount of salt are ground well and made it as a paste. This paste is mixed in one litre of cold water and fed to the affected animal. This should be given once a day.

Precautions While Administering the Medicine

Green fodder should be avoided.

Feasibility and Relative Advantages

All the ingredients are locally available. It is very easy to prepare the medicine. Similarly, the method of preparation is also very simple which doesn't involve cumbersome process. Herbal cure is relatively advantages over other types of medicines due to its nil side effects.

Beneficiaries

According to this Vaidhiyar, nearly 300 animals have been benefited by this treatment.

General Profile of the ITK Holder

Name & address : Th.Bellalappa, Bogasandram, Irudhu kottai (Post), Thenkanikottai taluk, Krishnagiri district, Educational status : Nil, Experience : 57 years.

Treatment for Poisoned Animals (Insects or Snakes)

Symptoms

- Body surface edematous
- Swollen eyelids
- Hypothermia
- Body surface chilled
- Discolouration of the skin or coat
- Sweating and rashes on the body

Treatment – 1

Ingredient

1. Poolai Keerai (*Aerva lanata*) – A fistful
2. Thumbai (*Leucas aspera*) – A fistful

Preparation and Administration

A fistful of Poolai keerai, a fistful of Thumbai and required quantity of water are ground well. The extract is obtained and two drops of it is added to the eyes and ears of the animal. After ten minutes of the treatment, the poison will come out of the affected animal through cow dung and urine.

Treatment – II

The ears of the animals bitten by poisonous insect or snakes. Make cut in a corner of the blood oozes out of the cut portion which is covered with Erukku (*Calotropis gigantea*). It is observed that poison comes out along with the blood drops.

These two treatments should be administered immediately after noticing the poisonous bite symptoms.

General Profile of the ITK Holder

Kampalathu Nayakkars are a unique category of people residing in isolated pockets of Kampalathupatty in Manapparai taluk of Trichy district. Hunting and agriculture are their main occupation. The whole community follows the same Ethno veterinary practices. The traditional knowledge on Ethno veterinary practices being practiced by them have been gathered from their ancestors. Thiru Kuppamuthu and Kannan were one among the respondents.

Vishakkadi (Poisonous Beetle Bite)

Method I

Symptoms

The specific symptoms for the poisonous beetle bite in the cattle are,

- The body surface of the animal becomes oedomatous.
- The eyelids become swollen.
- Hypothermia.
- Chilled body surface and it is felt by touching the abdomen of animal.
- Discoloration of the skin or coat.
- Sweating and rashes in the body.

Causes

There is no specific reason for the disorder. While the animal is grazing in the field there may be chances for beetle bite.

Ingredients

1. Aduthinnaapaalai leaves *(Aristolachia bracteolata)* - a plant with a root
2. Garlic (*Allium sativam*) - 1 bulblet
3. Pepper *(Piper nigrum)* - 40-50
4. Green Pirandai leaves *(Cissus quadrangularis)* - 200g.
5. Water - 50 ml.

Method of Preparation

The "aduthinnaappaalai", garlic, pepper and "pirandai" are to be taken, ground well with 50 ml. of water to make a bolus. The grinding is done by either grinder or "ammi"- (specially designed stone used for grinding-a rectangular slab stone with a roller for grinding). Thus 2-3 boluses will be prepared weighing 80-100g each.

Administration

The mode of application is internal. The bolus is to be administered as intake medicine to the affected animal by using hand.

Two boluses are sufficient for normal cattle and in the case of calf or heifer one such bolus is sufficient. The recommendation is that 2-3 boluses immediately after beetle bite and/or during evening hours for 2 days continuously. If swelling is noticed in the bitten portion, green "pirandai" leaf paste is to be smeared over the portion after grinding it with addition of water.

Mode of Action

The mode of action is that the ingredients namely, "aduthinnaappaalai", garlic, pepper and "pirandai" have the chemical components to prevent the poisonous effect of beetle bite in animals.

Conditions to be Followed

- After the administration of medicine, the animal is to be fed with dry paddy straw and cotton seed cake.
- Avoid roughages during the course of treatment since it has high concentration which may reduce the medicinal effect of contents.
- The cattle should be fed with hot water cooled for 3-5 hours instead of hot water as the ordinary cool water has carrying effect of germs.

Relative Advantages

- Raw materials are locally available in the petty shop and surrounding place
- The cost of entire medicine is less than Rs.25-30/-
- It hasn't any side effects.
- Easy preparation.

Profile of the Practioner

Address : Mr. Karupiah, Manakudi, Tirupathur tk., Sivagangai Dt. Ph : (pp) 04577-26607,

Age : 65, Education : Illiterate, Animal treated so far : 1000 cattle

Method II

Ingredients

1. Erukku (*Calotropis gigantea*)
2. Nanchilaiparichaan root (removing poison) - 1 (2-3") length.
3. Water - 100 ml.

Method of Preparation

The "Nanchilaiparichaan" root is to be taken, ground with addition of 100 ml. water to make the content into a paste. The grinding is done by either grinder or "ammi" (specially designed stone used for grinding. Then, a 100 gram of paste will be prepared.

Administration

The mode of action is internal. The paste is to be administered as intake medicine to the affected animal. The resultant 100g paste is to be given to the animal as intake medicine during morning and evening hours continuously for 3 days and/or immediately after the poison bite.

Mode of Action

The mode of action is that the "nanchilaiparichaan" root has the chemical components capable to reduce the venomous effect of the animal.

Conditions to be Followed

1. After the administration of medicine, the animal is to be fed with dry paddy straw and cotton seed cake.
2. Avoid roughages during the course of treatment since it has high concentration which may reduce the medicinal effect of the contents.

3. The cattle should be fed with hot water cooled for 3-5 hours instead of hot water as the ordinary cool water has the carrying effect of germs.

Relatve Advantages

- Raw materials are locally available in the surrounding place and the cost of preparation is nil.
- It hasn't any side effects.

General Profile

Address : Mr. Karupiah, Manakudi, Tirupathur tk., Sivagangai Dt. Ph : (pp) 04577-26607Age : 65, Education : Illiterate, Animal treated so far : 800-1000

Method III Naaikadi (Dog Bite)

Introduction

It is a common problem found among cattle. While grazing there may be a chance of dog bite to the animals.

Symptoms

The specific symptoms for the dog bite are,

- The animal often looks restless.
- Makes noise frequently and itching symptom noticed.
- While getting up the animal scratches its body on the side wall
- In severe cases and if unattended it leads to death of animal.

Ingredients

1. Thumbai leaves (*Leucas aspera*) - 25 g.
2. Thulasi leaves (*Ocimum tenufolium*) - 25 g.
3. Thaivalai leaves (*Cleome gynandra*) - 25 g.
4. Aindilai notchi leaves (*Vitex negundo*) - 25 g.
5. Thalaisuruli leaves (*Aristolachia indica*) - 25 g.
6. Veppaalai leaves (*Wrightia tinctoria*) - 25 g.
7. Sirugurinchaan leaves (*Gymnema sylvestre*) - 25 g.
8. Aduthinnapaalai leaves (*Aristolachia bracteolata*) - 25 g.
9. Hot water - 1.5 litr.

Method of Preparation

All the ingredients are to be taken, ground well either by using grinder or "ammi" and 1.5 litres of water is added to make the content into a 2 litre juice.

Administration

The mode of application is internal. The prepared juice is to be administered as intake medicine through "moongikottaan" (a cylindrical hollow bamboo tube having a length of 20-30 cm.) by using a hand.

The 2 litre juice is sufficient to cure the dog bite disorder. The oral intake is to be carried out immediately after the symptom gets noticed and/or during evening hours continuously for one week. If necessary, treatment is to be continued for one more week. So for, 400 animals are treated in this method.

Mode of Action

The leaves of the ingredients have chemical components able to cure the dog bite.

Conditions to be Followed

- The animal should not be washed immediately after administration of medicine.
- After taking the medicine, animal is to be fed with dry paddy straw only.
- Avoid roughages and green grasses during the course of treatment which may reduce the medicinal effect of the contents.
- The cattle should be fed with hot water cooled for 3-5 hours instead of hot water as the ordinary cool water has carrying effect of germs.

Relative Advantages

- As the raw materials are locally available in plenty it fetches no or little cost.
- It will not have any side effects.
- Easy preparation.

Kunthu (Anthrax)

Method I

Ingredients

1. Asafoetida (*Ferula asafoetida*) – 5 g
2. Betel vine (*piper betel*) – 4 numbers

Method of Preparation

Five gram Asafoetida and four betel vine leaves are taken and ground well. Paste is made out of it. Half litre of hot water is added to that paste and made it as a solution. Finally the solution is fed to the affected animal by using bamboo utensil called 'Kotton'. This should be fed to the animal for thrice. (I.e. first day two times and next day morning one time)

Method II

Ingredients

1. Pepper (*Piper nigrum*) - 10 numbers
2. Garlic (*Allium sativum*) -1
3. Lavangam (*Cinnamomum zeylonicum*) -5g
4. Raddish (*Raphanus sativus*) -1
5. Coriander (*Coriandrum sativum*) - required amount.

Treatment

Above said ingredients are taken, cleaned and ground well. Paste is made out of it. This paste is fed to the animal for thrice a day.

Feasibility and Relative Advantages

Since all the ingredients used are grocery items which are available in a house hold level. It is very easy to prepare the medicine. Similarly, the method of preparation is also very simple which doesn't involve cumbersome process. Moreover, while taking this treatment, the animal need not adhere to feed restrictions.

Limitation/Constraints

Instant/immediate cure is not possible. It has own time limitation for curing the disease.

Beneficiaries

According to this Vaidhiyar, nearly 400 animals have been cured by this treatment

General Profile of the ITK Holder

Th. Ramaiah, pasavapuram, irudhu kottai (post) thenkanikottai taluk, Krishnagiri district.

Education status : 10th standard.

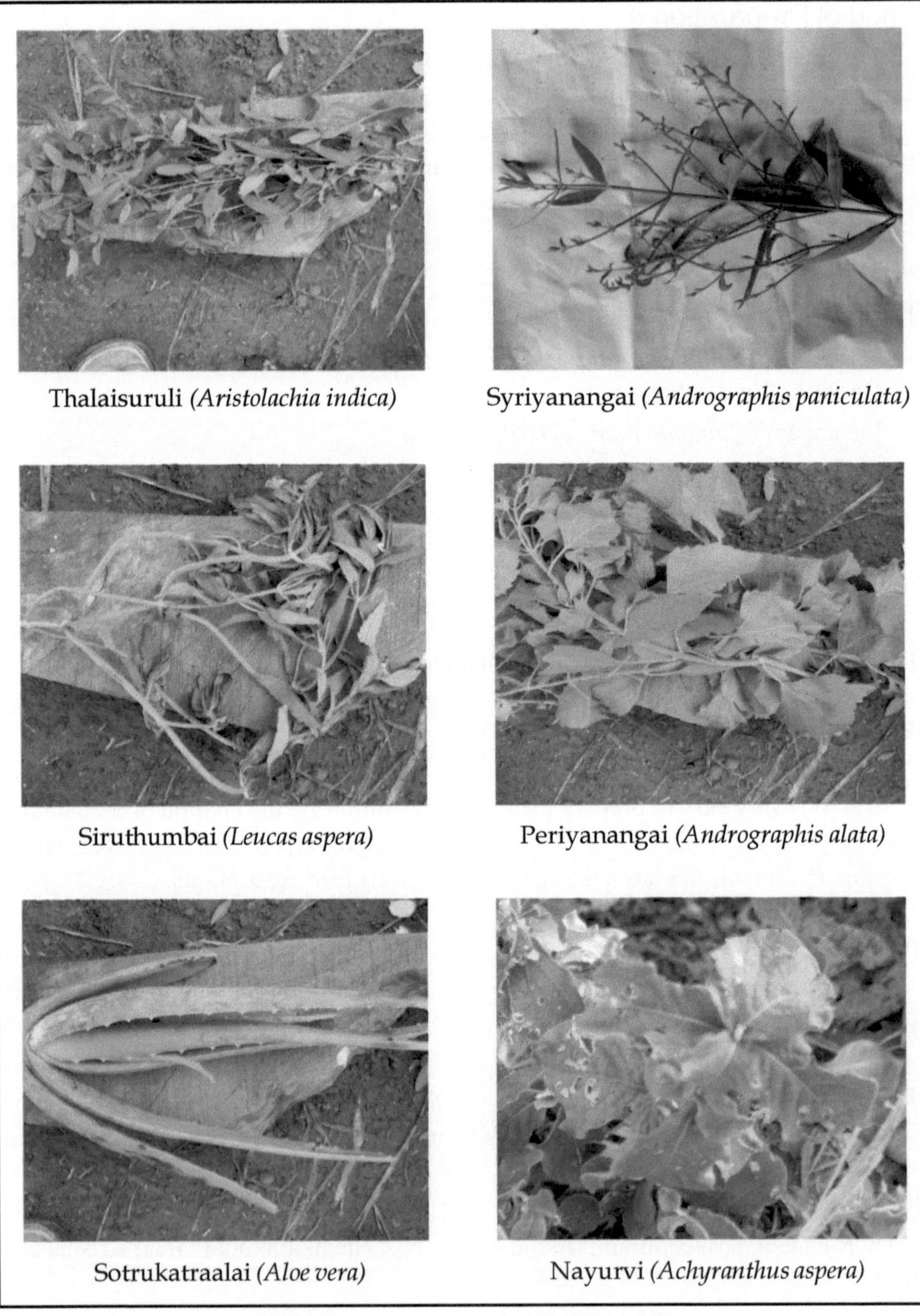

Thalaisuruli *(Aristolachia indica)*

Syriyanangai *(Andrographis paniculata)*

Siruthumbai *(Leucas aspera)*

Periyanangai *(Andrographis alata)*

Sotrukatraalai *(Aloe vera)*

Nayurvi *(Achyranthus aspera)*

Fig. 29. Ethno Veterinary Medicinal Plants

Method III

Ingredients

Veliparuthi (*Pergularia daemia)*
Kundumani (*Abrus precatorius)*
Small onion (*Allium cepa)*
Garlic bulb (*Allium sativum)*
Gingilly oil (*Seasamum niger)*

Medicinal Preparation

A fistful of veliparuthi, a fistful of kundumani leaves, 2-3 numbers of small onions, 4-5 garlic bulbils are ground well and made it as paste. Finally 10 ml of gingelly oil is added to it and it should be given as internal feed.

Dosage : Twice a day.

Method IV

Ingredients

- Kuppaimeni – (*Acalypa indica)*

Method of Preparation

A fistful of kuppaimeni leaves are collected, cleaned and squeezed well and juice is extracted. This extract is poured into ear, nose and eyes of the affected animal. This should be continued upto one week.

Dosage : Twice a day.

Method V

Ingredients

- Kuppaimeni – (*Acalypa indica)*
- Salt

Medicinal Preparation

The mixture of kuppaimeni leaves and small amount of salt is smeared over the body of the cattle. It should be administered till the cattle come to normal condition.

Dosage : Once in a day.

Conditions to be Followed

While administering the medicine the animal is to be fed with hot rice gruel.

Method VI

Ingredients

- Aaveli
- Cow urine

Medicinal Preparation

Handful of Aaveli leaves is soaked in cow urine for 15 minutes. It is ground well and the juice is extracted. It is fed to the nose and mouth of the affected animal for 3 days.

Dosage : Once in a day.

Vaitrupokku (Diarrhoea)

Symtomps

- Watery diarrhoea with foul odour
- Cattle will feel more thirsty
- Undigested feed will be found and mixed along with the dung
- Dried dung will be found adhered to tail and hind legs.
- Cattle become losing weight.

Medicines and its Ingredients

There are various treatments for curing diarrhoea. Four good methods are detailed here.

Method of Preparation

Method I

Ingredients

- Palm Gur and water.

250 g palm gur is added with 200ml of water and fed to cattle thrice a day. Similarly it is done for 3 days. In case of calf 100g of palm gur is added to 100ml of water, but the dosage is same.

Method II

Ingredients

- Castor oil and cotton seed

Cotton is made into three small balls (lemon size).These balls are dipped into castor oil and fed to cattle, thrice a day.

Method III

Ingredients

- Eggs, raw boiled

It is very simple method i.e. three eggs are fed thrice a day.

Care to be Taken While Preparing

In all methods, purity of the ingredients is the only factor to be checked i.e. gur purity in method one and cotton ball without any foreign materials in method two.

Conditions and Precautions

No special Conditions and precautions to be followed while or after administering the medicines.

Relative Advantage

In all the methods cost is the main advantage i.e., each method costs only Rs 5/. (Approximately)

General Profile of the ITK Holder

Name and Age : Thiru R.Osaladiyan,52 years Address :S/O Thiru .Rangasamy, Perahambi (post)

Manachanallur (TK) Trichy (DT) Educational status : 5[th]standard Occupational status : Farming

Method IV

Ingredients

- Agathi leaves – *(sesbania grandiflora)*

Preapation

A handful of Agathi leaves are collected, ground well and made it as paste. This has to be given to the affected animal for about three times a day.

Method V

1. Perumarunthu (*Aristolachia indica*) - A fistful
2. Shinni *(Acalypha fruiticosa)* - A fistful
3. Sirukurinchan *(Gymnema sylvestre)* - A fistful
4. Tobacco (*Nicotiana tobacum*) - A little quantity
5. Pepper *(piper sp)* - A little quantity
6. Chillies *(Capsicum annum)* - 3 Nos

Preparation

The abovesaid ingredients are ground well and prepare paste out of it. Make the paste into small balls. This has to be fed to affected animal for about 3 days, thrice a day.

General Profile of the Traditional Holder

Thiru. Ramalingam, Muniayathirayanpatti, Udayarpalayam (Post), Perambalur (Dt), Age 65 years

Educational status-5 th standard, Educational status : 45years, Source : Farm traditional medicinal books.

Patchai Narampu (Green Nerves)

Ingredients

1. Tamarind - *(Tamarindus indica)* A little quantity
2. salt - A little quantity

In summer season, the cattle body heat is increased. Due to the over heat, the green nerves are occurred in the lips. Green nerves are removed by using needle. And small amount of tamarind and salt is smeared over the tongue. Paddy straw is tied around the tongue. One time treatment is sufficient to cure the disease.

General Profile of the ITK Holder

Name :Raman, Age :80, Education :illiterate Occupation :farming , Farming experience :50 years, Land holding :2acres, Address :kotarakundru, Perambalur District.

Cold-Symptom (Respiratory Disease in Poultry Birds)

Symptoms

Cold is a common disorder found among poultry birds. The symptoms are,

- Affected birds show watery discharge from the nose.
- The eyes are watery.
- Loss of body weight or appetite.
- If unattended, it leads to death of a bird.

Causes

- Unhygienic bird management practices.
- Moist atmosphere
- Absence of required heat.

Ingredients

1. Small onion - 50 g.
2. Butter milk - 500 ml.

Method of Preparation

The onion is to be ground by using either grinder or "ammi" (specially designed stone used for grinding- a rectangular slab stone with a roller for grinding). It is to be mixed with butter milk and prepare a solvent.

Administration

The mode of administration is internal. The aforesaid prepared solvent is sufficient for 10 birds. The recommended rate for individual poultry bird is 10 ml per bird. The medicine is to be administered during any time by using a silver tumbler. One time administration is sufficient for a normal bird. In case of chicks the recommended rate is 5 ml per bird.

Mode of Action

The onion and butter milk acts as an immunization agent against cold disease in poultry birds. Prior application of this medicine before the occurrence of disease show best result.

Conditions to be Followed

- Immediately after administration of the medicine the feed should not be given as it affects the respiratory organ of bird.

Relative Advantages

1. The raw materials are locally available in the shop and the cost is less than Rs.10/- for 10 birds.
2. It is a simple method compared to vaccination.
3. It will not create any side effects.

General Profile

Birds treated so far : 200-300 own birds.

Mooku Adaippan (Sneezing in Poultry Birds)

Symptoms

Affected birds show watery discharge from nose, watery eyes, loss of body weight and difficulties in respiration. In severe cases, it may lead to death of a bird.

Causes

- Ectoparasites
- Unhygienic bird management practices
- Moist atmosphere

Ingredients

1. Small onion - 1kg
2. Water - 10 lit.

Method of Preparation

The onion is to be ground by using either grinder or "ammi" (stone mortar) it is to be dissolved in 10 litres of water to make a solvent. Ensure complete dissolving of onion paste in water.

Administration

The mode of application is internal, oral mouth feeding through a silver tumbler. The prepared solvent is sufficient for 100 birds. The recommended rate is10 ml for a bird and it can be administered at any time during the treatment a period. The duration is 3 days continuously. In case of chicks the recommended quantity is 5 ml per chick.

Mode of Action

The onion contains sulphur and act as an immunization agent against sneezing problem in poultry birds. Prior application of this medicine before the occurrence of disease has best effect.

Conditions to be Followed

- Immediately after administration of the medicine , the feed should not be given since it affects the digestive tract of the bird. The feed may be given 1-2 hours after the application of medicine.

Relative Advantages

- The raw materials are locally available in the shop and the cost is also lesser
- Approximately the cost is Rs.2 per bird
- It will not create any side effects
- Easy preparation of medicine.

General Profile

No. of birds treated so far : 1150 birds.

Serumal Kaichal (Cold and Fever)

Symptoms

- Shivering of body and limbs
- Erect hairs
- Watery eyes
- Recumbent and unable to get up
- Off feed and reduction in the milk yield.

Ingredients

1. Chitra moolam root - (*Plumbago zeylonica*) – a fistful
2. Garlic bulb - (*Allium sativum*) – ½ fistful
3. Pepper - (*Piper nigrum*) -½ fistful
4. Salt - required amount

Treatment

Chitra moolam root – a fistful garlic bulb – ½ fistful pepper-½ fistful, salt-required amount are ground well. This should be given twice a day.

Precautions While Administering the Medicine

Green fodder should be avoided.

General Profile of the ITK Holder

Name : Thiru Bellalappa, Bogasandiram, Irudhu kottai (post), Thenkanikottai Taluk, Krishnagiri District. Educational Status : Nil, Experience : 57years.

Kanai Noi (FMD-Foot and Mouth Disease)

Method I

Symptoms

- Wound lesions in the mouth and hooves.
- Ulcer in the mouth and hooves.
- Saliva secretion from the mouth and nose.
- Fever and it is felt by touching the abdomen of the affected animal.
- It won't feed normally and often resorted to off-feeding.
- The affected animal becomes difficult to walk.

Causes

There are several reasons for FMD. But, the following are the specific reasons for the disease.

1. Fever and exhaustion.
2. Infectious diseases.
3. Ecto parasite infection
4. Winter season is susceptible climate for the spread of disease.

Ingredients

Nalabalasthi – a handful

Treatment

A handful of nala balasthi leaves are collected, cleaned and boiled in ½ litre of water. The content is filtered. This filtrate is applied over the affected portion (leg and mouth) to cure the wounds. This should be done once a day and this should be continued for 3 days.

Conditions to be Followed

- The animal should not be washed during the course of treatment.
- After the administration of medicine the animal is to be fed with paddy straw, cotton seed cake, sesame cake and groundnut cake.
- Avoid roughages during the course of treatment since it has high concentration which may reduce the medicinal effect of the contents.
- While administering the medicine the animal is to be fed with hot water which is cooled for 3 hours.
- It is better not to allow the cattle to graze during the course of treatment.

Feasibility and Relative Advantages

- The raw materials are locally available.
- The cost of medicine is less than Rs.30/-
- It will not create any side effects.
- Easy preparation.

General Profile of the ITK Holder

Name : Thiru Bellalappa, Bogasandiram, Irudhu kottai (post), Thenkanikottai Taluk, Krishnagiri District. Educational Status : Nil, Experience : 57years.

Method II

Ingredients

1. Neem oil - 100ml
2. Raw rice gruel - 1/2 lit
3. Poovan banana - 10-15 nos
4. Cumbu koozh (pennisetum glacum)

Treatment

Neem oil 100 ml is applied externally and smeared on the mucilage layer.

As internal, oral feed, ½ lit raw rice gruel mixed with 10-15 Poovan bananas and fed to the cattle. This acts as a healing agent for ulcer in mouth. Cumbu is pounded, boiled and made into paste called "cumbu koozh".This paste is finally converted into balls which is fed to the animal until the wound heels. This can be fed thrice a day.

Conditions and Precautions While Feeding

Feed like roughages and fodders should be reduced in quantity.

Relative Advantages

The medicine for curing the disease is serving both as medicine as well as feed.

General Profile of the ITK Holder

Thiru : Osaladiyan, Age : 60 years, Perahambi Village, Manachanallur Taluk, Trichy district.

Method III

Ingredients

1. Neem oil - 100ml
2. Castor oil - 50 ml
3. Kalyanamurunkai (*Erythrina suberosa*) -A fist ful
4. Thuthuvalai (*Solanum trilobatum*) -A fist ful
5. Turmeric (*Curcumba longa*) -50g
6. Camphor - 5 nos

Preparation and Administering

A fist ful of Kalyanamurunkai, a fist ful of Thuthuvalai and 5 nos of Camphor are ground well and Neem oil, Castor oil and Turmeric are added to it one by one. Paste is made out of it and smeared over the affected portion. This should be continued until the wound heels.

General Profile of the ITK Holder

Tmt Rasathi, Perahambi village, Manachanallur taluk, Trichy district.

Method IV

Introduction

FMD is a common disease, prevalent among cattle. It is a severe disease as it makes the animal become weak and the affected animal can not be used for agricultural operations.

There is a web of causes for FMD. But, the following are the specific reasons for the disease.

- Infection from other animals.
- Unhygienic cattle management practices in the shed.
- Diseased atmosphere.
- Cattle exposed to heat for longer duration.

Method of Preparation

(For Foot Lesions)

Ingredients

1. Neem oil - 100ml.
2. Camphor - 10-15nos
3. Hot fine sand - 250-500g.

The camphor is to be boiled in 100 ml of neem oil for 10-15 minutes and cooled for 1 hour for smearing and/or the hot sand is to be smeared over the wounded portion using hands.

(For Mouth Lesions)

Ingredients

1. Country banana fruit - 10-15nos
2. Sesame oil - 100 ml.
3. Kasakasa (*Papavar somniferum*) - 100 ml.
4. Coconut milk - 300-500 ml.

The Country banana fruit and sesame oil are to be mixed thoroughly to make the content into a semi solid form.

Administration

(For Foot Lesions)

The mode of application is external. The hot fine sand is to be smeared over the wounded portion or the camphor, neem oil mixture is thoroughly smeared over the wounded portion.

(For Mouth Lesions)

The semi solid content is to be given orally into the mouth of animal.

Time of Application

(For Foot Lesions)

Smearing over the wounds is to be normally carried out during morning and evening hours continuously for 3 days.

(For Mouth Lesions)

The prepared mixture is to be given during morning hours continuously for 3 days.

Conditions / Precautions to be Followed

- Avoid tick and lice infestation over the wounds.
- Spray the cattle shed with bleaching powder to evade ectoparasites.
- The cattle should be fed with "kasakasa" *(Papavar somniferum)* + Coconut milk.
- Not to allow the cattle for grazing green grass.
- Not to allow the cattle to take bath during the course of treatment.
- The cattle should be fed with hot water cooled for 3-5 hours.

Relative Advantages

- Raw materials are locally available in petty shop and it fetches Rs.20-30/- only. But the clinical method costs Rs.100-150/-
- It will not create any side effects and other diseases.
- Simple preparation and easy application.

Animals treated so far : 1200

Method V

Introduction

Foot and Mouth disease is a common disease prevalent among cattle during winter period. It is an infectious disease and a precursor for other lethal diseases.

Causes

There are web of causes for FMD. But, the following are the specific reasons for the disease. They are,

- Fever and exhaustion.
- Other Infectious diseases like Ephemeral fever etc.,
- Ectoparasites.
- Winter season is the susceptible climate for the spread of disease.

Ingredients

1. Konrai tree leaves (*Cassia spp.*) - 25 g.
2. Vishamalli leaves (*Rhinocanthus nasuta*) - 25 g.
3. Kaduvanthalai leaves - 25 g.
4. Arilai notchi leaves (*vitex negunda*) - 25 g.
5. Erukku leaves (*Calotropis gigantea*) - 25 g.
6. Thirugukalli leaves (*Euphorbia tirucali*) - 25 g.
7. Water - 50 ml.

Or

1. Pomegranate fruit (punica granatum) - 1
2. Coconut - 1
3. Vidhathalai leaves (*Dichrostachys cinerea*) - 25 g.
4. Water - 100 ml.

Method of Preparation

The leaves of the aforesaid ingredients are to be ground well by using either a grinder or "ammi" adding 50ml of water to make the content into a bolus. It is sufficient for cattle and incase of half or heifer one such bolus is sufficient. The rate is two boluses per day per animal for 3 days continuously during morning and evening hours.

Or

Half litre of water is to be added to make the content into an ointment, to be smeared over the affected leg and mouth portion of the animal. Care should be taken while grinding to avoid excess addition of water as the content is to be applied in the form of "ointment".

Or

The pomegranate, coconut and "vidhathalai" are to be taken, ground well along with addition of water to make the content into a paste and is to be applied orally.

Administration

The mode of application is external or internal. The aforesaid prepared ointment is to be smeared over the animal as external medicine by hand or the aforesaid prepared paste is to be administered as intake medicine.

The 500 ml. ointment is sufficient to cure the well developed wounds. The smearing is to be normally carried out during morning or evening hours continuously for three days or the 100 ml. paste is to be administered to the affected animal as mouth feeding during morning and evening hours continuously for 3 days.

Method VI

Introduction

Foot and mouth disease is a common disease prevalent among the cattle during the winter period. It is an infectious disease and precursor for other lethal diseases.

Ingredients

1. Fish water - 500 ml.
2. Dried fish - 2-3 pieces

Method of Preparation and Administration

Here, the method of preparation is simple. The water obtained after washing the fish is used to clean the wounds of animal affected by FMD. Then, 2-3 dried fish is to be put in the fire generated near the cattle shed and the animal is to be in haled the smoke emanated by the dried fish.

This practice is to be normally carried out during early morning or evening hours continuously for 3 days or till the ailment gets cured. Thus, the mode of administration is external.

Mode of Action

It is believed that the specific odour of fish water and dried fish smoke curing had the capacity to cure the wounds caused by foot and mouth disease in cattle.

Animals treated so far : 10-20 times her own cattle.

Method VII

Introduction

Foot and Mouth disease is a common disease prevalent among cattle during the winter period. It is an infectious disease and a precursor for other deadliest diseases.

Ingredients

1. Boiled cumbu grain (*pennisetum glacum*) - 500 g.
2. Hot water - 1 litre ml.
3. Salt - a pinch.

Method of Preparation

Cumbu grain is to be boiled in a vessel by the addition of one litre of water the boiling is to be continued till the water is reduced to 200 ml. Add with a pinch of salt.

Administration

The mode of application is internal. The prepared medicine is to be administered as an intake feed to the affected animal by "moongikottan". The 500 g. of medicine is sufficient for the well developed wounds. The internal feeding is to be normally carried out during morning and evening hours continuously for 3 days.

Mode of Action

The salt and cumbu grains have the chemical compounds to cure FMD in cattle.

Animals treated so far : 200 times his own cattle.

Method VIII

Introduction

FMD is a common disorder prevalent among cattle. It is a severe disease as it makes the animal become weak and affects the economical value of animal (the affected animal can't be used for draught purpose)

Causes

There is a web of causes for FMD. But, the following are the specific reasons for the disease. They are,

- Infection from other animals.
- Unhygienic cattle shed and cattle management practices.
- Diseased atmosphere.
- Cattle exposed to heat for longer duration.

Ingredients

For Foot Lesions

1. Wild sp.of mushroom - 50 g.
2. Neem - 50 ml.

For Mouth Lesions

1. Hill banana - 2
2. Ghee collected from pig milk - 50 ml.

Method of Preparation

For Foot Lesions

Crush the wild mushroom with hand and put in neem oil to make the content into a semi solid form.

For Mouth Lesions

Hill banana and pig ghee are to be mixed thoroughly to make the content into semi solid form.

Administration

For Foot Lesions

The mode of application is external. The prepared semisolid content is to be smeared over the wounded portion by using a feather (hens' feather). During morning and evening hours continuously for 2 days.

For Mouth Lesions

The semi solid content is to be fed to the cattle. During morning and evening hours continuously for 2 days.

General Profile

Name : Mr.V.Panaiyan, Vaidhyar, S/o.Mr.Vellaikannu, S.V. Mangalam, Alagunatchiamman Tirupathur taluk, Sivagangai dt. Age : 60 Education : Illiterate Animals treated so far : 150 cattle.

Rattha Kalichal (Dysentry) or (Hemorrhagic Gastero Enteritis)

Method I

Symtomps

- Watery diarrhoea with foul odour
- Cattle will feel thirstier
- Undigested feed will be found and mixed along with the dung
- Dried dung will be found adhered to tail and hind legs
- Cattle become losing weight.

Causes

There is a nexus of reasons for dysentry. But, the following are the specific reasons for the disease.

- Excess body heat during summer
- Eating excess quantity of "karuvel" (*Acacia nilotica*) pods
- Drinking stagnant and contaminated water during summer and rainy seasons
- Unhygienic management practices in cattle shed
- Other infectious diseases like fever etc.,

Ingredients

1. Gur - 25 g
2. Ipomea (*Ipomea sepiaria*) - A handful

Treatment

A handful of Ipomea leaves and 25 g of gur are ground well. And the prepared paste is mixed with one litre of water and made it as solution. This solution is fed to the affected animal once a day. This should be continued until affected animal comes to normal condition.

Feasibility and Relative Advantages

Since all the ingredients are locally available, it is very easy to prepare the medicine. Similarly, the method of preparation is also very simple which doesn't involve cumbersome process. Herbal cure is relatively advantages over other types of medicines due to its nil side effects.

Limitation/Constraints

Instant/immediate cure is not possible. It has its own time limitation for curing the disease.

Beneficiaries

According to this Vaidhiyar, nearly 200 animals have been benefited by this treatment through Thiru Bellappa.

General Profile of the ITK Holder

Name : Thiru Bellalappa, Bogasandiram, Irudhu kottai (post), Thenkanikottai Taluk, Krishnagiri District. Educational Status : Nil, Experience : 57years.

Method II

Introduction

Hemorrhagic Enteritis is a specific disease affects the cattle. It is a highly infectious disease and prevalent among the cattle in southern districts of Tamil Nadu.

Symptoms

The specific symptoms of the disease are,

- Excretion of blood tinged dung
- It won't feed normally and often resorted to "off"-feeding"
- Affected animal becomes dull and inactive
- It frequently lies down and gets up and looks disturbed (called as staggering gait).

Causes

There is a web of causes for Gastero Enteritis. But, the following are the specific reasons for the disease. They are,

- Excess body heat during summer
- Eating of excess quantity "karuvel" (*Acacia nilotica*) pods
- Drinking stagnant and contaminated water during summer and rainy seasons
- Unhygienic management practices in cattle shed.

Ingredients

Seeragam (*Cuminum cyminum)*	-	250 g.
Coconut spadix or Kiluvai tree bark	-	250 g.
Water	-	50 ml.

Method of Preparation

The "seeragam" and coconut spadix or "kiluvai" tree bark are to be ground well with little addition of water (50 ml) to make the content into bolus. Thus, 2-3 boluses would be prepared weighing 100-120 gm each.

Administration

The mode of application is internal. The aforesaid bolus is to be administered as intake medicine to the affected animal.The 2-3 boluses are sufficient for a normal milching cow and bullock and in case of calf or heifer one such bolus is sufficient. The exact recommendation is that 2-3 boluses per day per animal for 3 days continuously during morning (between 6-7 a.m.) hours continuously.

Mode of Action

The mode of action is the ingredients, namely, seeragam & coconut spadix or "kiluvai" tree bark has the required fibre content to cure dysentery in animals.

Precautions / Conditions to be Followed

- The animal should not be fed with green grass or concentrated green legumes
- While administering the medicine the animal is to be fed with hot water which is cooled for 3 hours
- It is better not to allow the cattle for grazing during the course of treatment.

Relative Advantages

- The cost of preparation of medicine is less than Rs.80/-. The ingredients can be either purchased or collected.
- It will not create any effects.

General Profile

Address : Mr. O. M. Chinnapottu, S/o. Mr. Oyyaduthan, Kattapomman st., Kirungakottai, Singampunari via, Tirupathur taluk, Sivagangai dist. Age : 60, Education : Primary Occupation : Farming, Land own : 5 acres.

No. of cattle treated so far : 50-60 times his own cattle

Method III

Introduction

Gastero Enteritis is a specific disease which affects the calf or heifer. It is a highly infectious disease and is most prevalent among cattle in southern districts of Tamil Nadu.

Ingredients

1. "Seeragam" (*Cuminum cyminum*) - 10 g.
2. Tender pomegranate fruit (*Punica granatum*) - 1
3. Butter milk - 250 ml.

or

1. Cotton seed - 50 g.
2. Water - 250-500 ml.

Method of Preparation

The "seeragam" and tender pomegranate fruits are to be taken, ground well with addition of water to make a bolus. This is to be dissolved in 250 ml. of buttermilk to make the content into extract.

Administration

The mode of application is internal. The 250 ml. extract is to be administered as intake medicine by using silver tumbler to the affected calf or heifer. The recommendation is that 250 ml. per day per animal for 2 days continuously during early morning between 6-7 a.m.

Mode of Action

When compared to cattle the calf or heifer is lesser age (2-2.5 years). The digestive tract of this animal can be cured by tender pomegranate. Buttermilk is used to cool the whole body. "Seeragam" will induce the medicinal effect of the content.

Another Method

Method of Preparation

The cotton seed is to be boiled in 250-500 ml. of water for 30-40 minutes. The resultant extract is to be cooled and filtered and the water alone is to be collected.

Administration

The mode of application is internal. 200 ml. extract is to be administered as intake medicine by using silver tumbler to the affected calf or heifer. The recommendation is that 200 ml. per day per animal for 2 days continuously during morning or evening hours.

Mode of Action

The boiled cotton seed water is capable of purifing the digestive tract of the calf or heifer.

Conditions to be Followed

- The animal should not be fed with green grass or concentrated legumes
- The animal should be fed with milk instead of water
- It is better not to allow the cattle for grazing during the course of treatment.

Relative Advantages

- The entire cost of medicine is less than Rs.10/-
- It is cheaper compared to veterinary medicine costs nearly Rs.150-200/-
- It will not create any side effects.

No. of animals treated : 100 time his own cattle.

Kulambu Noi (Hoof Lesions)

Introduction

Foot lesion is the common disease prevalent among cattle especially used for cattle competition purpose. Though it is not a lethal disease it is a precursor for other lethal diseases.

Symptoms

- Affected animal shows ulcer and cleft in hooves
- Dropping of saliva from the mouth of affected animal
- The animal won't feed normally and often resorted to off-feeding
- Affected animal is unable to walk
- The animal looks feverish.

Causes

The causes are,

- Improper management practices for cattle.
- Infectious diseases.
- Ectoparasites.

Ingredients

1. "Kadukkaai" (*Terminalia* chebula) - 100g.
2. "Kathirimanjal" (*Curcuma aromatica*) - 50g.
3. "Karuncheeragam" (*Cuminum cyminum*) - 50g.
4. Small amla (*Emblica officinalis*) - 50g.
5. Sesame oil - 200ml.

Mode of Preparation

The first four ingredients are to be ground well and 200 ml. of sesame oil is added to make the content into an ointment to be smeared over the hoof lesions.

Administration

The mode of application is external. The prepared ointment is be smeared over the hoof lesions as external medicine by hand.

The 200-300 ml. ointment is sufficient to cure the well developed wounds. The smearing is to be normally carried out during morning or evening hours continuously for 3 days.

Mode of Action

The 'kadukkaai", "kathirimanjal", "Karuncheeragam", amla and sesame oil mixture have the chemical components capable to cure the foot lesions in cattle.

Conditions to be Followed

- Washing of animal should be avoided during the course of treatment
- After the administration of medicine the animal is to be fed with paddy straw, cotton seed cake, sesame cake and ground nut cake
- Avoid roughages during the course of treatment as its concentration may reduce the medicinal effect of the ointment
- While administering the medicine the animal is to be fed with hot water cooled for 3 hours
- It is better not to allow the cattle for grazing during the course of treatment.

Relative Advantages

- The raw materials are locally available.
- The cost of preparation of the mixture is less than Rs.25/-
- Easy preparation.

General Profile

Animals treated so far : 200

Pulippu Noi or Thullumari Noi (Struck)

Introduction

Struck is one of the common disorders found in the cattle. According to veterinary doctors the disorder caused by the veterinary pathogen namely, *Pasturella multicida.*

Symptoms

The specific symptoms are,

- Affected animal is unable to walk
- Shivering of body of the animal
- Constipation
- Animal faces difficulty in urination.

Causes

Though Web of causes is indicated for struck in cattle, the following is the specific reason as per veterinary for the disease.

- Feeding the cattle with highly concentrated green fodder.

Ingredients

1. Aduthinnaapaalai (*Aristolachia bracteolata*) - 25 g.
2. Sambiraani (*Canarium stricutum*) - 20 g.
3. Ramasimitti fruit - 2-3
4. Water - 50 ml.

Method of Preparation

Both the "aduthinnaapaalai" leaves and "ramasimitti" fruits are to be taken, ground well with little addition of water to make a bolus. Thus, two boluses are prepared weighing 80-100 g. each.

Administration

The mode of application is internal. The bolus is to be administered as intake medicine.

Two boluses are sufficient for normal cattle and in case of calf or heifer one such bolus is sufficient. The recommendation is that 2 boluses per day per animal for 2 days continuously during evening hours only. Along with this the affected animal is to be inhaled with the smoke of "sampiraani" for 2 days.

Precautions to be Followed

- The cattle should not be allowed for grazing green grass during the course of treatment
- While administering the medicine the animal has to be fed with hot water which is cooled for 3 hours
- Correct quantity of water is to be added while preparation of the medicine as the medicine is too administered as bolus.

Relative Advantages

- Raw materials are locally available
- It will not create any side effects.

General Profile

No. of Animals treated so far : 50 cattle.

Moochupididippu" (Traumatic Pericarditis)

Symptoms

Traumatic pericarditis is a common symptom prevalent among cattle. (Cow, buffaloe, and bullock). The symptoms of the disease are, respiration difficulty, animal is reluctant to walk, recumbency and even death may occur sometimes.

Causes

- Wound in the respiratory organ
- Drying of the nose tube
- Herding cattle consistently.

Ingredients

1. Milagaranai (*Toddalia asiatica*) root - 4-5 (2" length)
2. Wild onion bulb - 1
3. Water - 50 ml.

Method of Preparation

Both the "milagaranai" root and wild onion bulb are ground well with little addition of water (50 ml.) to make the content into bolus weighing 80-100 g. Thus, 2 boluses would be prepared.

Administration

The mode of application is internal. The boluses are to be administered into the mouth of animal as intake medicine.

Two boluses are sufficient for cattle and in case of calf or heifer one such bolus is sufficient. The recommendation is that 2 boluses per day per animal during morning or evening hours continuously for 3 days. If severe infection is noticed treatment for one more day is sufficient.

Mode of Action

The "milagaranai" root and wild onion bulb has the ability to cure the wounds in the respiratory tract of animal and cure the disease accordingly.

Precautions or Conditions to be Followed

- Correct quantity of water is to be added while preparation of the medicine as the medicine is to be administered as bolus.
- While administering the medicine the animal is to be fed with hot water which is cooled for 3 hours.
- It is better not to allow the cattle for grazing during the course of treatment.

Relative Advantages

- Raw materials are locally available in the surroundings and it fetches no cost
- It will not create any side effects
- Easy preparation.

General Profile

No. of animals treated so far : 100 cows.

Vaadhanoi (Anorexia)

Method I

Symptoms

The symptoms of the disease are

- High fever and it is felt by touching the abdomen of animal
- It won't feed normally and often resorted to off-feeding
- As the affected animal is off feeding in nature it will not excrete dung or urinate.

Causes

There are web of causes for the disease. But, the following are specific,

- Unhygienic cattle management practices
- Intake of indigestible feed
- Intake of contaminated water.

Ingredients

1. Veeli leaves (*Cadaba fruticosa*) - 250g.
2. Small onion (*Allium cepa*) - 250 g.
3. Seeragam (*Cuminum cyminum*) - 100g.
4. Buttermilk - 1 litre

Method of Preparation

The "Veeli" leaves, onion and "seeragam" are to be taken, ground well and 1 litre of butter milk is to be added to prepare the extract.

Administration

The mode of administration is internal. The prepared extract is to be given as intake feed by "moongikottan" The "moongikottan" is a traditional feeding structure to the animals used by the cattle growing population in the southern districts of Tamil Nadu. The 1.5 litre medicinal extract is to be administered during evening hours of animal continuously for 2-3 days. In case of half or heifer half the quantity is sufficient.

Mode of Action

The ingredients namely, "veeli leaves", small onion and 'seeragam" are made the animal more hungry by inducing the digestive tract of animal and buttermilk will reduce the fever.

Conditions or Precautions to be Followed

- After administering the medicine the animal should not be fed with green grass for 2 days as the concentrated nutrients and it will take more time for digestion and leads to another disease
- As the animal is not taking feed, it should be fed with hot water cooled for 3 hours.

Relative Advantages

- It is economical since the cost of preparation of entire medicine is below Rs.20/
- Raw materials are locally available
- Easy to prepare.

General Profile

Animals treated so far : 2000

Method II

Ingredients

1. Veppaalai leaves *(Wrightia tinctoria)* - 20 g.
2. Tender maavilam leaves *(Crateva adansonii)* - 20 g.
3. Erukku flower *(Calotrophis gigantea)* - 3
4. Betel leaves *(Piper betel)* - 2
5. Omam *(Trachysporum ammi)* - 10 g.
6. Pillaikilangu *(Acorus calamus)* - 20 g.

Method of Preparation

The contents are to be pounded well along with this one litre of hot water is to be added to dilute the content.

The resultant content is to be fed to the cattle by using "Moongikottan" (It is a hollow bamboo tube having 20-30 cm. length and cylindrical). The hooks in the bamboo are to be shaped well and it should not cause any injury to the animal while feeding. The "moongikottan" is a traditional feeding structure to the animals used by the cattle growing population in the southern districts of Tamil Nadu.

Administration

The mode of administration is internal. The prepared extract is to be given as intake feed by "moongikottan" for a single day during morning or evening hours.

Mode of Action

The ingredients namely, "veppaalai","maavilam" "erukku", Betel leaves, omam, "pillaikilangu" have the chemical components to cure the disorder in cattle.

Recommendation

The 1.5 litre medicinal extract is to be administered during morning or evening hours to the affected animal.

General Profile

Animals treated so far : 500.

Pullarithal (Ephemeral Fever)

Method I

Symptoms

The specific symptoms of the disease are,

- High fever and it is felt by touching the abdomen of animal
- Mucus secretion from the nose
- It won't feed normally and often resorted to off-feeding
- As it is off-feeding in nature it won't excrete dung or urinate.
- Shivering of body and limbs.
- Erect hairs
- Watery eyes.

Causes

There are web of caused for the disease. But, the following are the specific,

- Unhygienic management practices
- It is an infectious disease and spreads through dung and urine through maggots from the dung and urine of the affected animal
- Intake of indigestible feed
- Poor quality water feeding
- Complexity of other diseases like FMD etc.

Ingredients

1. Omam (*Trachysporum ammi*) - 10 g.
2. Betel leaves (*Piper betel*) - 2
3. Pillaikilangu (*Achorus calamus*) - 20 g.
4. Veppaalai thalai (*Wrightia tinctoria*) - 20 g.
5. Tender Maavilam leaves (*Crateva adansonii*) - 20 g.
6. Sukku (Dried ginger) - 20 g.

Method of Preparation

The contents are to be pounded well Along with these 750 ml. of water is to be added to dilute the content.The resultant content is to be fed to the cattle by using "Moongikottan" The hooks in the bamboo are to be shaped well and it should not cause any injury to the animal while feeding. The "moongikottan" is a traditional feeding structure to the animals used by the cattle growing population in the southern districts of Tamil Nadu.

Administration

The mode of administration is internal. The prepared liquid content is to be administered into the mouth of animal during early morning or evening hours for a single day. The dosage of medicine depends on the intensity of fever.

Conditions or Precautions to be Followed

- After administering the medicine the animal should not be fed with green grass for 2 days since it contains high concentration of nutrients will take more time for digestion and leads to another disease.
- As the disease spreads through maggots and germs, the animal should be fed with hot water cooled for 3 hours. Similarly, the animal should not wash for 3 days or during the course of treatment.

Relative Advantages

- Easy to prepare
- Raw materials are locally available
- It is economical since the cost of preparation of entire medicine is below Rs.20/ only.

Additional Information

The same medicine can also be used to cure FMD in cattle.

General Profile

Animals treated so far : 400

Method II

Introduction

Ephemeral fever is common disease found in cattle. It normally affects the cattle during winter. It is felt by touching the abdomen of animal.

Ingredients

1. Karuncheeragam (*Cuminum cyminum*) - 50g.
2. Kathrimanjal (*Curcuma longa*) - 50g.
3. White kungilium (*Shorea robusta*) - 50g.
4. Sampiraani (*Canarium strictum*) - 50g.
5. Omam (*Trachysporum ammi*) - 50g.

OR

1. " Aavaarai" (*Cassia auriculata*) - 250g.

Method of Preparation

The Ingredients are to be fed in the fire and the resultant smoke is to be inhaled by the animal.

Or

The green "Aavaarai' leaves alone is to be fed in the fire and the smoke is to be inhaled.

Administration

Inhaling of smoke by the affected cattle after put them in fire.

Mode of Action

The mode of action is that the ingredients namely, "aavaarai", "seeragam", "kathrimanjal", "kungilium" and "omam" have the capacity to purify the blood of animal and make the atmosphere hygienic. Thus, it able to cure the disorder of the animal.

General Profile

No.of animals treated so far : 500-600.

Method III

Ingredients

1. Vishamalli leaves (Rhinocanthus nasuta) - 25 g.
2. Aduthinnaapaalai leaves (*Aristolachia bracteolata*) - 25 g.

   Additional ingredients named as "kadaisarakku" and the contents are,

a. Vaalmilagu (*Piper cubeba*) - 25 g.
b. Karuncheeragam (*Cuminum cyminum*) - 25 g.
c. Vellaimilagu (*Piper sp.)* - 25 g.
d. Garlic (*Alium sativum*) - 25 g.
e. Kadukkaai (*Terminalia chebula*) - 25 g.
f. Thirumilagu (*Piper sp.*) - 25 g.
g. Vasambu (*Acorus calamus*) - 25 g.
h. Sukku (dried ginger) - 25 g.
i. Valamburikkai (*Helecteres isora*) - 25 g.
j. Karumilagu (*Piper nigrum*) - 25 g.

Method of Preparation

Initially, the "vishamalli" and aduthinnaapaalai" leaves are to be taken in banana leaf and the leaf is folded to facilitate internal feeding into the cattles' mouth. The banana leaf is used as a carrier for the medicine as it is easily digestible and also it will not reduce the medicinal properties of the ingredients.

The second ingredient is "kadaisarakku" (local name). It is a mixture of several ingredients used to be given to women immediately after delivery of a child. The contents of "kadaisarakku" have to be taken @ 25 g. each, which should be 250 g. in total. The content is be pounded well Along with this 750 ml. of water is to be added to dilute the content.

The Resultant content is to be fed to the cattle by using "moongikottan"

Administration

The mode of administration is internal. First, the "vishamalli" and "aduthinnaapaalai" leaves are to be internally fed the animal with the help of banana leaf then the aforesaid prepared "kadaisarakku" dilution is given to the animal as intake feed with the help of "moongikottan".

The dosage of medicine depends on the intensity of fever. The rate is 750 ml. per animal per day for 3 days continuously during morning and evening hours. In severe cases, the treatment will go up to one week.

Conditions or Precautions to be Followed

- After administering the medicine the animal should not be fed with green grass for 2 days since it contains high, concentration of nutrients will take more time for digestion and leads to another disease.
- As the disease spreads through maggots and germs, the animal should be fed with hot water cooled for 3 hours. Similarly, the animal should not be washed for 3 days or during the course of treatment.

Relative Advantages

- Raw materials are locally available in the surrounding place and in petty shop
- It is economical since the cost of preparation of entire medicine is below Rs.20/ only. But the chemical method costs nearly Rs.100/-

Additional Information

The same medicine is also used to cure FMD in cattle.

General Profile

Animals treated so far : 400

Method IV

Introduction

Ephemeral fever is a common disease found in cattle. It normally affects the cattle during winter. The fever is felt if touch the abdomen of the animal.

Medicine I

1. Thalaisuruli (*Aristolachia indica*) root - 10 g.
2. Erukku dried stick (*Calotrophis gigantea*) - 1 (20-50 cm length)

or

1. Wood ash - 250 g.

Medicine II

1. Asafoetida (*Ferula asafoetida*) - 50 g.
2. Betel vine leaves (*Piper betel*) - 50 g.
3. Omam (*Trachysporum ammi*) - 20 g.
4. Small onion (*Allium cepa*) - 5-6
5. Dried ginger - 2 (1-1.5" length)
6. Sidharathai (*Alpinia galanga*) - ½" length
7. Water - 100 ml.

Method of Preparation

1. The "Thalaisuruli" root is to be kept in the mouth of the animal and it is pushed into the mouth by using a dried sesame stick
2. Dried wood ash is to be smeared throughout the body of animal
3. Asafoetida, betelvine leaves, omam, onion, dried ginger and "sidharathai" are to be taken, ground well with addition of 100 ml of water to make a bolus. Thus, 2-3 boluses will be prepared weighing 80-100 g. each.

Administration

The mode of application is internal or external. The "Thalaisuruli" root is administered into the mouth with the help of dried sesame stick or the wood ash is to be smeared throughout the body of animal or the prepared 2-3 boluses is to be administered as intake medicine to the affected animal.

20 g. of "thalaisuruli" root is sufficient for normal cattle. It is recommended that during morning and evening hours continuously for 3 days or 250 g. of wood ash is to be smeared over the animal body during evening hours continuously for three days or 2-3 boluses of medicine is given orally during morning or evening hours continuously for three days.

Mode of Action

The mode of action is that the ingredient namely, "thalaisuruli", wood ash and bolus have the chemical components to cure the Ephemeral fever in cattle.

Conditions to be Followed

1. The animal is not to be washed immediately after administration
2. After the administration of medicine the animal is to be fed with paddy straw, cottons seed cake, sesame cake and ground nut cake
3. Avoid roughages during the course of treatment since it has high concentration which may reduce the medicinal effect of the contents
4. The cattle should be fed with hot water cooled for 3-5 hours instead of hot water as the ordinary cool water has carrying effect of germs.

Relative Advantages

1. The raw materials are locally available in the surrounding place and the cost is less than Rs.20/-
2. It will not create any side effects
3. Easy preparation.

General Profile

Animals treated so far : 60-100 times his own cattle.

Kaal Odidhal (Leg Fracture/Ethno Veterinary Bone Setting Treatment)

Method I

Introduction

Bone setting treatment is necessary for the animals that are participating in animal competitions (often called as "manjuvirattu" or "jallikattu"in tamil) as they might be affected by bone fracture.

Symptoms

The specific symptoms of the disorder are,

- Affected animal is unable to stand straight
- The cattle are not fit for agricultural operations
- It reduces the lifespan of bullock if it often attacks.

Causes

Improper management practices for bullocks participating in "manjuvirattu" and agricultural operations are the prime causes for the problem.

Ingredients

1. Tamarind seed - 100g.
2. Paddy grains - 100g.
3. Thesavaarai (*Asparacus racemosus*) bulb - 50g.
4. White kungilium (*Shorea robusta*) - 50 ml.
5. Cumbu grains *(pennisetum glacum)* - 100g.
6. Water - 500 ml.
7. White cotton bandage cloth - 1-2 m.

Method of Preparation

The ingredients are to be taken, ground well with little addition of water. The grinding is done then, the content is to be boiled by addition of 500 ml. of water it is to be cooled for 1-2 hours.

Administration

The mode of administration is external. The recommendation is that the aforesaid prepared ingredient is to be applied on the fractured bone of the animal and tied with a white cotton bandage cloth. One time treatment is sufficient for 18-20 days. Within the stipulated period the disorder gets cured. If severe injury is noticed, we can go for the treatment one more time.

Mode of Action

The mode of action is that the ingredients namely, tamarind seed, paddy grains, "thesavaarai", "kungilium" and cumbu have binding agents capable to joint the fractured bone of the animal.

Conditions to be Followed

- The cattle should not be washed during the course of treatment
- After administration of medicine the animal is to be fed with dry paddy straw
- Avoid roughages during the course of treatment since it has high concentration which may reduce the medicinal effect of the contents
- The cattle should be fed with hot water cooled for 3-5 hours.

Relative Advantages

- Easy preparation
- The cost of the preparation is less than Rs.20/- . But veterinary practice fetches Rs.100-150 for a single time
- The raw materials are locally available
- It will not create any side effects.

General Profile

Address : Mr. M.Chinnathambi servai, S/o. Mariappa servai, Ram nagar, Kurinchi tea stall,

Devakottai, Sivagangai dt. Age : 81 Occupation : Herbal healing for animals

Animals treated so far : 1000

Method II

Introduction

Bone setting treatment is necessary for the animals that are participating in Animal competitions (locally called as "Jallikattu") as they might be affected by bone fracture.

Ingredients

1. Sotrukatraalai (*Aloe vera*) - 2 No.
2. Egg (Country hen) white portion - 1
3. Black gram flour (*Vigno Mango*) - 100 g.

Method of Preparation

The sorus (pulp) is collected from the "sotrukatraalai" plants and it is to be tied with a white cotton cloth after washing the wounds with hot water during night. In the early next morning, unwind the joint with cotton bandage and wind again the wound with egg white portion and black gram flour mixture and cotton bandage. The procedure is to be repeated till the disease gets cured or two weeks whichever is earlier.

Administration

During evening hours tieing the wound with "sotrukatraalai" sorus and during early next morning wind the wound with egg-yellow portion and black gram flour mixture.

One egg and 100 g. black gram mixture is sufficient to joint the single leg of the animal along with 2 "sotrukatraalai" sorus. One time tieing is sufficient for 15 days and if needed second tieing of the medicine with cotton bandage is to be done. One month is sufficient time for the treatment.

Precautions / Conditions to be Followed

- During the course of treatment period the bullock should not be washed
- The bullock should not be used for any agricultural operations.

Relative Advantages

- The raw materials are locally available and the cost is also lesser than 15/-
- It will not create any side effects.

General Profile

Animals treated so far : 50

Method III

Introduction

Bone setting treatment is necessary for the animals that are participating in animal competitions (the animals participating in "Jallikattu") as they might be affected by bone fracture.

Ingredients

1. Coconut mesocarp (fibre) - 50 g.
2. Palm jaggery - 50 g.
3. Sunnaambu (Calcium hydroxide) - 50 g.
4. Cotton bandage cloth - 50 -100 cm. length.

Method of Preparation

The palm jaggery and calcium hydroxide are to be taken, ground well land it is to be mixed with coconut mesocarp thoroughly, to prepare ointment.

Administration

During evening hours the ointment is to be smeared over the wounded portion and tieing with white cotton bandage cloth would be carried out once a week.

The resultant 150g ointment is sufficient for one time application. It is to be carried out during evening hours. One time treatment is sufficient and if necessary, we can go for another round.

Mode of Action

The mode of action is that the calcium hydroxide, jaggery and coconut mesocarp have the chemical components to reduce the effect of wound injury.

Conditions to be Followed

- During the course of treatment, period the bullock should not be washed.
- The bullock should not be used for agricultural operations and racing if the treatment begins.

Relative Advantages

1. The raw materials are locally available and the cost is also lesser. The cost is less than Rs.20/-.
2. It will not create any side effects.

General Profile

Animal treated so far : 100 times his own cattle.

Veekkam (Boils / Swelling or Abrasions)

Method I

Introduction

Swelling is a specific disorder affects draught bullocks. It often called as physiological disorder.

Symptoms

The specific symptoms of the disorder are,

- Swelling of neck region
- The animal faces difficulties to bear the yoke while ploughing or any other agricultural operations.

Causes

This disorder is due to the web of causes. But, the following are the specific and they are given below,

- If the animal is engaged for long hours in agricultural operation during bright sunny hours
- Insect bite
- Yoke gall and naval illness

Ingredients

1. Avuri (*Cassia senna*) leaves - 100g.
2. Water - 100ml.

Method of Preparation

The "avuri" leaves is to be taken, ground well with addition of water to make the content into an ointment Thus, the 50 ml. of ointment is to be applied on the swelling. Care should be taken while grinding to avoid excess addition of water as the content is to be applied in the form of "ointment".

Administration

The mode of application is external. The aforesaid prepared ointment is to be smeared over the swelling as external medicine by using hand.

Mode of Action

The "avuri' leaves has the capacity to weeken the swelling.

Recommendation

The 50 ml. ointment is sufficient to cure the swelling of individual animal. The smearing is to be normally carried out during evening hours continuously for 3 days.

Three days treatment is sufficient to cure the disorder. If necessary, we can go for another three days continuously.

Precautions / Conditions to be Followed

- Immediate washed of animal after administration is avoided
- The animal is to be tied in ventilated place
- The affected animal should not be engaged for work during the course of treatment.

Relative Advantages

- Raw materials are locally available in the surrounding locality
- The cost of preparation of medicine is nil
- Easy preparation
- It will not create any side effects.

Profile of the Practioner

Address :Mr.O.M.Chinnapottu,S/o. Mr. Oyyaduthan, Kattapommanst., Kirungakottai, Singampunari via. Tirupathur taluk, Sivagangai dist.

Method II

Symptoms

The specific symptoms of the disease are,

- Yoke gall, naval illness
- In severe cases, if the wound is deeper it leads to maggot's infection

Causes

The following are the specific reasons for the disease. They are,

- Excess heat
- External parasites
- Engagement of bullocks in the field for longer period during bright sunny days.

Ingredients

1. Pirandai (*Cissus quadrangularis*) leaves - 100g.
2. Neem oil - 50ml.
3. Cotton white bandage cloth - 2-3' length.

Method of Preparation

The required quantity of pirandai leaves are to be fried in the neem oil for 10 minutes after this, the content is to be cooled for 30 minutes by keeping as such after boiling.

Administration

The mode of application is external. The fried "pirandai" leaves are to be smeared over the swelling portion and tied with a white cotton cloth once in 4 days. One time treatment is sufficient and if needed we can go for another round.

Mode of Action

The pirandai leaves and neem oil have the capacity to weaken swelling as it contains medicinal properties.

Precautions / Conditions to be Followed

- When treatment is in progress the animal should not be washed.
- The animal is to be feed with dry paddy straw.
- The cattle should be fed with water cooled for 3-5 hours instead of hot water

Relative Advantages

- The cost of preparation of he medicine is less than Rs.30/-
- It will not create any side effects.
- The method of application is simple.

Thalai Pungal/ Kayangal (Wounds /Abrasions in Head)

The animal when dashed on hard wood trees, walls rubbing the head on stones to avoid irritation. The symptoms are associated with of various diseases like fever, cold and cough.

Symptoms

- Swelling in the affected portion, horn joints
- Animal unable to stand. It results the animal always laid down
- Eyes get reddened and watery white liquid discharges from the eyes.

Medicine and its Ingredients

1. Thaivelai leaf – 100g
2. Kuppaimeni leag (*Acalypha indica*) – 100g
3. Dry ginger /chukku rhizome (*Zingiber officinale*) – 100g
4. Pepper fruit (*Piper nigrum*) – 100g

Method of Preparation

Freshly collected leaves of Thaivelai, kuppaimeni are taken in a bowl and washed with water. The leaves are crushed using a wooden pestle and stone mortar, juice is obtained. The extracted juice is filtered using nice cloth and kept separately. Dry ginger and pepper are kept under sun dried for complete removal of moisture. They are powdered, the sieved powder is added with a little of extracted juice and mixed well, kept for 2 hours. Finally, the mixture appears as paste which is to be used for application.

Administration

The mode of intake or method of application is external smearing on the affected portion. The dosage is not a fixed one and it varies depending on the affected portion. The medicine has to be applied early morning and evening for three days. The paste is applied over the wounds and allowed to dry.

Conditions and Precautions

The animal should not be fed with concentrate and roughage immediately after application of the paste. Instead rice gruel water or water can be given. There is a need for continuing of treatment if the wounds are not cured.

Relative Advantages

The preparation cost accounts only for chukku and pepper. It costs Rs.10/- and no other cost is incurred by the farmer.

Reasons for Severe Stages

The animal unable to lick the wounds on head and horn joints. If the farmer is unaware and do not take any control measure for the wounds, its severity increases.

Mutti Veekkam (Swelling of Joints)

Symptoms

Swelling of joints is a common disorder found in animals. The symptoms are,

- High temperature in the animal body and it is felt by touching the abdomen of animal by hand.
- Accumulation of fluid in the joining portion of legs.

Causes

The following are the specific reasons for the disease. They are,

- Engagement of bullocks/animals in the field for longer period during bright sunny days.
- Excess heat
- External parasites

Ingredients

1. Pirandai (*Cissus quadrangularis*) leaves - 50 g.
2. Camphor (*Cinnamomum Camphora*) - 5-6
3. Garlic bulblets (*Allium Sativam*) - 2
4. Lime fruit (Juice extracted) - 3

Method of Preparation

The "pirandai" leaves, camphor, garlic are to be taken, ground well with lime juice to make the content into an ointment. It is to be smeared over the bulging area.

Administration

The mode of administration is external. The prepared ointment is to be smeared over the animal as external medicine. The 50 ml ointment is sufficient to cure the disorder. The smearing is to be normally carried out during morning or evening hours continuously for threedays. It is sufficient to cure the disorder. If necessary we can go for another three days continuously.

Mode of Action

The "pirandai", garlic, camphor and limejuice have the chemical components to weaken the swelling of joints and additionally it can cure the wound.

Conditions to be Followed

- The animal should not be washed immediately after administration of medicine.
- The animal is to be fed with paddy straw instead of green grass.
- Avoid roughages during the course of treatment since it has high concentration which may reduce the medicinal effect of the contents.
- The cattle should be fed with hot water cooled for 3-5 hours instead of giving hot water as the ordinary cool water has carrying effect of germs.

Relative Advantages

- Raw materials are locally available and the cost of preparation is less than Rs.10/-
- It will not create any side effects
- Preparation is simple

General Profile

No. of animals treated so far : 500 cattle.

Madi Veekkam (Mastitis)

Symptoms

The specific symptoms of the disease are,

- Swollen and hot udder.
- Milk is discolored, saltish taste with pus and occurrence of thread like substances in the milk.
- Less or no milk secretion.
- Fever.

Causes

There are web of causes for the disease. But, the following may be the specific reasons and are given below,

- Less nutritious feed.
- Unhygienic cattle management practices often leads to infection.
- Off-feeding nature of animal.

Ingredients

1. Parboiled rice - 50g.
2. Sunnaambu (Calcium hydroxide) - 50g.
3. Kadukkaai (*Terminalia chebula*) - 50g.
4. Paakku (*Areca catechu*) - 50g.
5. Jathikkaai (*Myristica fragrans*) - 50g.

Method of Preparation

The ingredients are to be taken, ground well and 50 ml. of water is added to make the content into ointment. Care should be taken while grinding to avoid excess quantity of water as the content is to be applied in the form of "ointment".

Administration

The mode of application is external. The aforesaid prepared ointment is to be smeared over the swollen portion as external medicine by hand.

Mode of Action

The "kadukkaai", calcium hydroxide, "Jathikkaai" and "Paakku" has the chemical components which can drive away the maggots and cure the swelling.

Time of Application

The 100 ml. ointment is sufficient to cure the swelling of udder. The smearing is to be normally carried out during evening hours continuously for three days.

Three days treatment is sufficient to cure the disorder. If necessary, we can go for another 3 days continuously.

Conditions to be Followed

- The animal should not be washed during the course of treatment
- The animal is to be fed with paddy straw, cotton seed cake, sesame cake and ground nut cake

- Avoid roughages during the course of treatment since it has high concentration which may reduce the medicinal effects of the content
- Ensure hygienic cattle management practices
- The cattle should be fed with hot water which is cooled for 3 hours

Relative Advantages

- The raw materials are locally available in the surrounding place and in the shop.
- It is cheaper and the entire preparation cost is below Rs.50/-
- It will not create any side effects.

General Profile

No. of animals treated so far : 1000.

Kaluthu Veekam (Swelling of Neck)

Symptoms

The specific symptoms of the disease are,

1. Accumulation of fluid in the swelling portion
2. High fever and it is felt by touching the abdomen of animal
3. The animal won't feed normally and it resorted to "off-feeding"
4. The bullock can't able to do any agricultural operations and it is economically unproductive.

Causes

There are web of causes for the disease. But, the following are the specific. They are,

1. The animal is engaged in agricultural operation in sunny days for longer period.
2. The animal is engaged to pull heavy weight during ploughing.
3. Excess heat.
4. Other infectious diseases.

Ingredients

1. Ridge gourd leaves *(Luffa acutangula)* - 100g.
2. Salt - 100g.
3. Water - 500ml.

Method of Preparation

Both the salt and ridge gourd leaves are to be taken, ground well and it is to be boiled in 500 ml. water for 20-30 minutes till it attains ointment stage.

Administration

The mode of application is external. The prepared ointment is to be smeared over the swelling portion. The 300 ml. ointment is sufficient to cure the swelling. The smearing is to be normally carried out during morning or evening hours continuously for three days.

Three days treatment is sufficient to cure the disorder. If necessary we can go for another three days continuously.

Mode of Action

The "Ridge gourd" leaves and salt after boiling have the capacity to weaken the wound. As it is bitter taste it drives away maggots.

Conditions to be Followed

- The animal should not be washed during the course of treatment.
- After the administration of medicine the animal is to be fed with paddy straw, cotton seed cake, sesame cake and ground nut cake
- Avoid roughages during the course of treatment since it has high concentration which may reduce the medicinal effect of the contents
- The cattle should be fed with hot water cooled for 3-5 hours instead of hot water as the ordinary cool water has carrying effect of the germs.

Relative Advantages

- The raw materials are locally available in the surrounding place and it fetches minimum cost (less than Rs.20/-)
- It will not create any side effects.
- Easy preparation.

General Profile

No. of animals treated so far : 2000-2500.

Kaluthupun (Neck wounds)
Symptoms

Wounds in the neck region are common disorder found in animals. The symptoms are,

- Fever and high temperature in the body of affected animal and it is felt by touching the abdomen of animal
- In severe cases maggots would be found in the wounded portion
- The affected animal is unfit for agricultural operations and facing difficulties while taking feed.

Causes

There are web of causes for the disease. But, the following are specific for the disease. They are,

- Excess heat
- Engagement of bullocks/animals in the field for longer period during bright sunny days
- External parasite infection.

Ingredients

1. Tharaipugailai (wild tobacco) (*Lobelia nicotianaefolia*) - 50g.
2. Small onion (*Allium Cepa*) - 50g.
3. Sunnaambu (Calcium hydroxide) - 50g.
4. Veeram (*Nerium odorum*) & Pooram (*Bacopa monnieri*) - 50 g. total
5. Neem oil - 50 ml.

Method of Preparation

The "tharaipugailai", small onion, calcium hydroxide "veeram", and "pooram" are to be taken, ground well with little addition of water and mixed with neem oil to make the content into an ointment. It is to be smeared over the wounded portion.

Administration

The mode of application is external. The aforesaid prepared ointment is to be smeared over the animal as external medicine.The 50 ml. ointment is sufficient to cure the wound. The smearing is to be normally carried out during morning and evening hours continuously for three days.Three day's treatment is sufficient to cure the disorder. If necessary we can go for another 3 days continuously.

Mode of Action

The mode of action is that "tharaipugailai", small onion, calcium hydroxide "veeram", "pooram" and neem oil mixture ointment has the chemical components to cure the wounds and neem oil, the main ingredient fly away the external parasites.

Conditions to be Followed

- During the course of treatment the animal should not be washed
- It is better not to allow the cattle for grazing to avoid any external damage
- Avoid roughages during the course of treatment since it has high concentration which may reduce the medicinal effect of the content
- The cattle should be fed with hot water cooled for 3-5 hours.

Relative Advantages

- The cost of the entire medicine is below Rs.10/-
- Preparation is easy
- It will not create any side effects.

General Profile

Total no. of animals treated so far : 200

Kaal Pungal (Leg Wounds)

Animals standing in marshy places for a longer time get wounds on its hooves. While walking, the legs may hit on a stone which leads to bleeding. It continues and creates cut wounds in the affected portion. The infection spread through crows as it pecks an unaffected animal after pecking at wounded one. House flies also increases the wound intensity among animals.

Symptoms

- The shoes become weaker and the animal will unable to stand steadily
- Shivering of the legs results in difficulties while walking
- Animal always lies down
- Wounds are seen in the hoof partition and small abrasions are seen in the shoes and on the legs.

Neem Paste and its Ingredients

The disease can be cured with indigenously developed neem oil paste. Ingredients used in neem oil paste are;

1. Neem oil - 50ml
2. Hen's feather - one no.
3. Naphthalene balls - 4 nos.

Method of Preparation

- Neem oil is extracted from the dried neem seeds by crushing them in a wooden pestle and stone mortar
- After collecting the oil from mortar it is heated well till the water evaporates leaving only pure oil and used for the treatment
- The naphthalene balls are to be powdered
- Fresh neem oil extracted in a bowl is added
- Do not add water as they will lose the efficiency of medicine
- Preparation cost of the paste is very meager. Because only naphthalene balls costs two rupees (Rs.2/) and no other cost incurred.

Administration

The mode of application is external. Before applying the neem paste, the wound is cleaned with hot water. Already prepared mix is smeared over the affected portion with the help of birds/hen's feather for smoothness. The neem paste serves as a shield to prevent the entry of pathogens and cures the wound. Dosage of the paste mainly depends on the wound size. If the legs are severely wounded then the dosage will be more than the normal dose. The treatment is carried out twice in a day i.e. in early morning hours and evening. It takes at least three days to get the wounds cure it. If it does not show any improvement the treatment may be continued for another 3 more days.

Care to be Taken While Preparing Paste

If more number of naphthalene balls are mixed it will act as irritant and gives burning sensation to the wounds. As a result, the time required to heal the wounds extends them the expected period.

Relative Advantages

The preparation procedure is easier as well as the cost is very less. So it's widely followed by the local practitioners in kuzhithurai block of Kanyakumari district.

Additional Information

If the medicine is not available, the farmers used"kayamarunthuchedi" *Tridax procumbens* and onion *Allium cepa*. The above materials are grinded and the paste is applied over the wounds. This was the prior and alternate method to control leg wounds of animals.

Kaambu Adaithal (Teat Blockage)

Symptoms

Teat blockage is not a disease. It is a common phenomenon in milch animals. This will reduce the milk yield considerably and treatment of this disorder is economically important.

Causes

There are nexus of causes for this disorder. But, it is often termed as nutritional imbalance.

Administration

An experience medical practioner can effectively cure the disorder by using hen's feather. The smooth hen's feather is used to break the teat blockage without any detrimental effect to the animal.

Time of Application

Morning and evening hours are the suitable time for treatment of teat blockage. One time treatment is sufficient to cure the disorder.

Conditions to be Applied

- The animal may be fed with paddy straw, cotton seed cake, sesame cake and ground nut cake
- Avoid roughages during the course of treatment since it has high concentration
- The cattle should be fed with hot water cooled for 3-5 hours instead of hot water as the ordinary cool water has carrying effect of germs.

Relative Advantages

- It is a simple method.
- It would not create pain to the milching animal.
- Milk yield gets increased.

General profile

No. of animals treated so far : 2000 animals

Kombu odithal (Horn fracture)

Method I

Introduction

Horn itching is a common disorder found among the cattle. It can be easily treated by using locally available materials.

Symptoms

The specific symptoms of the disease are,

- Animal will scratch in between horns
- Scratch its head against tree
- Falling of horn(s)

Causes

There are several reasons for horn itching. But, the following are the specific reasons for the disease.

- Ecto parasites.
- Infectious diseases like fever etc.,

Ingredients

1. Palm sugar - 50g.
2. Sunnaambu or (Calcium hydroxide) - 50g.
3. Cotton bandage cloth - 50 -100 cm length.

Method of Preparation

The palm sugar and "sunnaambu" are to be taken, ground by using grinder.

Administration

The mode of application is external. The medicine is to be smeared over the wounded portion and tied with white cotton bandage cloth.

Mode of Action

The mode of action is that the Calcium hydroxide ($Ca(OH)_2$) has the capacity to cure the wound.

Recommendation

The prepared medicine is to be smeared over the wound and tied with white cotton bandage cloth once in 15 days. One time treatment is sufficient. If necessary we can go for another round of treatment.

Conditions to be Followed

- While tieing the cotton bandage cloth washing the animal is to be avoided
- While administering the medicine the animal has to be fed with hot water which is cooled for 3-5 hours.

Relative Advantage

- There is no specific medicine for horn itching in veterinary method.
- The cost of the medicine is less than Rs.50 – 60/-.

General Profile

Mr. O.M. Chinnapotu, S/o. Oyyaduthan, Kattapomman St., Kirungakottai, Singmpunari via., Tirupathur taluk, Sivagangai dt. Age : 61, Education : 5th standard, Land own : 5 acres

Crops : paddy, coconut and vegetables, Cattle own : 5

Cattle treated so far : 20 times his own cattle.

Method II

Causes

There are web of causes for horn itching. But, the following are the specific symptoms.

- Ectoparasite infection.
- Aged animal often faces this problem.
- Wounded injury in the root portion of horn.
- Infectious diseases like horn cancer etc.,

Ingredients

1. Ottadai (*cob webs*) - 50g.
2. Sunnaambu (*Calcium hydroxide*) - 50g.
3. Hair (ladies) - 2 pieces
4. Veeram (*Nerium odorum*) and "Pooram" (*Bacopa monnieri*) - 20 g. total

Method of Preparation

The "Ottadai", hair, "veeram" and "pooram are to be ground well this is to be mixed with Calcium hydroxide for smearing over the wounded horn.

Administration

The mode of application is external. The aforesaid prepared mixture is to be smeared over the wounded portion. The recommended rate is 50 g. of mixture for big cattle and 20-30 g. for calf or heifer during morning and evening hours for a single day. One day treatment is sufficient to cure horn cancer. If necessary we can go for one more day.

Mode of Action

The mode of action is that the "veeram", "pooram", "Ottadai", hair and Calcium hydroxide have the medicinal compounds to cure horn cancer.

Conditions to be Followed

- After administration of medicine the animal is to be fed with dried paddy straw.
- Avoid roughages during the course of treatment since it has high concentration which may reduce the medicinal effect of the content.
- The cattle should be fed with hot water cooled for 3-5 hours instead of hot water as the ordinary cool water has carrying effect of germs.
- The cattle should not be washed during the course of treatment.

Relative Advantages

- Raw materials are locally available in the surrounding place and the cost of preparation is below Rs.20/-.
- Easy preparation.

General Profile

Animal treated so far : 500 cattle

Method III

Introduction

Horn itching is a common disorder found among the cattle. It can be easily cured by using locally available materials.

Ingredients

1. Human hair - 20-50 g.
2. Palm sugar - 50 g.
3. Sunnaambu (Calcium hydroxide) - 50g.
4. Neem oil - 50 ml.

Method of Preparation

The human hair, palm sugar, Calcium hydroxide and neem oil are to be mixed thoroughly to prepare the content into an ointment. This ointment is used for smearing of the affected portion.

Administration

The mode of application is external. The ointment, thus prepared is to be smeared over the wound once in 15 days. One time treatment is sufficient. If necessary we can go for another round of treatment.

Mode of Action

The mode of action is that the Calcium hydroxide has the capacity to cure the wound and the neem oil flies away the external parasite. The whole ointment will be act as a binding agent to cure the disorder.

Conditions to be Followed

- Washing of animal is to be avoided the course of treatment
- While administering the medicine the animal is to be fed with hot water which is cooled for 3-5 hours instead of giving hot water.

Relative Advantages

- Raw materials are locally available
- The entire cost of preparation is less than Rs.20/-
- Simple preparation technique
- There will be no breakage of horn once this treatment is carried out.

General Profile

Animals treated so far : 20 times her own cattle.

Reduction in Milk Yield

Symptoms

The specific symptoms of the disorder are,

- Fever and it is felt by touching the abdomen of animal.
- Drastic reduction in milk yield.
- It won't feed normally and often resorted to "off-feeding".
- As it is off feeding in nature it won't urinate and excrete dung.

Causes

There are nexus of causes for the disorder. But, the following are the specific,

- Less nutritious feed.
- Fever and it yields blockage of teats.
- Aged animal often faces this type of problem.

Ingredients

1. Coconut kernel -1 full.
2. Parboiled rice - 250 g.

Method of Preparation

Both the coconut kernel and parboiled rice are to be ground well by using and added with little addition of water to prepare a mixture.

Administration

The mode of application is internal. The prepared mixture is to be given as intake orally during morning hours continuously for 20 days. Increase in milk yield would be noticed after 20 days of treatment. If necessary we can go for another one week.

Mode of Action

The parboiled rice and coconut milk has the capacity to induce milk yield as they stimulate milk secretion in milch animals.

Conditions to be Followed

- After the administration of medicine the animal is to be fed with paddy straw, cotton seed cake, sesame cake and ground nut cake.
- Avoid roughages during the course of treatment since it has high concentration which may reduce the medicinal effect of the contents.
- The cattle should be fed with hot water cooled for 3-5 hours.

Relative Advantages

- The raw materials are locally available and the cost of the medicine is less than Rs.25-30/-

General profile

Animals treated so far : 1000 animals.

Manjal Kaamaalai (Jaundice)

Introduction

Jaundice is a common disease prevalent among milching cow. The affected animal becomes unproductive as the milk is not suitable for consumption.

Symptoms

- Jaundice occurs in association with fever.
- The animal's body is too hot to touch.
- Watery discharge from eyes.
- It won't feed normally and often resorted to "off-feeding".
- Reduction in milk yield.
- Milk is unfit for consumption.

Causes

There is a web of causes for jaundice. But the following are the specific reasons,

- Fever and exhaustion.
- Imbalanced and untimely feed ration.
- Irregular feeding pattern (fed with only concentrates or roughages)
- Infectious diseases.

Ingredients

1. Cow urine - 1 litre.
2. Butter milk - 1 litre
3. Overnight soaked rice - 250 g.

or

1. Keelanelli (*Phyllanthus niruri*) - a single plant.
2. Notchi (*Vitex negundo*) leaves - 100 g.
3. Adathodai (*Adatoda zeylanica*) leaves - 100 g.
4. Sirukurinjaan (leaves (*Gymnema sylvestre*) - 100 g.
5. Water - 1litre

Method of Preparation

Normally, in cattle the urine will be dark yellow in colour after the delivery of calf .This is to be collected from another cow @ 250 ml and mixed with 1 litre of butter milk. The content is to be mixed with overnight soaked rice and administered to the affected animal as internal medicine by "moongikottan"

Or

The "keelanelli", "nothci", "adathodai" and 'sirukurinjaan" are to be taken, ground well with addition if 1 litre of water to make a juice. Thus 1 litre juice will be prepared and administered.

Administration

The mode of application is internal. The cow urine+butter milk+soaked rice mixture is to be administered as intake medicine through "moongikottan".

Or

"keelanelli", "notchi", "adathodai" and "sirukurinjaan" leave juice is to be administered into the mouth through "moongikottan" to the affected animal.

The recommendation is 250 ml. cow urine + butter milk + soaked rice mixture for an animal continuously during morning and evening hours for 3 days or till the disease gets cured.

Or

200 ml. leaves mixture for an animal continuously during morning and evening hours for 3 days or till the disease gets cured.

Conditions to be Followed

- Affected animal is to be separated from other animals
- The animal should not be washed immediately after administration of medicine
- After administration of medicine the animal is to be fed with dry paddy straw, cotton seed cake, sesame cake and ground nut cake
- Avoid roughages during the course of treatment since it has high concentration which may reduce the medicinal effect of the contents
- The cattle should be fed with hot water cooled for 3-5 hours.

Relative Advantages

- The raw materials are locally available in the surrounding place
- The cost of medicine is nil or less than Rs.10/-
- It will not create any side effects
- Easy preparation.

General profile

Animals treated so far : 200 -300 animals.

Sandhu Noi (Black Quarters)

Introduction

A black quarter is commonly seen in robust young heifers, and is prevalent during July and August months. Hence, the following measure is to be taken up in the month of May/June every year.

Symptoms

The specific symptoms of the disease are,

- Fever.
- Lameness.
- Crepitating sound in pressure.
- Swelling on hind quarters.
- It unnoticed it leads to sudden death of animal.

Ingredients

1. Marudhamtree (*Terminalia arjuna*) bark - 25 g.
2. Seemaikaruvai tree (*Acacia mangium*) bark - 25 g.
3. Udhian tree bark (*Lannea coramendelica*) - 25 g.
4. Karuvel tree bark (*Acacia nilotica*) - 25 g.

or

1. Sesame oil - 100 ml.
2. Thumbai (*Leucas aspera*) leaves - a handful (50 g.)
3. Thaivelai leaves - a handful (50 g.)

Method of Preparation

The "marudham" bark, "seemai karuvai" bark, "udhiyan tree bark" and "karuvai" bark are to be taken, ground well with addition of water to make the content into juice.

Administration

The mode of application is internal. The juice is to be administered as intake medicine to the affected animal through "moongikottan".The 1-1.5 litre juice is sufficient for adult cattle and in case of calf or heifer half the quantity is sufficient.

The recommendation is that 1.5 litre per day per animal for three days continuously during evening hours only.

Or

The "thumbai" leaves and "Thaivelai" leaves are to be taken, ground well with little addition of water to make bolus. Thus 2-3 boluses will be prepared weighing 80-100 g. each. The mode of application is internal. The bolus is to be administered as intake medicine along with sesame oil. It is sufficient for normal cattle and incase of half or heifer one such bolus is sufficient. The rate is two boluses per day per animal for 3 days continuously during morning and evening hours.

Mode of Action

The mode of action is that the ingredients namely, "marudham", "seemaikaruvai", "udhiyam" and "karuvai" bark have the chemical compounds to cure the disease.

Or

The ingredients namely "thaivelai" and "thumbai" leaves have the chemical compounds to arrest the disease. Sesame oil induces the medicinal effect of the ingredients.

Conditions to be Followed (for Both Methods)

- After administering the medicine the animal should be fed with paddy straw, cotton seed cake and sesame cake
- Avoid roughages during the course of treatment since it has high concentration which may reduce the medicinal effect of contents
- The cattle should be fed with hot water cooled for 3-5 hours.

Relative Advantages

- Raw materials are locally available in the surrounding place
- The cost of medicine is nil or less than Rs.15/-
- It will not create any side effects
- Easy preparation.

General Profile

Animals treated so far : 400 animals and 300 calves.

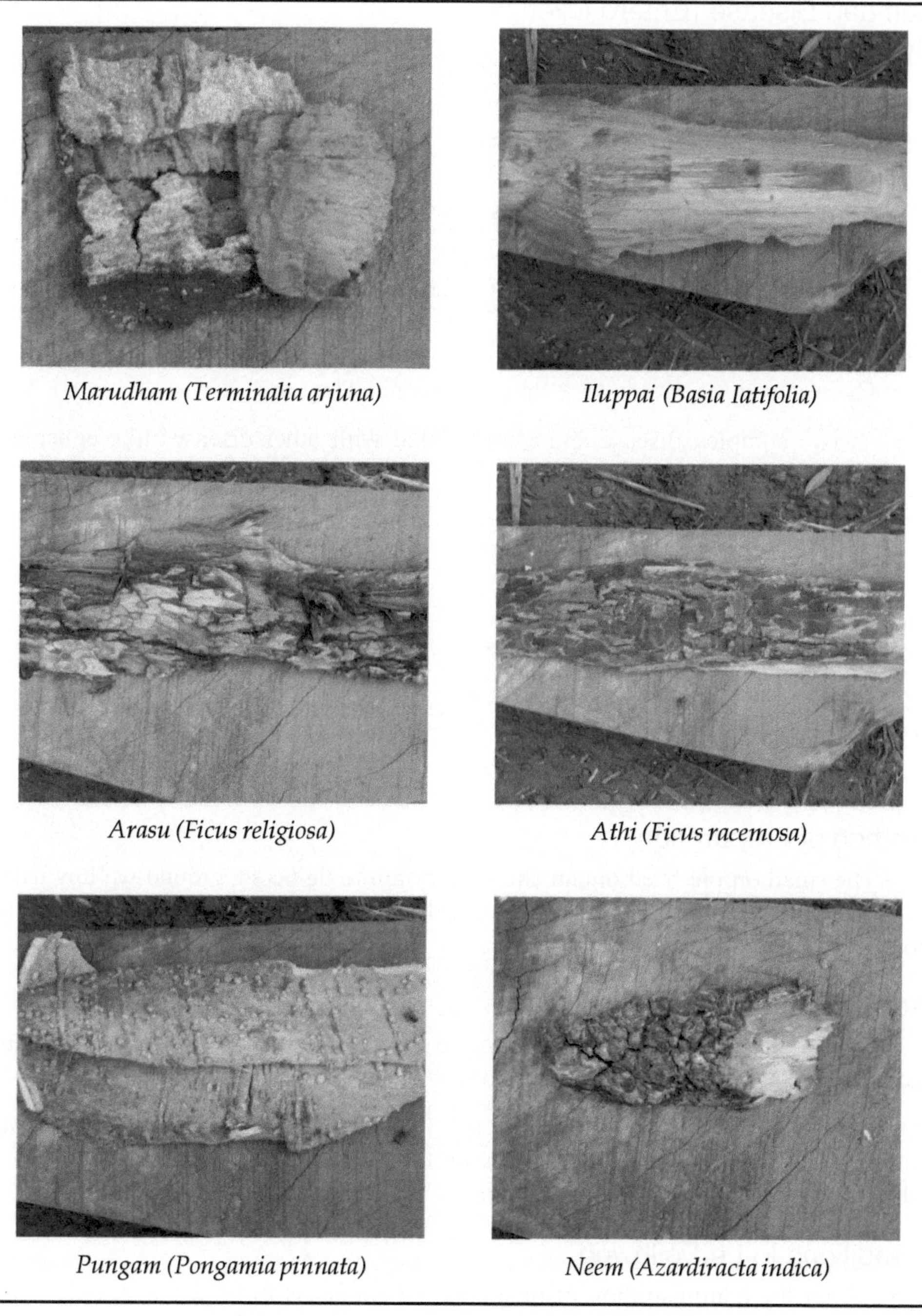

Marudham (Terminalia arjuna)

Iluppai (Basia Iatifolia)

Arasu (Ficus religiosa)

Athi (Ficus racemosa)

Pungam (Pongamia pinnata)

Neem (Azardiracta indica)

Fig. 30. Ethno Veterinary Medicinal Tree Barks

Komaari Disease (Entero toxemia)

Symptoms

The specific symptoms of the Intero toxemia in cattle are,

- Bleeding over the natural orifices
- Edma of throat
- Ropy secretion from the mouth of the affected cattle
- Hoof vesicles on the mouth and tongue
- If unnoticed, it leads to death of the cattle.

Causes

There is web of causes for "Entero toxemia. They are,

- It is a complex disease and often related with other diseases like ephemeral fever, dysentery etc.,
- Fever and Exhaustion
- External parasites.

Ingredients

1. Small onion (*Allium cepa*) - 250 g.
2. Omam (*Trachysporum ammi*) fried - 50 g.
3. Jaggery - 100 g.
4. Hot water - 1 litre.

Method of Preparation

The small onion, fried omam and jaggery are to be taken, ground well by using. The resultant mixture is to be again mixed with 1 litre of hot water to make the content into a semi solid form.

Administration

The mode of application is internal. The semi solid medicine is to be administered as intake medicine by using "moongikottan".

The recommendation is that the resultant 1.5 litre semisolid medicines are sufficient for cattle. In case of calf or heifer half the quantity is sufficient. It is to be administered during morning and evening hours continuously for three days.

Conditions to be Followed

- After the administration of medicine the animal is to be fed with only paddy straw, cotton seed cake, sesame cake and groundnut cake

- Avoid roughages during the course of treatment since it has high concentration which may reduce the medicinal effect of the contents.
- The cattle should be fed with hot water cooled for 3-5 hours.

Relative Advantages

- The raw materials are locally available and the cost of preparation is less than Rs.10-15/-
- It will not create any side effects
- Easy preparation technique.

General Profile

No. of animals treated so far : 300-400 animals

Elangodi Veliyearaamai (Retention of Placenta)

Introduction

Retention of placenta is prevalent disorder in milch cattle and buffaloes. Farmers have been practicing several indigenous techniques to treat this disorder in the southern districts of Tamil Nadu and the specific one is given below

Symptoms

The specific symptoms are,

- Non-shedding of placenta beyond 3-5 hours of post-phone.
- The animal will be looking uncomfortable.
- There is difficulty in expelling the placenta.

Causes

The causes are,

- Less nutritious feed or off-feed.

Ingredients

1. Sesame cake - 250g.
2. Palm sugar - 100g.
3. Pirandai leaves (*Cissus quadrangularis*) - 200 g. a lengthy plant
4. Water - 50 ml.

Method of Preparation

The sesame cake and palm sugar are to be taken, ground well with little addition of water to make a bolus. The content is to be mixed well by hand and the 2-3 boluses would be prepared weighing 80-100 g. each.

Administration

The bolus is to be administered as intake medicine to the affected animal by using hand further, if the placenta is not expelled out yet the affected, placenta is to be tied with 200 g. of "pirandai" vine and pulled out by using hand.

2-3 boluses are sufficient for a normal milching cow. The recommendation is that 2-3 boluses per day per animal for three days continuously.

Mode of Action

The main ingredient "sesamum" acts as fluxing agent (as it has high viscosity).

Recommendation

Conditions or precautions to be followed :

- While the administering the medicine the animal is to be fed with hot water which is cooled for 3 hours instead of giving hot water.
- It is better to feed the animal with dry paddy straw instead of green grass.
- Maintain a hygienic atmosphere by spraying bleaching powder in the cattle shed.

Relative advantages

- It is the easiest and best known practice followed in southern districts of Tamil Nadu
- The cost of the medicine is less than Rs.20/-
- It will not create any pain to the animal.

General Profile

No. of animals treated so far : 200 times his own cattle.

Sinaipidikkaamai (Oestrous Cycle Failure)

Introduction

Oestrous cycle failure is not a disease but, it reduces the economical value of cattle as this renders the animal unfit for insemination even after its maturity. To rectify this problem Mr. Chinnapottu has been using an indigenous medicine.

Symptoms

The animal will not produce the estrous sound even after it attained maturity.

Ingredients

1. Yaanai nerinjii (*Pedalium murex*) - 500g. (a whole plant)
2. Water - 200 ml.

Method of Preparation

The 'yaanai nerinjii" leaves and roots are to be taken, ground well with little addition of water to make a bolus. Thus, 2-3 boluses would be prepared weighing 80-100 g each.

Administration

The mode of application is internal. The prepared bolus is to be administered as intake medicine to the affected animal in its third age by using hand.

The recommendation is that 2-3 boluses per day per animal for two days continuously during morning around 6-7 a.m. After three days of this treatment the animal shows oestrous symptom. If it is not shown another round of treatment should be given.

Method of Preparation

The "yaanainerinjii" leaves have the chemical components to induce the reproductive organ of animal.

Conditions

There are no specific conditions for this treatment method.

Relative Advantages

- Artificial insemination takes longer duration and it is time bound in the veterinary hospital. But, the aforesaid practice has been traditionally followed by the people of the particular locality and it has very good results. Beside, it will not create any side effects.

General Profile

No. of animals treated : 50-60 times his own cattle.

Kannil Poochi Viludhal (Eye Infection Due to Insects)

Symptoms

The symptoms of the disease are,

- The eyes become red.
- The animal shows watery and purulent discharge from the eyes
- Frequent opening and closure of eyes
- Swelling of upper lid and cataract.

Causes

There are several reasons for the disease. But, the following are the specific.

- Screw flies infection in the eyes.
- Eye infection is due to nutrient(s) deficiency.
- Wounds and watery discharge is due to sharp objects and dust.

Ingredients

1. Kanchira leaves (*Spheorahthus amarantoicedes*) - 500 g.
2. Neem oil - 50 ml.

Method of Preparation

The "kaanchira" leaves are to be taken in a metal vessel and boiled in the neem oil for 15-20 minutes till it attains ghee stage. Then, it is to be cooled for 30-45 minutes and applied over the eye in the form of ointment.

Administration

The prepared ointment is to be smeared over the wounded eye as external medicine by hand. The 50 ml. ointment is sufficient to cure the well developed eye infection. The smearing is to be normally carried out during morning or evening hours continuously for three days. It is sufficient to cure the infection. If necessary we can go for another three days continuously.

Mode of Action

The "kaanchira" leaves and neem oil has the chemical components to cure the eye swelling and eliminate eye insects.

Conditions to be Followed

- The animal should not be washed immediately after application of medicine.
- Avoid roughages during the course of treatment since it has high concentration which may reduce the medicinal effect of the contents.
- The cattle should be fed with hot water cooled for 3-5 hours.

Relative Advantages

- The "kanchira" leaves are locally available and the cost of neem oil is below Rs.10-15/-.
- It will not create any side effects.
- Easy preparation and simple procedure for application

General Profile

Animal treated so far – 50 times his own animal

Serumal / Jaladhosham (Cold)

The disease usually affects the animal during rainy season. At first, muzzle gets dried up. The causative agent is virus and disease is air borne and water borne. When the attached animals are place in a hard the disease gets transmitted from affected one to healthier.

Symptoms

- Tracheal block is the primitive symptom which results the animal unable to breathe easily.
- The animal looks dull and drowsy, animal off feed, inactive and always laid down.
- The above symptoms accompany with fever, sunken eyes, running nose, phlegm comes out from nose and tears from eyes are continuous in nature.

Medicine and its Ingredients

1. Onion (*Allium cepa*) – 100g
2. Dry chilli (*Capsicum annum*) – 250g (*Capsicum annum*)

Method of Preparation

Outer skin of the onion is removed and it is dried for an hour. Dry chilli cut in to pieces is mixed with (onion skin) onion peel. A mud pot with fire balls is taken and kept under the front portion of the animal. The dried onion peel and chilli mix is added to fireballs taken in the mud pot. The smoke developed is pungent and irritant to animal. Instead the ingredients are roasted in a mud pot and smoke is made.

Administration

The treatment is externally done by allowing the animal to inhale the smoke. The smoke has pungent smell and cause irritation in the throat. After sometime, all the phlegm comes out through nose. Once the animal become free of cough and cold, it becomes normal. It is identified by the coolness of the body, erection of hairs through out the body. The animal is also able to stand freely. The treatment is continued for three days. If the cough is continued even after three days, treatment period may be extended for another three days.

Cost Effectiveness

Onion and dry chilli are available at every home, as they are regularly used for culinary purposes. Onion costs Rs.2 to Rs.5 for 100g.It varies depending on the season and the availability. Dry chilli costs Rs.10 for 250g. Total preparation cost is Rs.15/ - sometimes if the crops are raised in farmer's field no cost is incurred.

Care to be Taken While Preparing the Mixture

Drying of onion peel is important. If they are dried completely the pungency and irritant nature will be absent. So, onion peel is dried till optimum moisture is maintained. Chilli added to the dried onion peel is completely dried as they are capsaicin rich which is active on complete drying.

Conditions and Precautions

- When the mixture is added to the fireballs, the animal inhales and gets out mucilage from nose with tears on eyes.
- If the animal gets tears from its eyes continuously, treatment is stopped and continued later. Feeding roughages and green fodder should be avoided.
- Rice gruel water is given for three days as feed.

Relative Advantages

The ingredients used are locally available in farm and house. It is safer to use and the result of the mixture is good. The method is more economical and still existing in villages of kuzhithurai block of kanyakumari district.

Additional Information

An alternate method to cure cough and cold is to use pepper (*Piper nigrum*) and Narthangai (*Citrus medica*). Pepper 20 gm is to be ground and given in cow's milk orally. Also, narthangai 4 fruits each are cut into 4 pieces and feed once.

Ajeeranam (Indigestion)

The disease afflicts the animal which takes more of roughages consume which leads in lack of sufficient water and dehydration. Indigestion is associated with heat, stress, fever and exhaustion.

Symptoms

- Stomach are more than 2 lays with bulged abdomen this is associated with lack of defecation or pelletted hard faecal material with no rumination.
- Animals look dull, sick followed by lowered tail.
- Do not take food regularly.

Medicine and its Ingredients

1. Betel vine (*Piper betle*) -200g
2. Garlic (*Allium sativum*) -100g
3. Moringa bark (*Moringa tinctoria*) - 50g
4. Thaivelai -200g

Preparation

First the above said ingredients should be washed properly and ground well. The ground material is mixed together with little water. Finally the juice is extracted and filtered. The filtered juice is taken in a bowl.

The cost incurred is Rs.10/- for betel vine, Rs.5/- for garlic. Moringa bark is taken freshly and cut into pieces. Thaivelai is a medicinal plant locally available. The total population cost is Rs.25/-

Administration

The mode of administration is internal through oral feeding. The juice is fed to the animal with the help of "kottaan" Dosage of the mix is 100ml a time. Treatment is given twice a day for 3 days. The treatment can be stopped if the animal does not show any symptoms of indigestion.

Care to be Taken While Preparing Juice Mixture

The amount taken to prepare juice mix of betel vine, garlic should be at recommended dose only. Excess in the ingredients will result in irritation and increased infestation of the disease.

Conditions and Precautions

Do not give water immediately after intake and the juice. Roughage and grass can be given.

Relative Advantages

It is relatively cheaper when compared to allopathic medicines. The method of preparation is also easier and safer to the animal. Farmers were successful with this practice for more than 50 years.

Additional Information

An alternative to this method is castor seeds (100gm) is ground and give orally in to the animal with 250 ml of water.

Madi Noi (Teat Obstruction)

- The disease is caused by microbes.
- The teats get blocked and no milking is possible.
- This will be accompanied with fever as its associated symptom.

Symptoms

- Swollen and hot udder, milk discolored with saltish taste.
- Thread like structure in the milk.
- Sometimes lesser (or) no milk secretion.
- Udder becomes small in size and gets hardened.

Medicine and its Ingredients

1. Turmeric (Kasthuri) - *Curcuma longa* - 250g
2. Nannaari kizhangu (*Hemidesmus indicus*) - 100g
3. Kollankovai kizhangu (*Corollocarpus epigaeus*) - 100g
4. Seeni paahu - 100g
5. Palmgur - 50g
6. Rice gruel water - 200ml

Method of Preparation

There are two methods available.

Procedure I

Turmeric rhizome cobe ground well using while grinding a little water is poured. So that turmeric powder gets changes to paste. The paste is collected in a bowl and kept for further use.

Procedure II

Nannari kizhangu, kollankovai kizhangu are taken and cut into pieces and sun dried. The cut tuber is to be powdered .The powdered mix is added to seenipaahu and kept for some time. Palmgur is added to the mix and taken in banana leaf. The sooranam mix is given through oral feeding.

No cost is incurred to the farmer all the ingredients are available in farm and home. Palmgur alone costs Rs.5/-and turmeric 250 g costs Rs.20/-

Administration

Method 1 : Mode of application is external. The turmeric paste and the sooranam are smeared over the udder. Application has to be done daily in the morning. This is continued for 3 days and allowed to get dried. The rice gruel water is collected after cooking and salt is added to it. The water is sprinkled over the udder. After a day and night, the udder gets loosened, turns normal and gives milk as normal.

Method 2 : The soornam is given for 3 days at dose of 100g /day. Mode of application is internal i.e. given in to mouth as orally. Time of application is morning and evening time.

Care to be Taken While Preparing the Medicine

The turmeric paste prepared should not be dried, as dried turmeric do not have curing effect and are adhered to the udder. The dosage of sooranam is 100g and it should not exceed. When excess amount of the sooranam mix is given animal gets side effects. In the form of stomach ache, diarrhoea.

Relative Advantages

Turmeric is locally available in farmer's house, so no cost is incurred, also preparation procedure is easier, safer to use. The sooranam mix made from Nannaari kizhagu, Kollankovai kizhagu, Seena paahu are available in farm. The disease also gets quickly cured with these medicines.

Additional Information

An alternate method followed by the farmers of kuzhithurai block, Kanyakumari district is to use punnakku keerai (*Melochia corchorifolia*) leaves one handful, neem resin (*Azadirachta indica*) 10 gm, turmeric rhizome (*Curcuma longa*) 2 inch size. The above materials are to be ground and mixed with egg yolk of country fowl and made into a paste before applied over the udder region.

Koppulam (Vesicle)

Symptoms

The symptoms are,

- Maggots' infection in the external portion of wound.
- It won't feed normally and often resorted to off-feeding.
- The animal often emits sound due to deep wound injury.

Causes

There reasons are,

1. Unhygienic cattle management practices.
2. It is an infectious disease and infects through lice and cattle flies.
3. Excess heat.

Ingredients

1. "Ottadai" (cob webs) - 50 g.
2. Turmeric powder - 50 g.
3. Neem oil - 50 g.

Method of Preparation

The "ottadai" and turmeric powder are to be taken in a metal vessel and boiled in neem oil for 15-20 minutes by using a flame till it attains ghee stage. Then, it is to be cooled for 30-45 minutes and applied in the form of "ointment".

Administration

The mode of application is external. The aforesaid prepared ointment is to be smeared over the wounded portion as external medicine by using hand. The 50 ml prepared ointment is sufficient to cure the well developed wound. The smearing is to be normally carried out during morning or evening hours continuously for 3 days. It is sufficient to cure the infection. If necessary, we can go for three more days continuously.

Mode of Action

The "ottadai" will eliminate the insects and maggots. The neem oil and turmeric powder have the chemical components capable to cure the wound.

Precautions / Conditions to be Followed

- Immediate washing after administration to be avoided
- Avoid roughages during the course of treatment since it has high concentration which may reduce the medicinal effect of the contents
- The cattle should be fed with hot water cooled for 3-5 hours Relative advantages
- The raw materials are locally available
- The cost of preparation is less than Rs.20-30/-
- Easy preparation and application of medicine.

General Profile

Animals treated so far : 50-60 times his own animal.

Avoidance of Breeding in Goats

The neck of the male goat is tied with a thick rope and it is again connected with any one of the hind leg through the interval between the pair of front leg. This is done to avoid the movement of male goat nearer todow (female goat), in doing so we can avoid unnecessary breeding especially when the female goat is in gestation period.

Arecanut Juice (Dewormer)

Dried areca nuts are powdered and soaked in water up to 12 hours. Then the arecanut juice is collected and mixed with water .it can be given to chicks, hens, and cock, through drinking water which has got the medicinal power of controlling worms in poultry.

Indigenous Bullock (Kangayam Bull)

Origin

Origin of this bull is from kangayam block at Erode district of Coimbatore tract in Tamil Nadu.

Characteristics of Bullock

Naturally kangayam bull has strong and healthy body. The shape of the body is straight, legs are height and straight and four hoofs are equally flat in nature. Kangayam bull has attractive colour (Black with white, white with black fore head), tails are straight and long in nature; the end of the tail has black huge hair. Tail is the identification for good work efficiency of bull. After the age of 2 ½ years it is used for ploughing and transporting purpose. The work efficiency of this bull is 10 hours per day. The hump and horn are very beautiful when compared to other breed. Feeding habit of this bull is dried paddy straw, sorghum and groundnut stalk, normally it has the quantum of feeding is less but work efficiency is more. One day one animal give 5 kg of dung, it is used for manure purpose. On the 4th age of the bull have full capacity for work long time 8 hours per day. One pair of bull rate is Rs.30-35,000. Kangayam bulls are mostly used for agricultural purpose in Coimbatore, Erode and Salem district.

Kangayam Cow

Cow is normally used for multi purpose (Milching, Ploughing and transporting). This cow is naturally having good and strong health and also having more disease resistant power compared to exotic breed. Normally this cow will give milk upto 6 liter per day. Cow milk naturally has good flavor and taste.

Feeding Habit

During the milking period only green grass and feed material will be given. Apart from that watering with rice brawn, cotton seeds, groundnut cakes are increase given to the milk yield. Cow also having equal work capacity compared to kangayam bull.

Ramnad Sheep (Ram Aadu)

It is predominantly white in colour throughout the body and the edges have black colour. It is originated from Ramnad district but spread thorough out southern Tamilnadu state. The Ramnad male sheep is crossed with Kilakarisal female sheep for better body weight gain, because Kilakarisal sheep is comparatively lesser weight than the Ramnad white sheep. So the crossing between the Ramnad male sheep and Kilakarisal female sheep will yield better body weight kids when compare to original Kilakarisal kids. In a small unit of 10 or 20 female Kilakarisal sheep, one Ramnad male sheep is sufficient for the breeding purpose.

Salem Black or Pallai Aadu

These are medium sized animals and are originated from Salem District. Predominantly black in colour, some times a tinch of black colour may see in the extremities like tail, forelegs or hind legs. The disease resistance capacity of the animal is high. Twinning is the common characteristic feature of this animal. Normally 60 percent of the animals have the twins. Triplets and quadruplets are also seen. Some animals will deliver 5 kids in a single parturition as the exception. These animals are mainly meant for meat purpose.

Sandai Seval (Fighter Cock)

This desi breed poultry is derived from Asil poultry breed. These birds are reared in the areas of Cumbam, Theni, Tirunelveli, Madurai, Usilampatti, Sivagangai, Paramakudi and Ramanathapuram areas. These are large sized cocks and having the tendency to fight with fellow cock always. The cost of the cock is ranging from Rs. 2000 to Rs. 3000. It fetches good income to the owners. This bird is reared for ornamental as well as entertainment purpose.

Vempur Sheep

Vempur breed is famous for its thigh muscle. Comparatively this breed has larger proportion of thigh muscle than other breed of sheep. This breed is predominantly white in colour and having brown patches throughout the body. This breed has good adoptability in tropical areas. It is the native of the Tirunelveli District and now in Thoothukudi District.

Kilakarisal Sheep

Popular Sheep breed of then Tirunelveli District, now Kilakarisal village is in Thoothukudi District which is the native place of this breed. The animal is predominantly brown in colour with admixture of black & white colour through out the body. The animals have very good adoptability in tropical areas. The meat derived from this breed is tastier than other sheep breeds.

Indigenous Breed Burgur – Local

Th. Kariyan of Thamaraikarai village of Anthiyur taluk in Erode district stated that Burgur local breed is traditional to their locality.

Fig. 31. Local Breeds

Description

Burgur local breed is called 'sevalai' which is medium sized and slightly reddish brown in colour.

The specific feature of this breed is its local purpose nature tasty milk.

Other details : since it is less used the population is diminishing which is not a positive sign. So population conservation of this local breed is the need of the hour.

General Profile of the ITK Holder

Name and Address : M.Pomman, S/o Mathan, Malasjanoor Burgur (post), Bhavani (TK), Erode district. Age : 35 years.Educational status : 5th standard. Existence of the breed : More than 55 years.

Burgur Local Breed Sheep

Thamaraikarai is a village in Burgur Taluk of Erode district. It is a village in Western Ghats extension, where local Burgur breed sheep are found. It is a white and light brown colour breed which grows upto a height of 3 feet. It gives 2 lambs / year. Normally, people will not consume the milk of the sheep, with in one year it weighs about 15-20 kg which costs about Rs.2000.

General Profile of the ITK Holder

Name and Address : M.Pomman, S/o Mathan, Malasjanoor Burgur (post), Bhavani (TK), Erode district. Age : 35 years.Educational status : 5th standard. Existence of the breed : More than 55 years.

Manapparai Local

Manapparai, a place in Trichy district is predominant for several local breeds of cattle. The local breeds are used both as milch animals and draught animals.The same kampalathu Nayakkar groups still maintaining the local breed.

S.No	Name of the local breed	Colour of the local breed
1.	Mayilai	Black and white
2.	Pillai	Red and white
3.	Sevalai	Pure red
4.	Karukka mayilai / kari	Black
5.	Mavellai	Pure white

Special Features

- This local Manapparai breed cow serves dual purpose i.e. they are used as both milch animals and draught animals.

- Milk tastes good, but poor milk yielder i.e., 1 lit/day.
- No need for any special maintenance as like that of cross breed animals.
- Adaptable for all climatic condition.
- Regarding the feed available locally materials are fed to the animal(paddy straw, sorghum & cumbu stalks and Oil cakes).It is existing in that area for about 100 years.

Krishnagiri Local

Local breed cattle found in Krishnagiri area is used for dual purpose. It yields ½ - 4 liter of milk a day. It is in different colour like white, black, red and resistant to different diseases and needs less maintenance compared with cross breed cattle. It costs about Rs 12,000/ pair.

General profile of the ITK holder

Name &Address : Thiri Ramaiah, Pasavapuram, Irudhu kottai, Thenkanikottai taluk, Krishnagiri dt, Educatiomal Status : 10th std, Age : 55, Farming experience : 35years, Occuopation : farming, Land holding : 10ac.

Cheng Kanni Adu

Cheng Kanni Adu is the variety of Kanni Adu. It is the native of Kovilpatti, Sattur and Sankarankoil taluks of Thoothukudi, Virudhunagar and Tirunelveli Districts respectively. In total Kanni Adu population 75 % are Pal Kanni Adu and the remaining 25% only Cheng Kanni Adu. This variety is the tastiest meat producer in Tamil Nadu. Normally this breed delivers twins. 18% of Cheng Kanni does will deliver triplets. The average adult body weight is 36 kgs in male and 29 kgs in female animals. This breed is highly suitable for rain fed region.

FISHERIES

S.No	Particulars
1	**Suraa/Shark mattu (Fishing Net)** in Thoothur post of Vilavancode taluk of Kanyakumari district. It is used for fishing in sea water.
2	**Maayakkaa (Fish trapping device)** in Thoothur post of Vilavancode taluk of Kanyakumari district. It is used trapping the fish.
3	**Muthu** in Thoothur post, Vilavancode taluk of Kanyakumari district Used to catch the fish.
4	**Soorai thumbai** in Thoothur post of Vilavancode taluk at Kanyakumari district. It is used to trap the fish when the waves and wind velocity is more.

5	**Paamaram/paai maram** in Thoothur post, Vilavancode taluk of Kanyakumari district. It will protect the fisher man from high wind and waves.
6	**Catamaran Kattu Maram** Type I in Thoothur post of Vilavancode taluk of Kanyakumari district. This ensures zero pollution of sea water. **Catamaran** Type II in Thoothur post, Vilavancode taluk of Kanyakumari district. Fishing can be done upto 20kms into the sea. **Kattumaram** Type III in Alagatha valasai of Ramanadhapuram district. Tree used to design "Cattamaran". **Thuduppu (oar)** in Thoothur post, Vilavancode taluk of Kanyakumari district. It is a device used to running the catamaran. **Aniyam** in Thoothur post, Vilavancode taluk of Kanyakumari district. It will give protection to the fisher man during high wind and waves. **Kadiyaal** in Thoothur post, Vilavancode taluk of Kanyakumari district. Give protection to the fisher man when the strong wind is occur.
7	**Karaimadi** in Thoothur post, Vilavancode taluk of Kanyakumari district. Up to10 can be taken in to the sea for fishing.
8	**Nangooram/anchor** in Thoothur post, Vilavancode taluk of Kanyakumari district. Used during natural disasters like storm, heavy rain with thunder and lightning.
9	**Fishing with nanjukuru** in Thoothur post, Vilavancode taluk of Kanyakumari district. It is a herbal plant used for fish catching.
10	**Traditional Fishing Basket** **(i)Pari** in Agaraperunthottam, Sirkazhi taluk of Nagapattinam district. Used for fish collection. **Katcha** in Agaraperunthottam, Sirkazhi taluk of Nagapattinam district. Used for fishing in the rice field. **Ootha** in Agaraperunthottam, Sirkazhi taluk of Nagapattinam district. Used to capture the fishes in running water and stagnated water.
11	**Sorati (Crab collecting tool)** in Agaraperunthottam, Sirkazhi taluk of Nagapattinam district. Used to collect the crabs from nest in the rice field.
12	**Karuvadu Thayarithal (Dried Fish)** in Vellapallam, Keevalur taluk of Nagapattinam district. Preserved fish is very largely used by the poor because of its cheapness.
13	**Karaivalai** in Alagatha valasai of Ramanadhapuram district. Used for harvesting the fish.
14	**Dam Fishing** in Trichy district. Fish catching in local dam which is cheaper method.

Suraa/Shark Mattu (Fishing Net)

It is used in Vilavancode taluk of Kanyakumari district. It is a bigger net used to catch suraa/shark fish. The main component of this mattu is a thoondai, which is made of iron. Thoondai measures about 10cm in length, has a sharp edge at one end and it is bent inwards. With this thoondai, saalai, kalavai, saemeen fishes are caught. For this, the fishermen go deep in to sea from 6 am to 6 pm. If the mattu gets struck on to any rocks the time may extend.

Different Types and Their Uses

The 6no.mattu is less in weight and used to catch neimeen, sinna choorai, vaalai, ayilai whereas no.8 is to catch ketta kembu, saarai, chiravu, kerai. The larger one of no.12 is used to catch suraa/shark.

Special Features

The mattu/nets are being made by the women in the fisherman family. The material used is silk and nylon. They make the mattu manually; do not use any machines for it. It also adds an additional employment to the fishing community.

Maayakkaa_(Fish Trapping Device)

It is a trapping device created by the fishermen of Kanyakumari district especially in the areas of kolachel, thengai patnam, eraviputhan thurai. It is made up of iron and looks similar to fish. Its weight ranges from 150-200g.

Usage

Mayakkaa is made as per the requirements of the fishermen, vary in size and weight. It is connected to one end of tangusam / role, blue in colour. Tangusam is made of nylon, one part of role is kept in hand and other left under the vellam – fishermen boat.

Benefits

Maayakkaa invites by attracting the other fishes and acts as a prey to them. The fishes get trapped by this and struck on to the rope. With this nokkoli, saalai, kili, koduvaalai, ayilai fish are caught. Sometimes, the tangusam gets cut off when big fish are caught. To avoid this, experienced persons will get to know by the rope movement whether the fish is big or small. Immediately on seeing the movement of rope, pull out the tangusam by which they are freed off.

Muthu

It is a fish trapping device used in Kanyakumari district. They are smaller in size when compared to maayakkaa. It is used in many numbers of 5-6 for a single time. It is less weight and weighs about 25-100g.

Method of Use

The muthu is tied on the rope and the net is used to catch fish. Between two muthu one gap is left. After tying, the net and the rope they are left in to the sea under the vellam. It is mainly used to catch, 'kanmai'fish. When bigger muthu is used 'periya neimeen' is trapped.

Soorai Thumbai

This is used when wind velocity is more and waves are fast so that the fisherman unable to go deep in to sea. It also looks similar to fish and traps soorai, kerai, nemmeen. Some other small fish are also caught. For this, the role used must be thick and strong.

Paamaram/Paai Maram

Components

- Paai-thaamaa paai
- Wooden pole- thadi
- Rope
- Small wooden boat

Structure

The wooden boat at its bottom on which the wooden pole is kept and tied vertically. The rope is tied at an angle of 45^0 from the top. A triangular gap is created between the pole and the rope. Thaamaa paai is spread and tied in the gap.

Working Principle

The thaamaapaai gives protection when heavy winds occur. They move the boat in the direction the oar is swing. It mainly works with the oar and the wind blowing direction. With this upto 20kms fishing can be done.

Limitations

During rainy days, the thaamaapaai gets completely wet and add more weight to the boat. To overcome this nowadays, fisherman use paai made of plastics.

Catamaran

Kattu Maram- Type I

The coastal fishing communities in Sirkali Taluk, Nagapattinam district (Tamil Nadu, India), have survived by their catamaran fishing for centuries. Catamaran is a light watercraft, which was named from the Tamil language words "kattu" (tie) and "maram" (wood, tree) –meaning "two trees tied together". It was invented by the "Paravas", an aristocratic fishing community in the southern coast of Tamil Nadu, south India.

"Catamaran" is a 10 feet boat made out of timber .Timber obtained from "Matha" tree is specially used for constructing it. This timber is transported from Kerala where it is found abundantly. Generally, it is made by carpenters and bought by fishermen. The rate varies depending upon the size of the Catamaran .Normally, on an average one Catamaran costs about Rs. 3000/.Nearly 3-4 slightly curved wooden planks are tied together with rope, such that the entire structure forms a curve shape.

One end of the centre, timber is curved and lifted upwards to a height of nearly 1-1.5 feet. This facilitates the boat to adopt with wave movements, while roaring into the sea.

This kind of Catamaran can accommodate only one member. Whereas by making an adjustment even 3-4 persons can be accommodated i.e. when one or two more planks are added and tied together with a Catamaran more space is provided and so it enables more persons to go for fishing.

In Ramnad, Catamaran is made up wood of kalyana murungai (*Erythrina indica)* because it has good floating capacity in sea. The tree selected to make catamaran is cut down in "crescent moon" period. Then it is kept for 15 days. Thereafter it is shown in fire to remove outer barks. Kaatumaram is kept for 4-5 years based on usage. The larger woods are kept in outside, in the middle slightly bigger woods are kept in orderly manner. Finally these arranged woods are tied horizontally with wood and rope.

Cost Effectiveness

Timber is the main component that involves more expenditure. On an average a Catamaran cost up to Rs.3000/ (one member boat).When compared to other boats (like motor boats made out of fiber) the cost of this kind of Catamaran is cheaper, affordable by an ordinary fisher man.

Feasibility and Relative Advantage

Fishing by catamaran is a traditional way of harvesting marine life without damaging the marine environment and its biodiversity. Catamaran is highly preferable, because of its eco friendly usage. It is being rowed manually by using oar. Apart from that, entire structure is made out of wood. This ensures zero pollution of sea water. Since, the cost of construction is also cheaper, it is highly feasible and advantageous.

Constraints and Limitations

- Though it is cheaper and eco friendly, there are few constraints which limits its usage. It can be used only up to a distance of one mile (approximately) into the sea. So the distance covered is comparatively very less which effects in low fish catchment.
- It is difficult to rove the boat near marshy places, when the fishermen himself should get down and pull/push the boat. It happens near to the sea shore.
- Fish catchment is 7-10 kg
- Using catamaran for fishing is also threatened by the mechanized boats and technological developments that have been used by commercial fishing operators over the last few decades. Marine capture fisheries have been stagnating in recent years in spite of an increased fishing effort –an indication that over-fishing by commercial mechanized boats is already taking a toll. This is threatening the survival of the poor fishing communities that use traditional catamaran.
- The number of boats in operation is estimated at about 70,000. It has been indicated that the small country craft do not operate beyond a few miles from the shore and spend much of their time in going to and from the fishing grounds. Consequently production per unit of effort is low. Mechanization of fishing operations would enable the fishermen to reach areas outside the range of the existing craft and also to fish for longer hours.
- It consumes more time in roving rather than fishing.
- During cyclonic wind period it is not used for fishing, because of heavy wind.

Catamaran- Type II

Structure

It is a traditional fishing boat existing still today in the coastal areas Kolachel, Eravi puthanthurai of Kanyakumari district. It is a U shaped, having 12-15 feet length at the middle portion. The middle portion is lengthier one, which is slightly raised at both ends.

Making of Catamaran

The wood used is taken from locally available trees (Kalyanamurungai). The wood is cut to required size and shape. For a single catamaran, 6 pieces of wood with width of15cm, 12-15' length are taken and put it in to the sea by tying them to strong support preferably huge stones at sea shore. The process is continued for 3-5days. This is to make the wood get increase in weight, become stronger. After 5 days, they are taken out and tied together with rope to get U shape. Finally, kadiyaal and aniyam are kept and tied at front and back end respectively.

Use : Fishing can be done upto 20 kms into the sea.

Limitation

- Riskier than the karaimadi vellam
- Strong winds may lodge the boat in to sea.
- Only 2-3 persons can go at a single time, have to come back immediately on the next day itself

Kattumaram- Type III

Catamaran (Kattumaram) is traditionally used for fishing. It is made up of *Erythrina indica* because it has good floating capacity in sea. The tree is cut down in "crescent moon" period. It is kept for 15 days. Thereafter it is shown in fire to remove outer barks. Kaatumaram is kept for 4-5 years based on usage. The larger woods are placed in outside, in the middle slightly bigger woods are kept orderly manner, and finally tied horizontally with wood and rope. Fishing can be done in day hour's upto 7'0 clock using Kattumaram. Fisherman can catch 4 kg fishes by going 4 km range by using kattumaram. As the waves attack directly to fisher man under woods and also to the body, they can fish only for 2 hours.

a. Thuduppu (oar)

Local names : Tholavaai, Chukkaan

It is a supporting device, normally used for pushing the sea water to move a particular direction. This tool is used by the fishermen right from when fishing started. Hollow bamboo is used to make thuduppu/oar. The fully matured hollow bamboo of 7' is taken and split in to two half. Each half is used as a separate oar.

Method of Use

Oar is used by fishermen when they go in to sea with thadimaram. The oar is moved towards the direction which the fishermen want to move the boat. A limitation to the oar and thadimaram is fishing can be carried out only upto 10 kms.

b. Aniyam

It is part of catamaran which is used for fishing up to 10kms. Aniyam is attached at the back side of catamaran. Aniyam is a "Y" shaped structure, made of wood and has a bent at its two sides to hold on the main portion of catamaran. During heavy winds while fishing a wooden pole is tied to the one curved end of Aniyam. Similarly a rope is tied to the top of the pole. Paai is tied in the gap created between pole and the rope. This mainly provides protection to the fishermen. It is smaller than kadiyaal, the front portion of catamaran.

c. Kadiyaal

In catamaran, similar to aniyam a structure called kadiyaal is present at the front end. But, it is bigger than aniyam. This also serves as a protection when strong winds occur. It is connected with more than two pole can be kept and tied.

Karaimadi Vallam

It is an advanced form of catamaran. This fish catching device still exists in coastal areas of Kolachel, Thengaipatnam of Kanyakumari district.

Structure and Making of Karaimadi Vallam

It is about 20'in length, has a bottom which is made up of various wooden planks. To the flat bottom is attached a shallow wooden planks 5'length and 1' breadth at all its sides. They are connected to each other with thick ropes. For this holes are made in the wooden planks. At the front and backend they are closed to form a sharp edge. Inside the vallam, there can be kept wooden planks to sit while fishing.

Benefits over Catamaran

Karaimadivallam has a space to store food in the boat. So, fishing can be done for a week and carried upto 50 kms. Upto10 can be taken in to the sea for fishing.

Limitations : It requires more initial investment than catamaran.

Nangooram/Anchor

It is made of iron with a single main piece to which 4 bent arrows like structures are joined. It weighs about 5-10 kg, 1'length. The bent arrows are 15cm in length. At the bottom of the main piece, a hole is present to which rope is tied. If the boat wants to stay in a place without moving, the fisherman put the nangooram/anchor in to the sea. It gets firmly held at a point and do not move. This is mainly used during natural disasters like storm, heavy rain with thunder and lightning. It can also be used during heavy waves or winds.

Fishing with Nanjukuru

An herbal plant called nanjukuru, poisonous is collected. They are available in the Kulasekaram, Pechiparai hills of Kanyakumari district. The collected nanjukuru seeds are broken and made as paste. The paste is dissolved in unit volume of water. Pour the nanjukuru solution in to the pond, lake where ever fishes are available. The fishes get giddiness, unable to breathe. Now, a hand net is taken and the fish are collected. The water regains its originality and free of poison after 30 minutes.

Address of the ITK holder

Mr. Martin, Door no : 250, E.P.Thurai, Elanthai vilaham, Thoothur post-629176, Vilavancode taluk, Ph : 04651-293273

Traditional Fishing Basket

It is designed by the farmer Th.Rajendran Agaraperunthottam in Sirkazhi taluk of Nagapattinam district.

(i) Pari

It is an indigenous fishing basket which is made out of bamboo splits. Fresh bamboo splits are taken from the bamboo trees and cut into required length and width of 3feet and 2cm respectively. Pari has two openings. One is wide spaced top opening, the diameter of which is 1.5 feet and another one is bottom opening, whose $3/4^{th}$ portion is closed with a small hole. Width of the basket is gradually decreased from top to bottom, so it looks like a conical shaped basket. In four places, bamboo splits are tied with rope around the basket to obtain the conical shape. 30 cm from the top centre portion is blocked with small bamboo splits and two holes are provided with the diameter of 5cm for fish entry. After that it is divided into two halves. Y shaped stick is tied on the pari, which is buried in the ground in order to arrest movement of the basket along with running water. Stick is acting as an anchor for pari. Existence of pari in that is more than 100 years which costs about Rs.150.

Fishing Mechanism

Pari is used for fishing in the tail end of the running water like small canals. It is used for fishing during night time ie 7.00p.m-6.00a.m. Small hole in the bottom portion is blocked with paddy straw to prevent the fish to come out from the basket. It is kept on the tail end of the canal and anchored it. After that bund is made with muddy soil around two sides of the pari. It can be placed along the water flowing or running direction. Normally fish has the tendency to swim against the water flowing direction. So the fish is jumped into the basket and it will enter into the bottom portion through two small holes. Trapped fishes are collected through the bottom opening.

(ii) Katcha

Katcha is a traditional fishing basket used for fishing in the rice fields. It is also a conical shaped bamboo basket and looks like a Pari. But there is no division in the inner side of the basket. Fresh bamboo splits are taken from the bamboo trees and cut into required length and width of 3feet and 2cm respectively. Katcha has only one opening i.e. mouth for fish entry and there is no bottom side opening for fish collection. After completion of fish capturing fishes are collected from the basket through the broad top opening (mouth). Existence of Katcha in that area is more than100 years which costs about Rs.150/-

(iii) Ootha

Ootha is also conical shaped basket made out of thin bamboo sticks. Length of the bamboo stick is 3 feet. Fifteen cm iron ring is taken and smooth bamboo sticks are closely tied around the ring with the help of nylon rope. To obtain conical shape bamboo sticks are tied in 3-4 places. To strengthen the Ootha thick notchi stick is attached in a circular manner. Ootha has two openings such as broad opening and small round opening (iron ring). Ootha is used to capture the fishes in running water and also stagnated water bodies. Ootha is placed randomly in the water bodies in such a manner that broad opening facing downward and small iron ring opening facing upward. Ootha is hold tightly by left hand and captured fishes are collected by right hand through the small iron ring opening. Two kilo gram of fishes can be captured in an hour. It costs about Rs 200/-

Advantages

- Cost involved for making is very less.
- Less maintenance.
- Less weight and easy to handle.

Limitations

- It needs skilled persons for basket making.
- Fishing capacity is less.
- Only used to fish in the smaller areas.

General profile of the ITK holder

Th.Kasinathan, Agaraperunthottam, Sirkazhi Taluk, Nagappattinam District

Sorati (Crab Collecting Tool)

Sorati is used to collect the crabs from nest in the rice field. Two feet lengthy iron rod is inserted in the waste unused cycle peddle. Cycle peddle is used as handle. Iron rod is inserted into the crab nest and crabs will come out due to disturbance. Then crabs are collected manually for human consumption. One person can collect 40-50 crabs / day

Cost required to make this Sorati is Rs. 15/-

General profile of the ITK holder

Th.Rajendran, Agaraperunthottam, Sirkazhi Taluk, Nagappattinam District

Karuvadu Thayarithal (Dried Fish)

Preparation of Dried Fish

Excess unsold fresh fishes are used for making dried fish. Dried fish can be made out of all the species of fish. Fins and tail portion are cut by using knife and fish is split open. Inner portion is removed. Fish is washed without removing the scales (while drying the scales will fall off) and lined with knife, which facilitates better salt absorption. Salt is applied over the fleshy portion. It is dried in hot sunlight for a day. Next day the fish is washed and again salt is applied. Duration of drying varies from 3 days to 10 days, which depends upon the size of the fish. Women fisher folks are involved in this dried fish making. Dried fish is one of the major ingredients of the poultry feed.

Marketing

Captured fishes are sold in the local villages itself. Excess fishes are sold in the near by towns or far away places as ice preserved fishes. Price may vary. Normally price depends up on size and variety of the fish.

Only 20 per cent of the sea fish catch is marketed as fresh fish and the remaining 80 per cent is either preserved as sun-dried fish or as salted fish or converted into fish meal and manure. Preserved fish is very largely used by the poor because of its cheapness.

General profile of the ITK holder

Thiru Chellathambi, Vellapallam at Keevalur Taluk of Nagapattinam district.

Karaivalai

It is the traditional fishing net made out of coir consisting of three particles in it viz., boya, purai, stone. Boya is on the top of the net while fishing and stone is on lower of the net. The karaivalai can be used up to 1 km in sea and it can be kept up to 1 hour. The fisherman goes to sea in morning 8'o clock when they use karaivalai.

There is separate boat or thoni made from "nava", *(syzygium cumini)* "vahai" *(Albezia lebbeck)* trees for using Karai valai known as **"karaivalaithoni"**. "Marukkuvalai" or nylon net is used in large boat. In the large size "karaivalaithoni" 10 members may fish in the sea and the smaller in size only 5-6 members may fish in the sea. Karaivalai has large size holes and nylon net (marukku) has small size holes. Karaivalai can be used only for 3 years as it is made out of coir. If the fisher man catch fishes for Rs.1, 000/- and they give Rs.30 as labour charge to the assisting persons in fishing. Nylon net or “Marukkuvalai” is used for fishing up to 5 km sea distance. There is a separate thoni is used for catching fishes by using “marukkuvalai”. The fisherman weaves the “marukkuvalai” by adding certain portions such as boya, purai and stone. The boya used in the “marukkuvalai” (plastic made) is different from the “karaivalai”. The bag used to keep fishes is known as **kacha.** The karaivalai is being used to catch the fishes like nagarai, sundai, Mura, sheela, vazhai and ora

Dam Fishing

Traditional Fishing in Dam

This method of fishing involves traditional materials like bamboo splits, coir and net A square dimensioned normally 7 ftx7ft or a rectangular dimensioned (10x7ft) net is tied on opposite two sides (Top & bottom) with bamboo splits that serve as a frame. Bamboo splits are framed such that one of the sides i.e., bottom side is bow shaped. The straight portion of the split is tied to the net and the bended (bow shaped) split is left without tying. On the opposite side i.e. top portion frame is tied firmly with the net. On the four corners long coir ropes are tied.

Mode of Operation

The net is suspended at the shutter outlet such that the bow frames a broad base at the bottom. It is kept as such for 20-25 minutes (Time varies from season to season). For a day nearly 15-20 times they suspend the net for fish catching. Each time when suspended, fish catch ranges from 1 kg to 2 kg.

Principle

While fishes are forced by the flushing water through the out let, it tries to jump backward. Thus it is caught in the net which is suspended at the outlet and the bow shape frame facilitates in withholding the fish that have fallen in the net.

Cost Effectiveness

Since net, bamboo splits and the coir rope are locally available. When compared to sea fishing, fish catching requirement cost is very cheap in dam fishing.

Advantage

Necessary requirements are easily available. Making as well as handling is easier.

Disadvantage

It needs minimum two members to handle and the time it takes to get an average fish catch is comparatively more.

Fig. 32. Fishing Equipments

OTHERS

S.No	Particulars
1	**Paai Thayarithal (Traditional Mat Weaving)** in Thaikkal of Sirkazhli Taluk of Nagapattinan district. For easy transportation and drying.
2	**Cane (Moonkil)** ***(Calamus rotang)*** **Products** in Thaikkal of Sirkazhli Taluk of Nagapattinan district. Used to prepare wooden furniture's.
3	**Mattu Ladam Kattuthal (Cattle shoeing)** in K.N.Puram village of Salem district. Protect the animal shoe from cut wounds and other injuries.
4	**Chekku (Oil extracter)** in Chekkumedu village of Salem district. Traditional method of oil extraction.
5	**Pugai elai Pathapaduthuthal and Tharam Pirithal (Tobacco curing and manual grading)** in Panankkatture village of Salem district. To improve the quality of the tobacco leaves.
6	**Panai vellam/ Karuppatti (Palm candy Preparation)** in Aandipalayam village of Salem district. It is used as a sugary substances like Tea, Ginger tea and coffee.
7	**Pughai poduthal (Banana ripening)** in Devoore village of Salem district. Ripening of the matured green banana fruits at emergency period.
8	**Sorghum Vithai Pirithal (Post Harvest in Sorghum)** in V. Mettuppalayam village of Salem district. To remove the sorghum seeds from ear head manually by using a stick.
9	**Sarkkarai Thayarithal (Sugarcane Jaggery Preparation)** in Devoore village of Salem district. Cane candy / jaggery powder is prepared and it is used as a sugary substances.
10	**Manapalli Aruval (Implement Preparation)** in Manapalli village of Namakkal district. Tools and implements are used for cutting and harvesting the crop.
11	**Natural fruit ripening structure** in kalluvali, Pallithambam (via), Sivagangai taluk of sivagangai district. Traditional method of fruit ripening.

Traditional Mat Weaving

Thaikkal is a village in kollidam block of Nagapattinam district where majority of the people are involved in traditional mat weaving. In and around kollidam block korai *(Cyperus spp)* has been grown well in kollidam river banks, due to this reason traditional mat weaving is the major cottage industrial product in this area.

Korai is a weed plant, which is used for traditional mat weaving. Root tuber is the planting material for the cultivation of korai. It can be planted in all the season (or) through out the year. The field is ploughed well for about 3-4 times. All the other type of weeds available in the field is removed from the ploughed field. Then, the field is levelled by using wooden plank called "Parambu Palaki". Fifteen days after planting weeding can be done. Irrigation has to be given to the field for about once in three days. Six months after transplanting harvesting can be done. Flowers and Korai are cut separately upto 10-15 years ratooning can be done, but harvesting can be done once in six months. After harvesting stubbles are collected and burnt, then urea 75 kg is applied.

Processing

a. Post Harvest Practice

After harvest, harvested korai are separated/graded based on length ie 4ft, 5ft,3ft korai and splitted into 2 halves either by manually or by mechanically. Lower sheath in the lower part of korai is removed and then korai is ready for drying. It has to be dried for about three days. For easy transportation and drying, harvested korai are bundled which is locally called as "Allu" whose circumference is 18 inches. Wage for making bundle is Rs.3/bundle; one member can make 10 bundles per day. Korai are graded based on its length (i.e. 3 ft, 4 ft) and made it as big bundle (or) 4 allu=1 kattu. Cost of the big bundles-4ft korai=350Rs and 3ft korai=180Rs.

b. Dying

Common colours available are Violet, Rose and Green. Handful of colour powder is added into 30 liters of boiled water. Big bundled korai are immersed /dipped into the coloured water. Before dipping bundles are wet with water, to avoid breaking while dipped for drying. 'Y'shaped and hook like wooden sticks are used to lift the korai after dying. One member can dye for about 10-20 bundles of korai.

c. Drying

After dying, big bundles are bundled as small bundles for effective drying. Bundles are dried for about 3 days while drying first day one end of the bundle is tied other end is left free. Next day another end is tied; tied end is left free, so that even drying can be achieved. After the completion of 3 days drying and dying, again big bundles are made. Wage for making big bundles is Rs. 60.

d. Mat Weaving

Total measurement for the weaving structure is 12ft length and 5ft width. To mark the width and length bamboo poles are kept in four sides which is called as "Mullai Kazhi". Four bamboo poles are attached to the Mūllai Kazhi both length and widthwise. Then it looks like a rectangular structure.

Components of Weaving Structure

Annaippu- A Wooden plank its length is 4 ½ ft. It is movable.

Achu- It is also a 4 ½ ft lengthy wooden plank. It is immovable.

Vizhukaghi-It is also a 4 ½ ft wooden plank.

Mithipatti- It is also a 4 ½ ft and Right leg is kept while weaving one rope is attached in it, in which left leg is kept while weaving .

Manapalakai-Made out of wood, length is 5ft which is kept under thread, in which weaving person can sit on it and also it avoids thread breaking while weaving.

Kuchali- Kuchali is a wooden needle like structure which is used to hold the korai while weaving. Its length is 4 ½ ft.

Weaving Procedure

Thread is tied horizontally along the entire length of the 12 ft weaving structure. It looks two layers (upper & lower). Under this thread manapalakai is kept for sitting while weaving. Without manapalaki a person cannot sit and weave. Weaver can sit on the manapalaki kept the right leg on the mithipattai and kept the left leg in the rope which is tied on the tripod stand. While pressing the mithipalakai upper layer and lower layer will be separated. Then kuchali is used to hold korai. One korai is hold by the kuchali at a time.

Anaipu is used to arrange/ organize the korai tightly. Next mithipalaki is pressed to get the upper layer as lower layer and vice versa. Again the kuchali is used to hold korai again the same process continues until the mat weaving is finished. Manapalakai and tripod stand are moved forward while weaving. One member can weave one finished mat in a day. Cost is Rs. 150/mat .Different bird's design (peacock) and names are designed manually. 'Kappa korai' (thin) is used for filling and 'Motta korai' (thicker) is used for designing.

Cutting

After finishing the weaving edges/2 sides of the mat will be cut by using bigger size scissors. Four to five threads are tied together and knot is made to avoid splitting of mat. Thin cloth is stitched on the two corners of the mat.

Efficiency-80 mats/day

Wage- Rs 100 /day

General Profile of the ITK Holder

Th. Sekar,Thaikkal,Sirkazhi Taluk,Nagapattinam District

Cane (Moonkil) *(Calamus rotang)* Products

The raw material used for making the cane product is Pirambu. Based on the size, it is of two types.

- Long size pirambu (thick)
- Small size pirambu (thin)

The artisans got the long size pirambu from the wholesale market in Chennai and the small size pirambu is locally available in Manampattu, (Kollidam area) Nagapattinam Dist. Long sized one is used for outlining (designing) i.e. frame work and the small size one is used only for the binding, filling and joining work. The length of the pirambu varies from 11-17 feet.

Cost of the long size Pirambu : Rs. 3500-4000 / 100 Nos

Similarly the small size Pirambu : Rs. 400/100 Nos

Steps involved in Cane Product Making

a. Peeling

By using small knife, the outer skin is peeled off. The knife has specifically designed for this purpose.

b. Kanu Aadithal

The nodes of the pirambu are removed by using big size knife. Now, the outer skin of the pirambu will get smooth surface.

Labour charge : Rs.40/day Work done : 100 splits / day / men

c. Soaking and Drying

The pirambu are soaked in water for overnight. Next day morning, it is taken out from the water and dried in shade for one day.

d. Pirambu Valaithal (Bend the Cane)

Two iron rods are oppositely fixed on the ground. The length of the iron rod is 1-1.5 feet. The gap between two rods 20 cm. Pirambu piece is kept in between the iron rod and kerosene lamp is lighted and placed nearer to the pirambu piece where the pirambu is to be bent. Due to the heat, piece is melted slowly. At that time, the bending is done by manually. Then the piece is immediately soaked into the water for 15-30 minutes to retain curved or bent postition. Otherwise it will come to normal position.

e. Joining and Binding Work

Based on the shape of the products the pieces are designed separately. Next the pieces are atttached by small size pirambu pieces. The pieces are joined by using 2-2.5 inch nail and small hammer. The hammer is made out of iron. The length is 25-30cm. The extra pieces are cut by using small size knife.

f. Varnishing

The final step is varnishing. After completion of varnishing the product has shiny surface and attractive look.

Based on the products, labour charges and work done/day are varied. It is given in the tabulated form below.

Items	Labour required	Frame work (Rs)	Labour charge Designing (Rs)	Varnish work (Rs)	Work done / day (numbers)	Cost of the product (Rs)
Cot	3	250	200	15	2	2500
Sofa set	3	300	450	20	2	3500
Unnjel	2	26	40	3	20	450
Cradle	2	15	13	2	30	250
Stand unnjel	2	55	45	3	6	750
Delux racking chair	2	25	20	3	4	800
Lady rackin chair	2	30	20	3	4	800
Dinning table	3	40	30	10	5	2500
Shelf	2	80	150	15	2	1400

General Profile of the ITK Holder

Name and address : Th. Nagarajan, Thaikkal, Sirkazhi Taluk,Nagapattinam District

Cattle Shoeing

Required Materials

1. Shoe
2. Hammer
3. Sharpner
4. Wooden Board
5. Rope

Purpose

This method is traditionally followed in K.N.Puram village at Salem district. Normally cattle shoes are getting damaged during the time of ploughing and on road movement. In order to safeguard the hoofs from damage, shoeing is followed.

Mode of Shoeing

For shoeing the hoofs, three members are required. Four types of hoofs are used according to weight or hoof size of the cattle. Once in three months old shoes have to be changed for efficient in work. In this process, cattle is to be pulled down and the front leg has to be tied with rope. The top portions of the two leg hoofs are cleaned and old / damaged shoes are removed by using sharpner with the help of wooden board and hammer. Then turn the cattle for fixing iron shoes in another two legs. In horse also iron shoes used to be fixed for protecting the heels without pain during the time of running either early morning or late evening is the time suitable for fixing the iron shoes. Time taken should not be more than 20 minutes. If the time is exceeded to 20 minutes cattle will have problem in respiration. The next day itself animals can be used for any work.

Wages for Shoeing

Wages for shoeing one set of cattle is Rs.150, and for one horse is Rs.150.

Traditional Oil Extraction Chekku

Oil extraction is in existence of about 50 years in Chekkumedu village at Salem district.

Making

Chekku is made up of stone or wood and it' is covered with steel plate, to this large wooden pestle used for grinding purpose.

Process

Gingely and groundnut oils are extracted by this chekku. The capacity of the chekku is 20 kg. Time taken for extracting the oil from 20 kg of seeds are 1- 1 ¼ hours. Out of 20 kg of seeds 6 liters of oils can be extracted. In gingely oil extraction 1.5 kg of palm candy has to be added to increase the oil quality and sweetness. Candy is not needed for groundnut oil extraction. Time suitable for extraction of oil is morning 6-10 am and evening 3-5 pm. After completion of this process oil has to be collected in a small provision attached below the chekku. Extracted oil is collected in zinc sheet. One kg of gingelly oil is Rs.90/- and groundnut oil is Rs.65/- at the time of documentation. The quality of oil is high when compared to mechanical extraction of oil. This oil can be preserved upto 6-7 months. Finally oil cakes are collected which can be used for animals feed and some food preparation. One kg of groundnut cake is Rs.19/-, and the gingelly cake is Rs.10/- at the time of documentation. Two men power is required for extracting oils. Hundred kg of seeds can be extracted per day. Wages per day is Rs.100/person.

Traditional Tobacco Curing and Manual Grading

Tobacco is cultivated during January to February. After full maturation tobacco leaves are harvested and the leaves are exposed to sun drying for 10 days in the field itself. Once in three days tobacco leaves position will be changed for perfect drying. During the time of drying if there is rainfall; leaves are dried in thatched house with proper aeration. After drying the tobacco leaves, they are kept inside the house called "Param poduthal". Once in three days change the position of tobacco leaves which is essential to get good quality of tobacco leaves. This process will continue upto 10 days.

Manual Grading

After completion of tobacco curing, leaves are separated manually from the stem and graded into 3 different sizes. Graded leaves are bundled in different quality viz; Grade-1, G-2 and G-3. Each bundle has up to 100 leaves. The tobacco leaves are kept as bundles till it reaches ash colour. Then only the leaves are marketed. One kg of good quality tobacco leaves is Rs.40, second quality leaves are Rs.25. The yield obtained from one acre is 500 kg of good quality leaves and 200-300 kg of second quality leaves. Cost of cultivation in one acre of tobacco leaves is Rs.15000. Net profit is Rs.10, 000-15,000.

Palm Jaggery Preparation (Panai Vellam/ Karuppatti)

Purpose

Palm candy is prepared from neera.

Neera Collection

Neera is to be collected with a pot lined with slaked lime for preventing fermentation. Good time for collecting neera is morning between 5 and 10 am.

Process

After collecting the neera it is poured in to the palm kopparai(vessel). The diameter of the vessel is 1 meter, in this 54 liter of neera is transferred. Burn the stove upto 3 hours for neera getting a fit condition of treacle. Palmyra dried leaves and palmyra fronds are used as fuel. For getting fit condition of treacle, small amount of castor seeds juice is added. After getting fit condition, treacle can be transferred into big pot for cooling. Then it is mould or transferred into small sized pot (Mud Acchu/wooden acchu) for making palm jaggery. Treacle is preserved in Acchu and 3-4 hours is needed for getting jaggery. To prevent stickiness wet cloth are placed inside the mud pot. The weight of the each potted jaggery is around 2 kg. From 54 liter of palm neera, 10 kg of palm jaggery will be prepared. Each 2 kg of palm jaggery is Rs.50.

Uses

Palm jaggery is used for making "sukku kasayam" (dried ginger essence), tea etc, it will give energy to the human health.

Palm Sugar Crystal (Palm Gur)

For palm sugar crystal preparation no need to added caster seeds juice. Palm treacle is poured into mud pot and then kept upto 3-4 months. Then palm treacle is fixed inside the mud pot. It is consumed to reduce the intensity of small box disease.

Banana Ripening (Pughai Poduthal) - Smoke Champer

Purpose

Banana bunches after harvested has to exposed to smoke for ripening. This method is called as pughai poduthal.

Preparation Room for Ripening

In air tight and compact room banana bunches are arranged uniformly. For smoking a mud pot is used. Inside this pot banana leaves and charcoal are used for making smoke. Mud pot should be kept in the corner of the room. The room should be closed without any gap by plastering with cow dung slurry. In this process bunches are kept for one day. Next day the banana turns it colour. It is one of the indications for banana ripening. Then it can be used for sale.

Precaution

When the concentration of smoke increases it affects the quality of banana by showing black spots. This might be a problem in fixing the rate and mostly farmers won't prefer such type of bananas.

Post Harvest in Sorghum (Manual Threshing)

Drying the Ear Heads

After harvesting the sorghum, ear heads are sun dried for 1-2 days and then threshing is done. Some times immediately after harvest threshing is done by using sticks.

Separate the Grains

Ear heads are spread over the threshing floor and beaten with sticks for separating the grains from ear heads. After this husk and dust (kongu, sonai, pottu) are removed by using Muram or hand winnower. The good quality seeds are collected and sun dried for 4-5 days, so that the germination percentage can be maintained.

Special Feature

1. These seeds can be used upto one year, when stored in gunny bags.
2. The grains can be used as food material (chola choru, chola kali, chola thosa). It gives good energy to our body.
3. The manual threshed seeds possess high germination percentage while comparing with machine threshed seeds.

Traditional Chekku
Used for Oil Extraction

Traditional Oil Pestle-Ulakkai

Formation of Oil Cake
During Oil-extraction

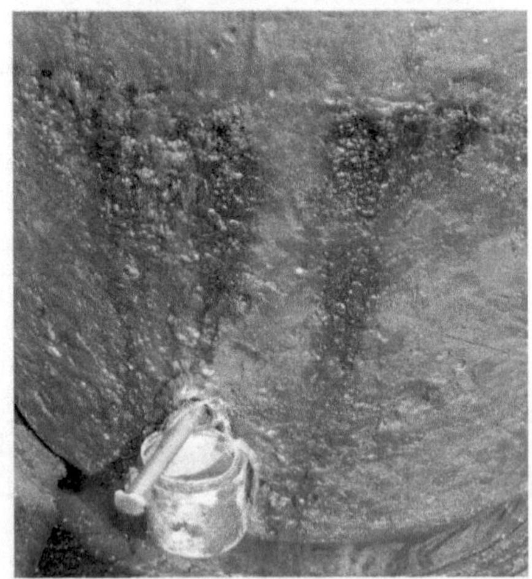

Oil Outlet in Chekku
After Extraction

Fig. 33. Traditional Oil Extraction Chekku

Indigenous Tobacco
Curing Process

Tobacco Grading

Graded Tobacco
Grade 1, Grade 2, Grade 3

Removed Tobacco Sticks
During Grading

Fig. 34. Indigenous Tobacco Curing Process

Sugarcane Jaggery Preparation

Purpose

Cane jaggey preparation

Mode of Preparation

On full maturation cane is harvested using Kodari (Axe). The harvested cane is transported to the crushing yard. Juice is extracted from cane by using meachanical crusher. Crushed juice is transferred through pipe line, and is collected in an iron tank, which has the capacity of 700- 800 litres. From this, the juice is transferred to the kopparai (container) through plastic hose. Then heating process is continued upto 1 hour, sugarcane baggases are used as a burning material. A small quantity of "soda" is added the juice for removing the dust from it, for removing dust "Mattipattai" is used. Stirr the treacle upto one hour using "Pagu Aathi" continuously without any gap, which help in avoid sticking of jaggery to the kopparai. After getting fit condition (Treacle) it is transferred to another kopparai, again soda is added to get thickness of treacle. Then the solid Jaggery is cooled for half an hour by using "Kuruppi" and "Muttu kattai". Three members are required for this process, jaggery clots are removed by using "Muttu kattai". From 800 litres of cane juice 90 kg of jaggery can be prepared. 900-1000 kg of jaggery can be prepared per day. Burned ash is used for manure mixed with cow dung.

Implement Preparation (Manapalli Aruval)-Sickle

Materials Required

1. Cast iron (car, jeep, lorry,)
2. Big sized hammer
3. Sharp iron
4. Mud air store.

Preparation

Old iron materials are purchased from local mechanical shop. One kg of iron cost is Rs.15-20 /-. Initially it is made into a straight blade and then it is shaped to the form in which we need to make implements. Then iron rod is placed in extreme heat for sharpening the implements. Heat is intensified giving cycle wheel roller to mud air stove. After it gets full reddish colour, the rod is beaten with big sized hammer according to the requirement of size and shape. It is heated again and then it is dipped in coconut oil or water. After that it is cleaned with coconut fibre. This process is done 3to 4 times. For making one cutting knife (vettu aruval) it will take half an hour the handle of the implements are also made in this unit itself. It cost around Rs. 20- 40/- according to the quality of the wood (guava and teak) and artistic work. Manapalli knife is famous in Karnataka, Andra Pradesh, and Tamil Nadu. Agricultural and home need implements are prepared here. The cost and time taken for each implements is given below.

S.No	Name of the implements	Time taken for made	Cost (Rs)
1	Aruval manai	3 hours	200
2	God knife	8 hours	1000
3	Chaffing knief	8 hours	1000-1200
4	Axe	2 hours	150
5	Small knife (kodikkal kathi)	2 hours	100
6	Koduval (Tree cutter)	30 min	150
7	Kathir arrival (Harvesting cereals)	30 min	50
8	Gardening knife	30 min	400

Natural Fruit Ripening Structure

Introduction

Normally, carbide stone is used in artificial chamber for fruit ripening. This is used for mango, sapota and jack. But it fetches poor market price and make the fruits into bitter in taste. To overcome these problems Mr. Abraham of Devakottai, Sivaganga district of Tamil Nadu has developed a natural fruit ripening structure by using a local carpenter.

Description of the Structure

The size of the structure is 1x1x1 m. It is made by using clay soil and covered by a wooden shield. The entire structure is smeared with cattle dung. The mini stove is attached with a whole tube so that to supply a smoke by firing farm waste. The capacity of the container is 15 kg.

By burning wood ash and natural waste the smoke is supplied into the hole. Thus, the smoke curing technique ripens the fruits.

Advantages

- The duration for fruit ripening is 2-3 hours
- This method does not ripen the tender fruits and prevent poor quality, low price in market
- The fruits ripened by this technique are sweet in taste and free from foul odour normally occurred during inorganic ripening method

Sugarcane Juice

Traditional Dust Removing Implements-mattipattai

Sugarcane Extracts Boiled for Jaggery Making

Traditional Jaggery Making Implement - Pagu Aath

Formation of Jaggery in Kopparai

Collection of Jaggery After Powedered Form

Fig. 35. Traditional Sugarcane Jaggery Making Process

- The cost of preparation of the structure is less than Rs.200/-. But the stony method of fruit ripening requires Rs.200/- for single attempt
- It is a permanent structure and is not affected by termites since is smeared by cow dung
- Consumption of the fruit ripened through this method would not cause any health hazards.

4

Rationality on Indigenous Technical Practices Traditional Varities

TRADITIONAL VARIETIES

S. No.	Documented ITK Practices	Rational
1.	**Puzhuthi Kar Nel – Dry Land Rice** Puzhuthi Kar Nel is being cultivated in hilly tracks of Western Ghats bounded in Erode district. It is suitable for rain fed and garden land conditions; villages lying in this hilly track. Owing to its suitability to rain fed condition it is direct sown crop. The duration of the crop is 125-130 days and the average height of the crop is 130 cm that makes it susceptible for lodging. It yields upto 1.5 tones/ac and features of the grains are bold and red coloured, which is preferred by the hill people for consumption. Keeping quality of this dry land rice is good.	Based on experience • Suited for soil brought to fine tilth. • Medium duration variety.

2.	**Kalurundai** It is being cultivated in Nagapattinam district. Late Samba (Purattasi – Thai i.e., Sep 15 – Feb 14) and Navarai (Margali – Masi i.e., Dec 15-Mar14) are the best-suited season for cultivating this variety. It requires clay loam soil for better growth. It is direct sown crop with duration of 120 days. It grows upto a height of 120 cm (4 feet) which makes it susceptible for lodging. It is resistant to drought, pest and disease as well as salinity. During maturity stage, colour of the grain is yellow with black colour faded stripes. However, the rice is bold sized with pale yellow colour. Average length of a grain is 22.5 cm and the 1000 grain weight is 32g.Due to its bold sized quality, the grain weight is comparatively more than other varieties. Straw obtained from this rice variety is used for thatching roof. It exists for more than 80 years and at present, it is cultivated in all coastal villages of Nagai District.	Based on scientific reason • May be used for developing drought tolerant varieties in the breeding programme.
3.	**Soolai kuruvai** It is cultivated in coastal villages in Sempodai, Periyakuthagai, Thopputhurai at Nagai District. Late samba (Purattasi-Thai) i.e., Sep 15-Feb 14 and Kuruvai (Mid Vaigasi – Mid Aavani i.e., June 1-Aug 31) are the suitable season for cultivating this variety. Clay loam and coastal saline soils are well suitable for cultivating this variety. It is cultivated in both irrigated as well as rain fed conditions. In irrigated system, when water is available in plenty transplanting method is followed, whereas in rain fed system, direct sowing is practiced. Since it is a drought tolerant variety, it suits even in rain fed areas. The duration of this variety is 130-140 days and grows upto a height of 110-120 cm. The feature of this variety is motta or bold and pale dirty brown color. It yields about 15 bags/ac (75kg/bag) i.e., 1125 kg/ac. This variety is characterized by the special feature like the straw yield will be more. Hence it is used as cattle feed and also used for thatching roof.	Based on experience Traditional cultivar higher straw yield benefit the faming system having cattle as component.

DSIR/TPDU/TIF 056 and E-28-GO-GOI Research Scheme on "Documentation and Preservation of Agricultural Traditional Knowledge through Farmer Participatory Approach"

4.	**Kuzhi Vedichan** This variety is being cultivated in Vellapalam and Keevalur taluk of Nagapattinam district. Samba (Aadi-Margazhi i.e., July 15 – Jan 14) is the apt season for	Based on experience Highly drought tolerant. The root biomass and length show the varieties ability for

cultivating kuzhi vedichan. It grows well in sandy clay soil and saline soil, so that it is very well suited for coastal areas. This can be raised by both direct sowing method as well as transplanting methods. It is resistant to salinity, drought and pest and diseases. It is a short duration crop and the cropping cycle is 110 days. On an average, this variety yields upto 950 kg/ac which is lesser than other local varieties. The average height of the crop is 2½ feet and it will leads to lodging. Size of the grain is bold in nature with pale dirty white colour. This variety is suitable for idly and dosai preparation. The specific character of this variety is induced milk secretion. Even when the field is flooded and the crop lodges, it has an unique feature of tillering again. This is secret behind its name called "Kuzhi vedichan" in Tamil, it means "Coming out of the pit.

drought tolerant. The variety keeps the greener ever in high drought condition.

5. **Val Sivappu**

It is also being cultivated in Vellapalam and Keevalur taluk of Nagapattinam district. This variety is well suited for late samba (Purattasi-Thai i.e., Sep 15-Feb 14). Sandy clay and saline soils are most favourable soil type for better growth. This feature facilitates coastal farmers, especially in periyakuthagai, Thopputhurai, Sempodai villages of Nagapattinam district to cultivate val sivappu. It is saline as well as drought tolerant variety but susceptible for lodging. The duration of this crop is 145-150 days and grows upto 160cm in height. The character of the rice will be red in colour and also good taste. The grain is coarse in nature and used for idly and dosai preparation. It is a low yielding variety and yields upto 900 kg/ac (12 bags/ac; 75 kg/bag). The identification character and special feature of this variety is the posterior end of each grain has a single bristle like structure which looks like a small bird's tail. So, the tail like structure and its colour (red-sivappu in Tamil) both contributed to the naming of the variety as "Val sivappu". It seems to exist even before 150 years.

Based on experience

- Awns may be present in the grains. And the awns may be red in colour as the name denotes.

6. **Vellai Kuruvi Kar**

This variety is also being cultivated in Vellapalam and Keevalur taluk of Nagapattinam district .Samba (Aadi-Margazhi ie. July15-Jan 14) and late samba (Purattasi – Thai i.e., Sep 15-Feb 14) are the best suited seasons for

Based on scientific reason

- The name indicates the colour of the grain as well as the duration of the variety.

this variety. It exists for more than 90 years. It prefers sandy clay and saline soil condition which is a special feature of this variety. Total duration of the crop is 125-135 days out of which 25-30 days are nursery period. In case of upland conditions, direct sowing can be followed. This variety grows upto 100-120 cm. Both the grains as well as straw are pale yellow to dirty white colour. During tillering stage, rat damage will be more because the stem is tastier.

7.	**Seeraga Samba** This variety is also being cultivated in Vellapalam and Keevalur taluk at Nagapattinam district .The suitable season for growing this variety is samba season. The cultivation method is transplanting system. The duration of the crop is 125-130 days. The nursery period is 30-35 days. The plant grows upto a height of 80.5 cm. It yields about 20 bags (1500 kg). The straw yield will be 1050 kg. The grain size is also very small and fine. It is highly used for making briyani on the special occasion	Based on Scientific reason • The grain has good aroma and shape. The variety can be used in a breeding programme but with reduced duration (upto 110 days).
8.	**Pichavari** This variety is also being cultivated in Vellapalam and Keevalur taluk of Nagapattinam district. The special feature of this variety is tolerant to flood and drought. The best season for cultivating this variety is Thaladi season. Direct sowing and transplanting system are used for cultivation of this variety. The duration is 110-115 days of which 21-25 days is nursery period. The plant grows upto a height of 100 cm. It yields about 4800 kg/ha. The straw yield is about 180 kg.	Based on experience • Short duration helps for direct seeding.
9.	**Suran Kuruvai** This variety is also being cultivated in Vellapalam and Keevalur taluk of Nagapattinam district. The suitable season for growing this variety is samba and late samba period. The cultivation method is transplantation system. The crop duration is 130-135 days and the nursery period is 30-35 days. The grain yield is about 1200 kg per acre. The straw yield will also 50 bundles (1200 kg) .The colour of the rice is dark brown and coarse in texture. This is highly suitable for making idli and dosai.	Based on experience • Traditional cultivar for choice purpose.

10. **Vai Kunda**

This variety is also being cultivated in Keevalur taluk of Nagapattinam district. Direct sowing and transplanting system is used for cultivating this variety. The cropping cycle is 145-150 days. The nursery period is 35-40 days. The plant grows to a height of 121 cm. It yields upto 18 bags (one bag weighs about 75 kg). The straw yield will be 180 kg. The special feature of this variety is tolerant to flood and drought.

Based on experience
• The grain is slender and similar variety by name vaiyakundan existed in old ramnad district, now extinct. Rice grain is white in color with good taste.

11. **Kichedi samba**

This variety is also being cultivated in Keevalur taluk of Nagapattinam district. Samba season is best suitable season for cultivating this variety. The cropping cycle is 140 days. The seed rate is 24 kg / ac. The plant grows to a height of 102 cm and it has a lodging habit. It yields upto 15 bags (1125 kg).The straw yield will be 1050 kg. The grain colour is pale white in colour. The special feature of this variety is resistant to pest and diseases.

Based on experience
Low Nitrogen requiring variety suitable for organic farming.

12. **Chinna samba / Sada samba**

This variety is also being cultivated in Keevalur taluk of Nagapattinam district. Samba season is very well suitable season for sowing (Aadi-Margazhi i.e., July 15-Jan 14).Total duration is 140 days in which 30 days are nursery period. Average height of the crop is 4½ - 5 feet and susceptible for lodging. Grain is light green in colour during panicle initiation and milky stage, whereas in maturity stage it becomes pale yellow colour.

Based on experience
• The panicle may possess more number of grains and this concept so called sada samba.

13. **Mara nel**

It is being cultivated in Thenkanikottai taluk of Krishnagiri district. The duration of the crop is 120-125 days. Late samba and Navarai (Summer) season is best season for cultivating this variety. Transplanting method is used for cultivation.The rice is red in colour and coarse in texture. It is ideal for preparing tiffin like idly, dosai. The special feature of the grains of this variety has a tough seed coat. So the grains do not germinate even if there is heavy rainfall at the time of harvest. It has another feature i.e resistant to storage pest because of the hard seed coat.

Based on experience
• This variety may be including for the further improvement (breeding programme) as the grains to not germinate on seeding which is highly essential in kuruvai season. Threshability of the grain is to be studied because hard seed coat may correlate with hard threshing which farmer do not prefer.

14. **Kallimadayaan**

It is being cultivated in Iyerpalayam village of Perambalur district. The duration of the crop is 6 months. Plant is tall in nature with lodging habit. It has good tillering ability with a plant height ranging from 110-120cm.The panicle length is 30 cm. It yields bold grains and has the potential to give 2 tonnes per acre under suitable conditions prevailing during crop management. The colour of the grain is white. It has good cooking quality. It is resistant to stem borer, ear head bug and also blight disease. Farmer's opinion about this variety is the starchy watering nature influenced the grain, to keep them fresh and energetic. The seed rate is 35 kg/ ac .The nursery period is 35-60 days. Now a day, only few farmers are cultivating this rice variety in Iyerpalayam because most of the people don't prefer bold rice which reduces the market price of this variety. Straw obtained from this rice variety is used for thatching roof.

Based on scientific reason

• The variety is a long duration variety which may not suit todays cultivation.

15. **Karuppukavuni**

It is being cultivated in Anumandhakudi village in Sivagangai district. The duration of the crop is 150-170 days. The best suitable season for growing this variety is September – January. It is suitable to direct sowing. The length of individual grain is 1cm. It needs only organic manures such as green and green leaf manure. It is quick growing crop. It does not require chemical pesticides.

Salient Features:

• Non-lodging habit.

• Profuse tillering.

• The colour of the grain is black and it is only used for sweet making along with coconut milk and it is not suitable for meals preparation.

• As it has profuse tillering habit, the paddy straw yield is 150 per cent higher than any other paddy variety.

• Tillers are moist in nature and it fetches good weight. Grain yield is 1800 kg per acre and straw yield is 700 kg per acre.

• Price of grain is Rs.7/- per kg. But the rate for ordinary variety is Rs.3.50/- only.

It requires lesser quantity of water comparatively and suitable for black cotton soil and red soil in dry lands.

Based on experience

• Kavuni is being consumed by chettinadu villages. Kavuni rice is mixed with jaggery or palm sugar and used for as a sweet consumed during marriages and other functions.

16.	**Chandikaar** It is being cultivated in Anumandhakudi village in Sivagangai district. The best season for growing this variety is September – January. It yields about 800 kg per acre and straw yield also 1800 kg per acre. It is suitable for edible purpose in the villages and organic in nature. It is quick growing in nature. The people often cut out the stem after three months of sowing to prevent lodging. It is suitable for black cotton soil, red soil and saline soil. It is suitable for direct sowing and transplanting methods. Generally, it does need not require pesticides and any nitrogenous fertilizers and it is resistant to paddy blight disease. The length of individual plant is upto 6 feet. The variety has dense crop canopy, the field often filled with weed and it needs frequent weeding. The weed growth suppressed the plant population and it will leads to reduced grain yield.	Based on experience Application of higher dose of fertilizers will lead to lodging. Farmers cultivate this variety because of straw yield for fodder.
17.	**Varappukkodanchan** It is being cultivated in Chatrakudi village in Ramnad district. This variety is suitable for rain fed condition. The outer skin is black in colour and the inner rice is red in colour. It is the drought resistant and tall variety. It is good in taste and cooked rice is kept upto 2- 3days. The rice cruel (kanji) of this variety is very palatable. The seed rate is 40-50 kg per acre for direct sowing .Farmers believe it gives more energy compared to other fine rice varieties for working people and the straw yield is very higher (3 tonnes /acre: 1 ton-RS.1000/-). Generally it is resistant to pests. Compared to all varieties, the cost of cultivation is less than Rs.1500/-. The duration of the variety is three months .It is sown normally in the middle of Puratasi (September) and harvested in the month of Tai (January).The yield is 1000kg-1200 kg/acre and it fetches higher price in Kerala state(Rs.7.50/kg)	Based on scientific reason • The variety grows very tall. The plant height is about 130-140 cm which is equivalent to the height of the bunds. Normally farmers prepare higher bunds with the aim of storing rain water which is received during the North East monsoon.
18.	**Kuliparichan** It is being cultivated in Pallabachery village in Ramnad district. This variety is suitable for coastal sandy soils. It is drought resistant and tall variety. The outer skin is slightly dark. The inner grain is bold and round. It is good in taste and cooked rice is kept upto three days. The rice kanji of this variety is good in taste than other fine rice varieties. The seed rate is 40-50 kg per acre	Based on experience • It is the drought tolerant variety. The other names of the variety are Kuliadichan or Kulipidichan.

for direct sowing. Farmers believe, it gives more energy than other fine rice for working people and they gave higher straw yield (3 tones / acre: 1 ton-Rs.1000/-). This variety is resistant to pests, generally except in very less water periods. It is affected by leaf folder in stress conditions. The cost of cultivation is less than Rs.1500.They applying only urea to the crop and one weeding is performed during season. The duration of the variety is three months .It is sown normally in the middle of Puratasi (September) and harvested in Tai month (January).The yield is 1000kg/acre and it fetches higher price in Kerala state (Rs.7.50/kg) than Tamil Nadu (Rs.4.50/kg) because they like red rice variety more than fine rice varieties.

19. **Sivappu Chitraikar**

"Sivappu chitraikar" is the traditional red rice variety grown in Paramakudi tracts. The outer skin is slightly black in colour and the inner skin is red in colour. It is the drought resistant and tall variety. The duration of the crop is 110 days and it is normally sown in the month of "Aavani" and harvested in the month of "Thai". The seed rate is 40-50 kg per acre for direct sowing. The yield is 1500-1800 kg per acre in normal monsoon period and in the failure of monsoon at least 600-900 kg per acre grains are harvested. So the farmers mainly rely on these traditional varieties rather than other varieties. The price of each bag is determined based on demand in market (Rs.300 to Rs.450). In the very lesser rain fed period there is small blight like formation found in the crop and ultimately lead to "Soorainookadu" disease. One or two weeding is performed in the field. "Rice kanji" of sivappu chitraikar is very tastier than other red rice varieties. It is good for preparation of dosai and idli. The seeds of this variety are stored by farmers from the previous harvest.

Based on experience

• The base of the tillers are red or purple in colour as the name of the variety denotes.

20. **Murungaikar**

Murungaikar is the traditional local red rice variety grown in Ramnad areas and it is suitable for even very low rainfall periods. The outer skin is slightly black in colour and the inner rice is red in colour. It is the drought resistant and tall variety. The rice kanji of this variety is very good taste in nature. The seed rate is 40 kg for

Based on scientific reason

• The variety may not be suited as it possesses loose panicle and lesser number of grains.

direct sowing. The yield of murungaikar is 1000-1300 kg per acre in normal rain fed conditions (15-20 bags 1 bag 65 kg). Even when failure of monsoon, the crop can withstand drought conditions. The seeds of this variety are stored by the farmer in the previous year. The murungaikar variety is not affected by pests and diseases and no foliar sprays of pesticide have been applied so far. As the crop grows tall, it will lodge at the time of harvest. Comparing to "Nootripathu" variety, Murungaikar has slightly lesser number of grains, but the straw yield is same (1ton per acre) as that of varappukodanchan and Nootripathu. The price of rice is fluctuating one, based on demand (Rs.350 to Rs.500). The crop is sown normally in the month of Aavani (August) and harvested in Markazhi (January). Goat manure and Urea alone are applied generally to the Murungaikar field. Only two weedings are performed in the cropping cycle but in larger land areas only one weeding is enough.

21. **Nootripathu**

Nootripathu is the traditional red rice variety grown in Ramnad district. The outer skin is white in colour and the inner coat is red in colour. It is the drought resistant and tall variety. The rice kanji of this variety is good in taste. It is sown in "Aavani" and harvested in "Thai". The seed rate is 40-50 kg per acre for direct sowing. One or two weedings are carried out during cropping period. The farmers observe "Senthalai Nookadu" disease in the paddy field in waterless condition. The total cost for cultivation of the paddy crop is Rs.2000 in rain fed condition. The grain yields about 1500-1800 kg per acre and straw yield is one ton per acre. Both black soil and red soil is found in nearby Paramakudi will lead to lesser growth of paddy, but in red soil more amount of rainfall lead to better rice growth. In some of the fields nearer to vaigai river basin, clay soil is found. The rain fed red rice varieties grow very well in clay soils rather than black and red soils. All the sivappu chitraikar, kuruvaikalnchiam and nootrypathu varieties fetch good price (Rs.6.50 per kg.) in market. The seeds of all varieties are stored by farmers from the previous harvest.

Based on scientific reason

• Any variety which matures in 110 days is called Nootripathu now so in original nootripathu variety, the purity is gone and it is cultivated and carried over to generations as populations.

22. **Ariyaan**

There are 4 types of Ariyaan rice varieties grown in sandy soils in Regunathapuram areas at Ramnad district. Among "Vellai Ariyaan", Karuppu Ariyaan, Sivappu Ariyaan, and Vazhai Ariyaan. All Ariyaan varieties have horn like structure. They grow very tall upto 5.5 to 6.5 feet. The duration of all the Ariyaan varieties is 120 days. The seed rate is 40-50 kg per acre for direct sowing. They are sown in the month of Ayppasi and harvested in the month of Thai. This variety is suitable for coastal sandy soils. It is the drought resistant and tall variety. The outer skin is dark. The inner grain is bold and round. The rice kanji of the variety is good in taste. It prevents hunger for several hours. It is resistant to pests and diseases, generally except leaf folder in very water less periods. It needs at least one rainfall shower in first 3 months for higher yield. The harvesting is done in three stages viz. top portion cut at first, middle portion cut at thereafter and bottom portion is cut at last. It gives higher yield of 2000-2100 kg per acre. (1 bag = 60 kg.) The total cost of cultivation for an acre is around Rs.1500 – 2000. All the Ariyaan varieties are maximum straw yielder upto 1000 kg/ acre. Among four Ariyaan varieties, karuppu ariyaan gave highest straw yield of 1-1.5 tones per acre. It is affected by leaf folder and stem borer. The seeds are stored by farmer from previous harvest.

Based on scientific reason

• All ariyaan varieties are long duration variety sown in areas near lakes and ponds and sown before rainfall and harvested after the water is received during January month.

23. **Poongar**

It is being cultivated in Raghunathapuram village at Ramnad district. If the season fails in "Puratasi" (September) another traditional early duration red rice variety of Poongkar is sown in the month of "Karthigai" (October) and it is harvested in Thai (January) but the yield is lower than kuliparichan and varappukodanchan. This variety is suitable for coastal sandy soils. It is the drought resistant and tall variety. The seed rate is 40-50 kg per acre for direct sowing. The outer skin is slightly dark. The inner grain is bold and round. The rice kanji of this variety is very good taste than other fine varieties. It prevents hunger for one hour. The crop can be able to grow in very less rainfall condition and it is drought tolerant than kuliparichan and varppukodanchan. This variety is not attacked by pests generally except in very less water periods. The cost of

Based on experience

• The variety actually matures in 70-80 days., the most early duration. It is extinct in most places, panicle will be very loose.

cultivation is less than Rs.1500. They applying only urea to the crop and one weeding are performed during season. It fetches higher price than other red rice varieties (RS.10/kg).The yield is around 800 kg per ha. It is a very good straw yielder (1 tone per acre).

24. **Kuruvai kalanchium**

"Kuruivaikalanchium" is the red rice variety grown generally in Ramnad district. The outer skin is slightly black in colour and the inner grain is red in colour. It is the drought resistant and tall variety. The rice kanji of this variety is good in taste. The duration of the variety is 110 days and it is sown normally in the month of Aavani. The seed rate is 40-50 kg per acre for direct sowing. One or two weedings are carried out during cropping season. The crop is harvested in the month of Thai. The paddy straw yield from an acre is one tonnes. There is no need of rainfall after milking stage. The total cost of cultivation for growing this variety is Rs.2000 in rain fed condition. The yield is 1500-2000 kg per acre in normal monsoon period. Rice kanji of Kuruvaikalanchiam is different than "sivappu chitraikar". The cooked rice is very good in taste even after two days. The seeds are stored by farmers from the previous harvest.

Based on scientific reason
• Most drought tolerant variety.

25. **Norungan**

It is one of the red rice varieties grown in Ramnad district with duration of 110 days. It is sown in rain fed condition in sandy soils. The seed rate is 40-50 kg per acre for direct sowing. The outer skin is dark. The inner grain is bold and round. The rice kanji of this variety is good in taste. It prevents hunger for several hours. It is sown in the month of Ayppasi (October) as rain fed crop and it is harvested in the month of Thai (January). The grains are bold type than other Ariyaan varieties. The crop yields about 2100 kg per acre. The straw yield from one acre is 1000 kg and it is 10 days shorter duration red rice variety than "varappu kodanchan". It is affected by leaf folder and stem borer, if there is no sufficient rain water. The crop is not affected by pest and diseases normally. The seeds are stored by farmer from previous harvest. After the end of the cropping season, gingelly or green gram is sown.

Based on scientific reason
• Drought tolerant variety. The straw is rough with hairy like structures which makes it suited to drought conditions may be recommended for breeding programme.

26.	**Kalurundaikaar** Traditional red rice variety growing in Paramakudi taluk of Ramnad district. The rice grain weight of kallurandaikaar is slightly more than kuliparichaan/ varappukodainchan varieties. It is the drought resistant and tall variety. The outer skin is slightly dark. The inner grain is bold and round. The rice kanji of this variety is good in taste. The seed rate of Kalurundaikar variety of an acre is 45 kg for direct sowing. This variety prevents hunger for more time than other fine rice. The yield is 1500-1800 kg per acre. Paddy straw yield is approximately three tons per acre. In some field, rice and kudiraivali (in border areas) system is followed. Kudiraivali yield is 600 kg per acre and it fetches higher price in market. (Rs.6 per kg.). The rice variety is not generally affected by pest and diseases. The seeds are stored by farmers themselves from the previous harvest. In some fields, 3 types of sorghum are grown as border crop. They are red, white and black.	Based on scientific reason • It has mostly drought tolerant nature.
27.	**Thanga Samba** It is being cultivated in Kanchipuram district. Late samba and samba period is the best season for cultivating this variety. The cultivation method is transplantation method with the duration of 160-165 days out of which 40 days is nursery period. The grain is golden colour and fine in texture. The grain yields about 1500 kg per acre and the straw yield is around 60 bundles (1800 kg). **Special features** • The ear head of this variety is very long. • This variety of rice is extremely fine and long it is used for the preparation of special dishes like pula.	Based on experience • By experience this variety is traditionally cultivated. These type cultivars are low nitrogen requiring used for special preparation.
28.	**Neelan samba** It is being cultivated in Sukkan kollai village at Kanchipuram district. Samba season is best suitable season for cultivating this variety with duration of 175-180 days. The rice has red in colour and coarse in texture. It yields about 1500 kg /ac and the straw yield will be 1800 kg. **Special features** • Highly suitable for water logging condition. • Resistant to brown plant hopper and ear head bug • Suitable for lactating women.	Based on experience • The name is related to season. This has the medicinal purpose, yield levels are low.

29.	**Vadan Samba** It is also being cultivated in Sukkan kollai village at Kanchipuram district. Samba season is best suitable season for cultivating this variety with duration of 155-160 days. Direct sown and transplanting system is used for growing this variety. The rice has dull white in colour and coarse in texture. It yields about 1125 kg/ac and the straw yield will be 1200 kg. **Special features** • Drought resistant • Resistant to pest and disease Since the grains shatter, it is important to harvest them when they are still green.	Based on experience • Kanchipuram district having tank irrigation, this local traits are preferred for direct seeding. This type of resistant character can be utilized for plant breeding purpose.
30.	**Kaliyan samba** It is also being cultivated in Sukkan kollai village at Kanchipuram district. Samba season is best suitable season for cultivating this variety with duration of 155-160 days. Out of which, 40-45 days will be nursery period. Transplanting system is used for growing this variety. The seed rate is 35 kg per acre. The rice has red in colour. It yields about 1200 kg /ac and the straw yield will be 1600 kg/ac. **Special features** • It grows very tall upto the height of 5.5 feet. • It can be grown in water logging and low lying area. • Highly resistant to pest and disease.	Based on experience • Because of the long duration, suitable for aged transplanting. Straw is valuable one because of its height, straw can be used for thatching purpose.
31.	**Samba Mozsanam** It is also being cultivated in Sukkan kollai village at Kanchipuram district. Samba season is best suitable season for cultivating this variety with duration of 160-165 days. Out of which, 40-45 days will be nursery period. Dry sown and transplanting system is used for growing this variety. The rice has red in colour and coarse in texture. It yields about 1800 kg /ac and the straw yield will be 65 bundles (1950 kg/ac). **Special features** This variety is also called Puzudikal, Erinel and Maduvu muzangi. It is suitable for growing in the vicinity of lakes. It is said that people traveled by boats and harvested the 'Samba Mosanam' in the lakes. The variety is good for preparing Aval (Flattened rice), idly and dosa.	Based on experience • Suitable for water logged condition. Choice types are suitable for special preparation.

32.	**Kadikazhuthan** It is also being cultivated in Sukkan kollai village at Kanchipuram district. Samba season is best suitable season for cultivating this variety with duration of 165 days. Out of which, 40-45 days will be nursery period. Transplanting system is used for growing this variety. The rice has white in colour and having the fine quality. It yields about 1350 kg /ac and the straw yield will be 40 bundles (1200 kg/ac). **Special features** • It is also known as 'Garudan samba'. The appearance of the grains is like the neck of the Brahmin kite (Garudan) i.e. it has a white ring and so called this name. • It is suitable for making a specialty dish called 'Puttu' of South India. It grows well in areas which have good drainage facilities.	Based on experience • Local trait used for special dishes.
33.	**Pisini** It is also being cultivated in Sukkan kollai village of Chengalpet taluk at Kanchipuram district. Samba season is best season for cultivating this variety with duration of 120 days. In the cropping period, 40-45 days will be nursery period. The rice has red in colour and having the coarse quality. It yields about 1500 kg /ac and the straw yield will be 50 bundles (1500 kg/ac). **Special features** • Highly suitable for making dosa, porridge and aval. • Highly resistant to drought. Since the ear heads are found with awn it is important to take extra care while removing the grains.	Based on experience • Locally adopted. Awn nature shows its drought tolerance ability.
34.	**Thuyamalli** It is also being cultivated in Sukkan kollai village at Kanchipuram district. Samba season is best suitable season for sowing this variety with duration of 135-140 days. Out of which, 30-35 days will be nursery period. Direct sown and transplanting system is used for growing this variety. The rice has white in colour and having the fine quality. It yields about 1125 kg /ac and the straw yield will be 35 bundles (1050 kg/ac).	Based on experience • The name of the variety based on its character fine quality. Low nitrogen requiring cultivar.

Special features

• During the flowering stage the ear heads look like flowers.

• Highly resistant to pest and disease. In Tamil, 'thuyamalli' literally means pure jasmine. Since the rice of this variety is white in colour like that jasmine, so that it is known by this name.

35. **Kalar palai**

It is also being cultivated in Sukkan kollai village of Chengalpet taluk at Kanchipuram district. Navarai season is best season for cultivating this variety with duration of 120 days. Transplanting system is used for growing this variety. The rice has brown in colour and having the coarse quality. It yields about 1200-1300 kg / ac and the straw yield will be 1300 kg/ac.

Special features

• Resistant to pest and disease.

• Suitable for kalar nilam i.e. saline soil. It is the reason behind this name.

Based on experience

• The name indicates the suitability of the variety to saline soil. Because of the shorter duration used for navarai season.

36. **Kudaivazhai**

It is also being cultivated in Sukkan kollai village of Chengalpet taluk at Kanchipuram district. Late Samba season is best season for cultivating this variety with duration of 120-125 days. In the cropping period, 25-28 days will be nursery period. Transplanting system is used for growing this variety. The rice has red in colour and having the coarse quality. It yields about 1400 kg / ac and the straw yield will be 50 bundles (1500 kg/ac).

Special features

· The ear heads of this variety appear like open umbrellas and hence the name 'kudaivazhai'.This variety is highly suitable for making dishes such as idly and Dosa.

· It grows upto 5 feet,the stem is 2.5 cm thick and the straw is used as roofing material.

· This variety is highly suitable for areas like water logging.

This is a short duration variety; hence it can even be cultivated in thrice a year

Based on experience

• The short duration helps to plant late samba season. Lengthy straw benefits for roofing.

37.	**Kumvalai** It is also being cultivated in Sukkan kollai village of Chengalpet taluk at Kanchipuram district. Samba season is best season for cultivating this variety with duration of 128-130 days. Direct sown and transplanting system is used for growing this variety. The rice has brown in colour and having the coarse quality. It yields about 1350 kg /ac and the straw yield will be 1800 kg/ac. **Special features** • This variety grown upto the height of 4.5 - 5 feet. • Straw is particularly preferred as a roofing material It is suitable for sandy soil and also water logging area.	Based on experience • Local trait. Tall growing helps for water logged areas.
38.	**Muttakar** It is also being cultivated in Edaiyur village of Thirukazhukundram taluk at Kanchipuram district. Navarai season is best season for cultivating this variety with duration of 120 days in which 30 days will be the nursery period. The seed rate is 45 kg per acre. Wet sown and transplanting system is used for growing this variety. The plant grows upto a height of 4.5 feet and it has lodging habit. The rice has red in colour and having the coarse quality. It yields about 1350 kg /ac and the straw yield will be 40 bundles i.e. 1200 kg/ac.	Based on experience • Short duration cultivar suitable for wet sown in navarai.
39.	**Sempalai** It is also being cultivated in Edaiyur village of Thirukazhukundram taluk at Kanchipuram district. Samba season is best season for cultivating this variety with duration of 100-105 days out of which 21-25 days are nursery period. Direct sown and transplanting system is used for growing this variety. The rice has dullish white in colour and having the coarse quality. It yields about 1125 kg /ac and the straw yield will be 45 bundles (1350 kg/ac). **Special features** • It is highly suitable for making 'Pori' (puffed rice)	Based on experience • Shorter duration, direct seeded condition. Special preparation like puffed rice. Short duration helps for direct seeding.
40.	**Kappakar** It is also being cultivated in Thakandarapuram village of Vandavasi taluk at Thiruvannamalai district. Samba season is best season for cultivating this variety with duration of 150-160 days in which 40 days will be the nursery period. The seed rate is 50 kg per acre. Dry sown and transplanting system is used for growing this	Based on experience • Local cultivar. Used for special preparation.

No.	Description	Basis
	variety. The rice has brown in colour. It yields about 1350 kg /ac and the straw yield will be 60 bundles i.e. 1800 kg/ac. **Special features** • This variety is highly suitable for making idly and dosa. • It gives high yield in clayey soil. • It is suitable for growing in water logging areas as well as drought prone areas. • It is suitable for preparing Aval.	
41.	**Perunkar** It is also being cultivated in Thakandarapuram village of Vandavasi taluk at Thiruvannamalai district. The duration of the crop is 120 days. The seed rate for direct sown crop is 35 kg per acre and 40 kg per acre is used for transplanting system. It yields about 1.4 tonnes /ac and the straw yield will be 1500 kg/ac. The plant grows to a height of 4.5 feet. **Special feature** • Tolerant to stem borer and ear head bug. Suitable for preparing idly and dosa	Based on experience • Local trait used for special dishes.
42.	**Karikattai Ragi** It is Suitable for hilly Region. It is traditional ragi variety which is being grown in Thamarakarai village of Erode district. As per their perception, it is hardy in nature. It is suitable to prepare ethnic food called Kali, Muruku, Koozh etc. It is harvested along with straw and dried for about 15 days. Threshing can be done manually or by using cattle. Whereas the other traditional varieties only the ear heads are reaped. The duration of the crop is three months. The season for growing this variety is in the month of Aadi. This variety is existence over 125 years.	Based on experience • Traditional cultivar, flour has the preferably making of cool with good keeping quality.
43.	**Kattucumbu- pearl millet** This traditional cumbu is cultivated in many areas of Trichy District, particularly Kampalathupatti village of Manapparai taluk, says that this traditional variety exists for more than 100 years. The apt season for cultivating this cumbu is Aadi Pattam (July-August). It is well suited for garden land and dry land eco system. Since it is a rain fed crop, direct sowing is the suitable method of sowing. The total duration of this crop is six	Based on experience • The name indicates the suitability of the variety for rainfed condition. Long duration also helps for drought tolerance.

months. So when it is sown during Aadi pattam and harvesting can be done in the month of Margazhi (January).Manual harvesting is done by using sickle, where the ear head alone is cut from the stem and collected. These ear heads are sun dried for 2-3 days. It is then threshed manually by striking it with sticks or even bullocks are allowed to stamp on the ear head and thus the grains are separated. The separated grains are winnowed and cleaned. Grains are elliptical shape with pointed nose at the point of attachment with the rachis. They are faded green in colour with milky taste. Normally, it yields about 600-700 kg/ac.
After cleaning process, it is stored in traditional storage structure called 'Thaniya kuthir'. When utilized for consumption, it is pounded well into course powdery form and consumed. It can be fed to both human and cattles. It contains minerals and other nutrients. Normally, an ethnic food called "cumbu koozh" is prepared by using the coarse powder made out of cumbu grains. This ethnic food controls fever and blood pressure

44. **Thattai varagu**
It is also being cultivated in Veppanthatai taluk at Perambalur district. The duration of the crop is 5 ½ months. The seeds are sown directly in the field. For increasing the yield, one traditional practice is followed by the farmer since 20 years i.e. 45 days after sowing, the grazing is done in the field by allowing the sheep for one day. This practice is called "kedai aadithal" and technically it is called sheep penning. For one acre, 100 sheep's are used. Sheep's are grazed the entire field and leave only the root of the crop. After that, the side tillers are developed well. Side tillers give more number of ear heads. It gives yield upto 10-12 qtls / ac.

Based on experience
• Tillering ability shows good yield. Long duration is beneficial for drought tolerance ability also.

45. **Puvadan Keazhvaragu –Ragi**
It is also being cultivated in Veppanthatai taluk at Perambalur district. The durationof the crop is three months. The seed rate is 3.5 kg/ac. The apt season for cultivating this cumbu is Karthigai. Direct sowing and Transplanting system is used for growing this variety. It yields about seven qtl/ac. This crop is sown at the time of "Pogi" (Puvadan Noonbu) i.e 30 th of Margahazi. This is the reason behind the name Puvadan Keazhvaragu.

Based on experience
• This crop utilizes the residual soil moisture. And hence more drought tolerance ability also.

46. **Sorghum (Kona Cholam)**
This sorghum is cultivated as an irrigated system. All type of soil is suitable for this sorghum cultivation. After first ploughing, Farm Yard Manure is applied as a basal. After that 2nd ploughing is done, during the 3rd ploughing seeds are sown. Seed rate is 3kg/acre. Season for cultivating kona cholam is Jan-Feb. Duration of this crop is 90 days. Irrigation is given once in 10 days interval .No need to done weeding. Grains yield in one acre of land is 2500kg. One kg of seed Rs. 12-15 for cultivating one acre of land Rs 3000/-is required.

Based on experience
• The local cultivar probably the sowing time during the pongal festival.

47. **Chittan Samai**
Chittan samai is cultivated traditionally in Alangayam, Javathu malai village at vellore district. Samai is cultivated during the month of June-August. Land is well ploughed and seeds are sown through hand sowing. In one acre of land, 12kg seeds are required. The crop duration is 90 days. One weeding is done in DAS 30th day. For weeding, 10 labourers are required. One acre of land wages for weeding is Rs 30/ for women labour and Rs70/day for men labour. Crops are ready to harvest after 90 days. Plants are harvested by using sickle. Grains are separated by threshing. The straw yield from one acre of land is 800kg. The market rate of samai grain is Rs 10-12/kg.

Special feature
• Very good taste and flavor compared to other varieties.
• Storage capacity is more than 4-5 years compared to other recommended varieties.
It is used for biscuit preparation.

Based on experience
• The crop is cultivated during the south west monsoon season. Short duration with high market rate for the grains benefits the farmer.

48. **Sen Cholam (Wild Sorghum)**
This is one of the traditional varieties in Salem district. The duration of the crop is 120 days and suitable for all the season. The color of the grain is dark reddish, ear heads are medium sized and oval in shape. The crop is mainly cultivated for cattle feed. In olden days, sorghum grains are used as a food material, but now it is not preferred. Farmers are eager to cultivate this variety while compared to other varieties. Green fodder is mainly used for milch animals to induce the milk production. Twenty kg of green fodder is fed to one animal /day (At milking time especially for cow). Dry fodders are preserved and given for all the animals. After maturation, the ear heads are harvested by using

Based on experience
• Fodder type, palatability is high. Both green and dry fodder is used.

knife and it is sun dried upto 2-3 days. The grains are separated manually by using sticks. This seeds can be used for next sowing after three months.

49. **Kakacholam**

Traditional cholam varieties are grown in Kothali village at Thoothukudi district. It is a black seeded variety. Duration of the crop is 6 months. They are cultivating this crop as rain fed and irrigated system. The seed rate is 30 kg /acre. It may be sown in the month of September-October and to be harvested in the month of January - February. The seeds are kept for next sowing by farmers themselves. This variety is suitable for livestock as fodder. It can be used as inter crop along with any other crops.

Based on experience

• Long duration. Rainfed cultivation during north east monsoon season.

50. **Black Gingelly**

Traditional Gingelly crop is grown in Ramnad and Paramakudi taluks. It is black seeded variety than normal white seeded type. It has more oil content than the other varieties and the duration of the crop is longer (i.e.90 days) than the other varieties (75 days). They are cultivated as rain fed crop and the yield is 400 kg per acre and the total cost of cultivation for rain fed gingelly is Rs.500/-. The traditional gingelly has more branches and vigorous growth than any other gingelly varieties. The seeds are kept for next sowing by farmers themselves. There is no permanent season for gingelly cultivation. It may be sown in the month of Chitrai, vaikasi or Adi (April – July) based on need and demand of farmer. It is not suitable for cultivating in rainy season because it will cause certain problems.

Based on experience

• Dark colour is the indication of high oil content and drought tolerant ability.

51. **Red Gingelly**

It is being cultivated in Raghunathapuram village at Ramnad district. Gingelly is sown in month of December and harvested in March. The gingelly is red in colour and it has good oil content. The rain fed traditional gingelly yield is 320 kg/acre (totally 4 bags: 1 bag – 80 Kg per acre). It fetches good price in market (Rs.18 per kg.).The seed rate is 1.5 kg/acre and the seeds are colleted from the previous year grains. The oil content is more than other varieties. The duration (90 days) is more than other varieties (70 days). It has more branches than other varieties. The seeds are collected from the harvested grains and it is kept upto one year without addition of any chemicals.

Based on experience

• Traditional cultivar.

52.	**Kadalaikai- Traditional Groundnut** It is being cultivated in Velankadu village at Vellore district. Land is ploughed 3-4 times before sowing by using country plough or cultivator. Five tones of farm yard manure are applied as a basal. The suitable season for growing this groundnut is June-August (it is a rain fed crop). The duration of the crop is 90 days. 50 kg of seeds are required for one acre of land. Seed are sown behind the country plough. 30th days after planting, 1st weeding is done .60th day second weeding is done and for each weeding 20 women's labourers required. For harvesting groundnut, farmers are given only ground nut as a wages for each 20 padi, 4 padi has given. 90th days after planting, ground nut are ready to harvesting. For one acre of land 20-25 women's labour is required / day. The yield is about. 5000 kg/acre. One kg of ground nut is Rs10/-. The expenditure for cultivating groundnut in one acre of land is Rs 5000/-.	Based on experience • This traditional variety is suitable for table purpose.
53.	**Kulithakali** It is being cultivated in Kanchipuram district. The season for growing this variety is throughout the year and also the bearing season is also throughout the year. The seed rate is 100g /ac and 10 plant / cent .The size is also small and more tartaric acid content. It is resistant to pest and disease. It is suitable for dry land cultivation. Root slip is enough for re-emergence (after drying of above ground portion i.e. stem and leaves). After two months, the plant will start bearing.	Based on scientific reason • The particular area of soil contains high amount of calcium. This nutrition is leads to develop tough skin of fruit. By this way back ground the local culture is resistance and quality of fruit.
54.	**Vazhuthunangai *(Solanum melongena)* - Egg plant/ Brinjal** It is another important tropical vegetable believed to be a native of India. It is being cultivated in Vettuvenni village of Marthandum Taluk at Kanyakumari district. Local preference varies considerably from region to region and even district to district. The suitable season for growing this crop is June to October. The duration of the crop is 90-110 days. The variety is well adopted to well drain deep soil for its growth. It is mainly used for table purpose. The variety is preferred more by kanyakumari people as it is a traditional part of their daily food. Little leaf disease is a major problem caused by Mycoplasma like organisms but the variety is resistance to this disease. First harvest starts from 55-60 days after transplanting. The fruits are lengthier than normal varieties and measures 12-15 cm height.	Based on scientific reason • Now this is selection made as imported selection and released as PPI-1 Brinjal from HRS Pechiparai.

The fruit is cylindrical in shape. Harvesting can be done by once in 5 days. Yield ranges between 20-25 tones/ha.

55. **Kannadi katthari- Wild Brinjal**

Kannadi kathari is traditionally cultivated in Devoor village at Salem district.The required seed rate is 200-300 grams/acre. Seeds are collected from well ripened, pest and diseased free fruits. Seeds are coated with ash and Dithane powder to protect the seeds from stored pests. It can be stored up to 2-3 years. Cultivation starts during the month of August and January and the duration of this crop is 6-7 months. This variety is mostly preferred in Erode market and has good market value while compared to other varieties. Cow dung, goat manure and poultry waste were used as manures. This variety possesses high pest and disease resistant, so that the farmers used to prefer to cultivate this variety. Plant height is 4-5 feet; high rainfall condition is not suitable for this crop because the root rot incidence is more in that time. After first harvest, the plants are left with 1 foot height from the ground level, so that other harvest can be done. The yield from this crop is upto 30-40 bags in 3 months. The market rat is Rs.15 per kg. So that, the farmer can fetches net profit of upto Rs. 30,000-50,000.During harvesting period, intercropping is done by using "kottai Pudalai' (Snake gourd) seeds are sown in brinjal field. The Snake gourd plant uses the wild brinjal plant as a support for its growth. The farmer generated the additional income of Rs.20, 000/- for growing kottai pudalai as inter crop.

Based on experience

• It is suitable for rationing as well as fruits used for making dry brinjal (vathal)

56. **Ramnad Gundu Chillies**

In Ramnad areas, the prevailing conditions are favoured to grow chillies well. The traditional Ramnad gundu variety is most preferably cultivated because of its yield is higher than the other samba type. When the amount of rainfall increases, the yield of chillies is drastically reduced. It is cultivated only in less amount of rainfall. The farmers get more profit in "Ramnad Gundu" chillies. The cost of cultivation of Chillies is Rs.10, 000 per acre and the profit is Rs.30, 000 per acre. It is sown in the month of Aavani / Karthigai and harvested in the month of Chittrai / Vaikasi. Cotton is used as intercrop in chillies field. Some of the farmers using coriander as intercrop in chillies field. The yield of chillies is 1000 kg per acre (50 bags: 1 bag = 20 kg).

Based on scientific reason

• It is specifically cultivated as direct seedling rainfed chillies and well adopted to saline and dry tracts of Ramnad. It is also cultivated same area at Sivagangai and Virudhunagar. This culture is high pungent and dark red in color

57.	**Bhendi /Okra** – ***Abelmoschus esculentus*** – **Red Bhendi ("Sivappu Vendai")** This is commonly known as lady's finger. It is being cultivated in Panavilai village of Vilavancode taluk at Kanyakumari district. It is a tropical direct sown vegetable crop with duration of 100-120 days. The optimum season is June-August. Sowing can be extended upto July but harvested before October. This is mainly to prevent the crop from North East Monsoon. Deep well drained loamy soils with undulating pattern are most suitable and also available in larger area. The variety is highly useful for culinary purpose. The plant grows to a height of 4 feet. Fruits are red colored and so called red bhendi. It is rich in carotene. 20th day onwards, the plants start yielding. After the emergence of fifth leaf, flowering occurs. Main stem is multiple branched, each branch gives fruits. The variety is resistant to yellow mosaic virus spread by vector white fly-*Bemisia tabaci*. Harvesting is done on alternate days, since the pods get matured. Quickly, the selection of pods with maximum size but still tender is taken. A total number of 15-20 harvests can be made. The yield varies from 12-15 tones/ha.	Based on scientific reason • Because of red colour, the vegetable marketability is very difficult. Our people always need green color bhendi. The red bhendi is imported from Australia. The same was released from TNAU as CO-1.
58.	**Covaikai (Coccina)** Th Crop can be cultivated in villurupuram district. The name of the variety is Padappai. The method of propagation is by stem cutting from pistil late plants. Crop spacing is 2x2m. The crop needs a special practice like erection of pandal or trellis and also trained the crop. Harvesting starts from 6 months after planting and continues upto through out the year. The yield varies from 6-8 tonnes/acre	Based on scientific reason • It is low very familiar in Kerala and Tamil Nadu. Because of it has medicinal value of anti-sugar and reducing the obesity (body sleemer). Now an improved clone is selected from Padapai of Kanchipuram. And now this Kovaikai is cultivated in pandal system at Theni district and produce are marketed at Kerala and Karnataka.
59.	**Gandhari chilli** – ***Capsicum frutescence*** **("Golkonda chilli")** Chilli is an important vegetable cum condiment called Red Pepper or hot pepper. It is being cultivated in Vettuvanni village of Marthandum taluk at kanyakumari district. The pungent principle is called 'capsaicin' which is mainly present in the central placenta of the fruit, as well as seed and to some extends	Based on scientific reason • It is perennial plant is occurred in the sub- tropical forest areas. It contains high pungency with maximum. It also has dark red colour. The leaves also used as greens.

in the ovary wall also. Generally, the capsaicin content varies from 0.2-0.4 %. Vitamin C is present in appreciable quantities. Spicy vegetables like chilli add taste and flavour to our food. The added advantage of these spicy vegetables is their protective nature against diseases. The duration of the crop is 100-120 days. Deep fertile loamy soils are best suited for cultivation. Gandhari chilli is more pungent than other varieties since their capsaicin content more than 0.5%. It helps in fat control, blood purification as a first aid by preventing blood clotting. The juice extracted from the ripe fruits is used for consumption. If excess of juice is consumed, it leads to ulcer. The fruits grow upwards, small in size. Yellowish or greenish colour at first stage, later on it turns to red on maturity. The fruits are also used in control of ear head bug along with garlic and Neem oil as paste. The paste is diluted in 10litre of water. From it 50ml/tank is taken and sprayed. The yield varies from 2.3-2.5 tones/ha. The variety has not been popular in other places of Tamil Nadu, because of its high pungency in nature. So that, the farmers don't prefer to grow this variety.

This special feature of the variety is used for breeding purpose.

60. **Manattakkali** ***Solanum nigrum***

It is being cultivated in Malavilai village at kanyakumari district. The common name is Black Night –Shade. The whole plant is used because of its medicinal properties. It cures asthma, general disability, heart diseases and hiccough.

Based on scientific reason

• Leaf is rich source of riboflavin, nicotinic acid and vitamin-c, b-carotene and citric acid. Fruits contain glucose, fructose, vitamin-c, and b-carotene. Immature green fruit contains four steroidal glycol alkaloids *viz.,* solamargin A and B, solanigrine and solasodine. Diosgenin is also present in leaves, stem and fruits.

61. **Blue chilli: (Neela milahai)**

It is being cultivated in Panavilai village of Vilavancode taluk at Kanyakumari district. The variety resembles all the characters of gandhari chilli except the fruit size, and colour. The flower appears blue along with blue leaf. It looks like an ornamental plant and attract observers. The fruits are round shaped and turn red on maturity. Fruits grow in bunches used for cooking, also eaten with tapioca.

Based on experience

• These genotypes also make it use for breeding purposes. It also perennial in nature.

62.	**Ther kothavarai-Local cluster bean variety** This traditional cluster bean variety is popularly cultivated in coastal districts, especially in Nagapattinam District. This variety exists beyond 75 years. June-July is the apt season for cultivating this variety. Ridges and furrows method is followed to sow the seeds, where 3 seeds are dibbled in one hill in the ridges. 30 days after sowing final harvesting can be done. It is harvested manually by plucking method. Time interval is set, once in 4 days harvesting can be done. On an average, it yields about 50kg/week and fetch about Rs5/kg of fresh bean. Normally in the tracks, 'Ther kothavarai' is cultivated organically by using farm yard manure, vermin compost and locally available backyard manures. Thus it finds own market value. Apart from that, the taste of this variety is also good which makes it highly preferable. It is capable of curing stomach ulcer and mouth ulcer. It yields about 1800 kg/ac.	Based on experience • This local culture is highly suitable for green vegetables. It contains high amount of fibre and vitamin B content.
63.	**Karuppu Avarai -Black Beans** This is suited only for rain fed or upland cultivation. It is grown in Valeppur village of Namakkal district. The whole bean as well as the pod alone can be separated and used as vegetable. The method of cultivation is more similar to that of the ordinary bean cultivation. Generally, this crop can be raised thrice a year during May-June (Vaigasi) September-October (Purattasi) and January-February. It comes to bearing on third month onwards and the total duration of the crop is four months with an average yields about 2-3 kg. The major difference between black bean and white bean is the taste and energy which is comparatively higher in the earlier than the later.	Based on experience • This genotype having high amount of calories and proteins.
64.	**Nattu Avarai** It is predominantly cultivated in Vallathodu village of Kanchipuram district. The suitable season for growing this variety is Aadi pattam (July). The crop will start bearing at 45 days after sowing. After six months, it starts bearing. Harvesting can be done in weekly interval. It yields about 3-4 kg/cent/week. If it is matured plant, ash is sprinkled whereas it is young one; cow dung is sprayed over the crop to control the pest and disease.	Based on scientific reason • This genotype is selected and released as variety from TNAU.

65. **Nattu Pusani- Pumpki**

The traditional pumpkin is sown in rainfed condition and grown in Ramnad district. The matured crop weighs about 20 kg. It is sown in the month of Aadi (July and August) and harvested in the month of Thai (January). Each pit is 2.5 feet depth. Thereafter thinning operations can be performed after growth of more seedlings. The seeds are kept by farmer themselves in previous year harvest

Based on experience

• It is very old pumpkin grown in this district. Fruits are very big in size. Storability is even six month periods. Highly tolerance to salinity, alkalinity and high temperature. Mostly cultivated in Pandiyur, Thirupullani blocks of Ramnad district.

66. **Curry leaf**

The curry leaf crop is grown as mixed crop in Thangachimadam areas (10cents in one acre) and 2000 trees can be grown. They are picked in every months and the yield is 3000 kg/ac (The rate is Rs.5/ per kg.) and they are transported to Paramakudi and Rameshwaram markets. It is suitable for coastal sandy soils. Goat manure is used for this crop. They are resistant to pest and diseases. Normally the soil is very fine coastal sandy in nature, the crop comes very well because of the availability of water at lower depth.

Based on experience

• It is green colored type. The multiplication is through own seeds of the grownup trees. It has good aroma and smell and high oil gland in the leaves. Well adapted to salinity, alkalinity and water logged conditions.

67. **Betel vine-Vethalai**

The traditional "betel vine" variety is grown along with "Velimurungai" which gives anchorage to betel vines. It is suitable crop for coastal sandy soils prevailing in Ramnad district. In the border areas, banana trees are grown and it gives higher crop intensity and also generates additional income. They are 150 banana trees surrounding around the border area of the 2 acres land and most of them are border crops in betel vine field. The field is irrigated by three wells and the water is available at 10 feet depth.

Based on experience

• Banana is planted to providing temporary to the betel vine. Because the coastal parts getting high temperature is due to sandy soil.

68. **Jaffna Moringa**

The traditional 40 years old Jaffna Moringa has been grown in Thangachimadam areas and the cutting was brought from Jaffna, Sri lanka during the cargo service between India and Sri lanka. It comes very well in coastal sandy soils. The 40 years old tree gives 1500 moringa sticks daily between March-June. They apply only organic manures such as goat manure to the tree. It is not usually affected by pests and diseases. The

Based on experience

• This perennial moringa is very popular and famous of every house at Ramanathapuram district. Especially in coastal belts. It is highly tolerant under saline condition, high temperature and water

	length of individual fruit is upto 12 feet and it fetches Rs.3-5/- based on demand. The rainy season of October, November and December is the off-season for moringa tree.	logging condition. The leaves also used for cattle feed. Leafs, pods are making use of water purifier.
69.	**Nattu Maravalli kizhangu (Vellai Bontha)** Vellai bontha is traditionally cultivated in Salem district. The duration of this crop is 8 months (August- March). 4500- 5000 setts are required for planting in one acre. Weeding should be done once in a month, so that yield could be increased. One plant can yield about 4-5 kg of tubers. In one acre we can get 10 tonnes of tubers. The market price for 1 kg of tapioca is Rs.5/-. Intercropping is possible in the period of first three months. Black gram and groundnut are cultivated as inter crops. Coconut tree is grown as a border crop. White fly incidence is more in this variety at the end of the harvest. **Special feature** • It is mainly used for culinary purpose. The colour of the skin is pink and some times white in colour and nattu variety tubers have more flour content.	Based on experience • This vellai bontha is used as pedigree of mulluvadi tapioca variety. The local is drought tolerant one.
70.	**Mottaipayaru** It is grown in Vadavathure village of Namakkal district and the seeds are sown in December. The yield is 6 bags per acre (one bag is 50 kg). The seed rate is 1 kg/ acre. As the soil is sandy in nature, water will percolate into deeper layers. So only rain fed sowing is possible in this crop. It has twining quality, so it is not used as intercrop with any other crops. The seeds are collected from the harvested grains and it is kept upto one year without addition of any chemicals. The crop is mainly used as cattle feed	Based on experience • The lab lab is containing more amounts of carbohydrates. And is better feed for cattle population. These plants can tolerant high temperature. Once germinated, the plant gets to develop and give minimum yield under unfavorable climate condition.
71.	**Kariyilai poriyan** As the name indicates, the tubers are black in colour. The variety has duration of eight months and cultivated in the tracks of kanyakumari district. It is a good and energetic table food also used as poultry feed. It possesses less bitterness in taste than other varieties. The tuber is sun dried. After cutting into pieces and the pieces are dried in naturally. It can be kept stored for 3 months. In pest and diseases, this is highly resistant to Tobacco Mosaic Virus and brown leaf spot respectively.	Based on experience • It contains more amount HCN content and it leads to bitterness.

72. **Adukku muttan**

It is predominantly seen in Kanyakumari district with the cropping cycle of six months. Tubers are small with increased starch and possess good cooking quality. Processed tubers can be stored upto 3 months after dehydration (processed one). It is suitable for making food items like "kizhangu puttu"- steamed flour sweetened or spiced. After processing powder obtained from the tuber is given to children. For both the varieties described above, the best suited soil is light sandy loam. Planting can be taken up in May-June. The crop does not grow well in other districts of Tamil Nadu because the soil type, required seed material is not suitable and available at the place. So care should be taken in selection of varieties and their season. Tapioca is an important subsidiary food and used in different forms. Yield of the crop varies from 30-35 tonnes/ha.

Based on experience

• It is one of the local line is available in particularly kanyakumari district. It contains rich in carbohydrate content. This starch is utilized for children food supply more calories energy. Also making chips, starch, gelli and semiya.

73. **Nattu seeai kizhangu –Yams – *Dioscorea spp***

The tubers are underground stem rich in starch and are usually single with various size and shape. Tubers are rich source of sucrose. A deep rich friable well drained sandy loam soil is ideal for development of tubers and it is prevailing in Malavlai village of kanyakumari district. The proximal end of tubers are cut into pieces and dipped in cow dung slurry allowed to dry in shade. The cropping cycle is 8-10 months. The yield varies from 20-25 tonnes/ ha. Yellowing and drying of leaves is the indication for harvesting the tubers.

Based on experience

• It supply rich source of starch and high energy vegetables. It contains more amount of calcium oxalate in matured stages. This chemical is responsible for acrid zing nature of this crop. The tubers should be dried same days and the calcium oxalate content will be reduced.

74. **Seeni kizhangu – Local sweet potato variety**

It has been cultivating for more than 15 years. It is suitable for both rain fed and irrigated condition seen in Irudhukottai village of Krishnagiri district with duration of three months. Leaf cuttings are best suited method for propagating this crop. It is all season crop. Furrows are made in the field with two feet length and the leaf cuttings are planted in that furrow. For cultivating one acre, 10 cent area is used for transplanting purpose. This crop is not affected by any pest and disease. Based on the water availability, the kizhangu size will vary.

Based on experience

• This local sweet potato line is resistant against pest and disease especially tuber borer (*Cylas* sp). The tuber borer is worst problem in sweet potato cultivation. This local line can be used for future breeding programme to evolve resistant against *cylas* sp.

75. **Traditional Jathi Mullai**

In one acre of land, 800 plants are planted. All soils are suitable for this crop *viz*., red, clay, and clay loamy and grown in Vellore district. After field preparation, plants are planted at a spacing of 5x5 feet interval. For line planting, rope is used. Furrows are made for plant to plant. Irrigation is given to the field once in two weeks. Flowering will start six months days after planting. Pruning is done in the month of July (yearly once). Flowering is start during the month of October-February. The yield will be given upto 5 months per/day. The total yield per plant is 100 kg. The market rate is Rs.15-20/kg. Wages for picking process is Rs.4/kg. The expenditure for one acre of land is Rs 5000/- .Jathi mullai is mainly used for panner preparation.

Based on experience

• This is local land race and adapted to the hot climate. The plant is having nature of to produce more number and lengthy roots. By which water will absorbed from long distance. This land race contains more amount of essential oil and leads to scented ness.

76. **Traditional Mullai**

It is grown in Vellore district and similar to that of traditional Jathi mullai. The market rate is Rs 20/kg. Mullai is mainly used for ornamental purpose.

Based on experience

• This land race is less scented ness. But big size with pure white in colour flowers having scentedness.

77. **Ramnad Gundumalli**

Ramnad gundumalli can able to grow well in sandy soils with less irrigation i.e. two irrigations per week. The name itself it is grown in Ramnad district. The crop withstands for 4 years and is sown at 2.5 x 2.5m interval. Usually marginal and small farmers are used 20 cents or 30 cents area for raising the crop. As December-January is the rainy season, no pickings are done. In January, all leaves are completely removed to grow new leaves. From February to June, flower pickings are done and mainly 8 kg flowers are plucked in every day during season time and are exported to Ramnad & Madurai. The market rate is Rs.40/-. The cuttings are done in thrice a year. In one time, 15000 cuttings are cut down. (1 cutting = Rs. 0.80 to 1). Only goat manure is applied in the field. Initially cuttings are kept inside coconut leaves, but recent times cuttings are grown small greenhouse chamber (inner Plastic /outer green sheet) and irrigation is done to maintain heat and coolingness.

Based on experience

• This local land race having its own characters of big size flowers, high scentedness, storabiling and tolerant to hot climate and salinity. From this place, it is introduced to many parts of India *viz*, Andhra Pradesh, Karnataka and Maharastra.

78. **Nattu vellai and senkoiya**

There are two types of guava varieties. Among these, one is white coloured and the other one is pink coloured. The main difference between the varieties released and the traditional varieties with reference to guava are;

• Taste – traditional varieties are tastier than other varieties.

• Skin – traditional varieties have a light skin when compared to other varieties which have a thick skin that makes it hard to eat.

Even within the traditional varieties, especially in red guava there are two forms, one which has a pointed end with less taste and the other which is round shaped with good taste. One basket containing 10-25 kg of fruit. The cost of one basket is Rs.30-40/-.

The reason for existence of such traditional guava variety still at kolli hills is because; Government variety doesn't suit to the climatic condition prevailing at kolli hills. Even if a same tree bears fruit, due to the harder skin, it has less or no market preference. So obviously, it loses its position in cultivation

Based on experience

• This local land races are adapted to particular climate. When gravity under hot climate and salinity area, the quality of the fruits will be enhanced.

79. **Pineapple (Annachi) - Naadan /local - *Ananus comosus***

It is being cultivated in kanyakumari district. A well drained loamy soil is ideal for pineapple cultivation. On 7-8 months after planting pine apple starts fruiting. It takes 3 months time to mature and marked with in a day or two. In maturity stage, the fruits are harvested in the month of April-July. Fruits obtained during peak summer (May) are faster compared to other varieties. If the fruits are harvested in rainy season, their sourness will be more. In early stage, the fruit becomes red in colour and later it turns to yellow. It weighs upto 2 kg and bigger in size. From each plant, 1-2 fruits are harvested per year. When planting is done in hill slopes, the mounds and channels should be formed across the slope and planting also should be done across the slope which will prevent soil erosion and conservation of rainwater. It is mainly used as a live fence because of its sourness in taste and more number of red thorns.

Based on experience

• The local land race is containing more fibres in fruit and Bromelic acid. This acid leads to acidity and sourness in fruits. This type is resistance against mealy bug.

80. **Ram seetha (*Annona reticulata*) – Local Anona variety**

Fruits are edible one and unripe fruits considered as anathematic; bark is a powerful astringent; leaves and seeds are insecticidal use. It cures asthmatic disorder.

Based on experience

• The fruit can also be used for anti-diabetic purpose. The local land races will be

Generally rural people use this plant as health tonic and farmers used the leaves for controlling pest. The colour of the fruit is yellowish orange. The single fruit weighs about ½ - ¾ kg. The maximum yield is 200 fruits/ year. The Cost of the single fruit is Rs. 2-3/-.

utilized for future breeding programme for developing resistant varieties in Anona.

81. **Chengavarkai**

The mangoes of kanyakumari district are off season, a variety is considered to be the best as a whole. It is reddish in colour with round shape. The fruit is devoid of fibre content and possess 'karkandu' taste (candy). It weights about 150-200g a single fruit. Fruits are not used for pickles and cooking because it possesses thick pulp (flesh).The shelf life of the fruits is very good because of thick skin. Mostly the mango varieties of kanyakumari district to harvest in summer. "chenkavarkai", sonachi to some extend it may be attended upto July. Mango can be grown on a wide range of soils, but kanyakumari possess deep well drained soils, clay loam are some places. The varieties have not been popularized in other places of Tamil Nadu because of seed scarcity, preference by other people and varied climatic conditions. The traditional varieties are now under extinction, the seed material is kept for next crop stored as germplasm at concerned research stations.

Based on experience

• The local line is high pulp content and will be made for processing purpose. Instead of going seed propagation, soft wood grafting will be beneficial get more plants in a short period as well as to produce true to type saplings.

82. **Chetti ottu**

The mangoes possess thick skin, bigger fruits, and good for consumption. The availability is extended throughout the year. But the yields are less when compared to others. The fully ripe fruits are used for preparations like jam, jelly, squash, syrups and nectars. The raw fruits before ripening can be used in preparations like pickles, chutney. Each variety has its own distinct taste, flavour, pulp consistency and yield potential. Because of this, the production of traditional varieties has been linked.

Based on experience

• The local line is gifted one for mango cultivation. Because, the fruit availability is throughout the year. The plants should be multiplied through vegetative methods and by which spreading of this line with parts of Tamil Nadu also.

83. **Parukkan pala**

There are two types of jack fruit called 'Parukkan pala' and 'koolam pala'.The varieties are differentiate by two major properties. The main difference between both of these traditional varieties is the earlier one is used for both consumption as well as cooking purpose, whereas the later one is not fit for consumption but the seeds are

Based on experience

• Parukkan is less sweet with thick and crisp carpels and highly suitable for table and processing purpose. The coola is less crispy in nature, more sweet, more moisture,

only used for culinary purpose. Another difference is the method of harvest, the Parukkan pala can be harvested only by using knife and fetches more market prices whereas the koolam pala can be plucked by hands and doesn't fetch more prices. Apart from these differences, the cultivation procedures are similar.
Normally, it starts flowering during February-March (Masi) and starts bearing during August-September (Aavani season).Mostly it is affected by pest especially a pest called 'Nurai poochi' which makes the fruit watery and water droplets drops from the leaves and seems as raining. In each tree on an average bearing 100 fruits and each fruit contain nearly 75-150 flakes (based on the size).If the flakes size is bigger then the number of flakes is less and vice-versa. The cost of each fruits is around Rs.10-20/-.

firm flesh and big size seeds. The immature fruits of coola can be utilized for vegetables. The seeds also used as culinary purpose and preparation of starch foods. Coola seeds are mainly used for seedling production and culinary purpose and starch preparation.

84. **Varikkai chakkai pala**

The variety thrives well in deep rich alluvial and loamy soils prevailing in Kanyakumari district. It belongs to firm flesh group and the carpel's are thick and tightly arranged. It is mainly used as a table fruit. Jack is a popular fruit with the house wife for making pickles. In a year, the tree bears fruits in two seasons, viz. April-June and November- December. Normally it starts fruiting from 7th – 8th year onwards. The yield ranges from 60-80 fruits/tree. The fruit weight varies from 12-18 kg containing 115-120 flakes. Flakes are pale yellow in colour. The shelf life is more when compared to other varieties.

Based on experience
• This local line could be used for future breeding programme.

85. **Koozhan chakkai pala**

The variety belongs to soft flesh group. Trees are medium tall and maximum bearing in tree trunk. Flakes are loosely arranged and poor storage life because moisture content is more when compared to other varieties. Carpels are sweet and tasty with pleasant aroma. It is suitable for commercial planting.

Based on experience
• This local line will be assessed the yield and quality characters and will be utilized for further breeding programme.

86. **"Mulagumoodu" local pala**

It's having less moisture content, specific to the place "mulagumoodu' of vilavancode Taluk, kanyakumari district. It is used as a parent in PPI -1(pechiparai-1). The fruits are firm fleshed and medium in size weighing about 7-10 kg. Patch budding is done in 3-5 months old

Based on experience
• Local cultivar with less moisture content

rootstock seedlings during January-February and October-November was found to be very successful. In all the varieties of jack during the pre bearing age, pulses are raised as intercrop and dried leaves are spread below the trees to serve as mulch for moisture conservation.
The type and bearing habit is called "cauliflorus". The fruit buds should be protected from any positive damage by rubbing the body of the stray cattle such as buffaloes through spreading dried thorny bushes. This is very essential especially when there is no proper fence in the fields.

87. **Palayankodan**
This is the foremost commercial banana variety in Tamil Nadu. The plant is tall, hard and grows vigorously under the ratooning system. The fruits are small to medium in size, yellow skinned, firm fleshed with a sub-acid taste and individual finger with prominent nipple. The duration varies from 11-14 months. The average bunch weight is about 15 kg having 8-12 hands. In each hand with 11-18 fingers.
Other names: Erode Poovan, Kadali, Fill basket, Adukku namarai.

Based on experience
• This type is suitable for perennial cultivation of banana.

88. **Nendran: *Syn*: Ethankai & Plantain**
The fruits of this variety are eater as raw and used for preparing various dishes especially chips. The pseudostem has a distinct shade of pink coloration. Leaves and young plants inclined towards ground unlike other varieties. The fruits are relatively longer and thicker than other banana varieties. The bunch is not compact like other varieties. Fruit with three prominent ridges strongly hold. Ripe fruits are yellow colored and pulp is flesh and tough. Duration of this variety is 11-12 months.

Based on experience
• This variety is suitable for making chips as well as culinary purpose. In Kerala, this fruit is used for making pajji or vadai. This variety is very demand for making chips especially during Onam festival period. The fruit also have long storage life.

89. **Sevvazhai: syn: Red banana**
This variety has a free suckering habit grown in Kanyakumari district. The colour of the pseudo stem, petiole, mid rib and fruit rind is purplish red. The fruit is of good size, slightly curved with a blunt apex. The bunch weight is alone 20-25 kg with over 100 fruits. It is a long duration variety and takes about 18 months from planting to harvest.

Based on experience
• Red banana is highly suitable for perennial cultivation. Growing of this variety under Kanyakumari area, fruit quality is high and size also bigger.

90.	**Matti** Fruits are very small but are very good in taste. The fruits having the medicinal value to cure dysentery and suitable for giving to small children. The fruit smell is similar to that of pineapple and no flour ness, bunch is compactly arranged.	Based on experience • This is a diploid genome variety in banana. The chromosome number is 2n=2x=22. Fruits are small in size due to diploid nature.
91.	**Traditional Spices varieties** **"Kudampuli" camboge –** ***carcinia cambogia*** "Kudampuli"is a large sized, long-lived, evergreen tall tree with a spreading crown. The pulp is edible, used as souring agent for squashes, chutney in beverages, and in general cooking. Ideal soil type is deep loamy soil with adequate moisture. This condition prevails in Kanyakumari district. Flowering is seen in September-October. Fully riped fruits are come to harvest in March-April. The yield ranges from 120-150 kg of pods/ear/ tree. Acidity of the fruits is lesser than *Tamarindus indicus.* The fruits are similar to orange, which has flesh in many numbers of edible juicy flakes.	Based on experience • This fruits can be replaced with tamarind for making rasam in household preparation. It contains more amount of acidity. It mainly used to reduce obesity in body.
92.	**Manjal – Turmeric** One such crop is turmeric, which is completely cultivated in organic method. Local variety is used with less seed rate of about 8 quintals /ac and it yields about 5000 qtl/ ac that is the special features of this variety. The apt season for growing the crop is Puratasi, Iypasi (October-November). This variety has the advantage of withstanding in more rainfall and mitigating from root rot / wilt diseases. It needs about 20 loads/ac of farmyard manure for best result in field. During Thai (January), the leaves will turn yellow which is the indication for time of harvest. Thereafter, during Vaikasi (May) it is harvested. An implement called 'Manjal kothi' does harvest the manjal. It is similar to hand hoe, but the handle is comparatively longer. Leaves are cut by using sickle and then, this manjal kothi is used and rhizomes are dug out. Harvested rhizomes are dried in shade for 2-3 days and stored. Such post harvest practice enhances the quality of the rhizome. This field is interrupted with onion during chithirai (May). Onion crop is grown in the furrows, and turmeric on ridges. If onion is inter cropped, then turmeric is harvested during Masi (February).	Based on experience • Local variety with high yield, resistant to root rot and wilt disease.

93. **Nattu milagu- Pepper**

Nattu milagu is small when compared to normal milagu but the spiciness is more than the later. Though the yield is less, its quality is good. It is cultivated through twines in Kolli hills of Namakkal district. The twines are cut into small pieces and planted near the support (preferably a tree). Generally it sets flower during may-June (Vaigasi-Aadi season) and starts bearing during March. Harvesting is done by plucking the berry bunches. From each tree, nearly 10 kg of pepper is obtained.

i) White pepper

The whole berry is soaked in water for one day completely and the outer skin is peeled off carefully. After that, it is sun dried and the resultant product is called white pepper. This fetches more market prices. However, it consumes more labour and time, which are the limitation.

ii) Black pepper

Berries are harvested and it is stamped by foot. Then they are boiled in hot water so that the black colour is obtained. In case of black pepper, the intensity of colour and the size determines the market price and care is given in selection of berries, colour processing, and while boiling. 1 kg of black pepper costs around Rs.80-100/-

Based on experience

• Nattu milagu is cultivated and trained with dadap and Jack tree in Kolli hills. The soil and climate is conducive for quality berries.

5

Rationality on Indigenous Technical Practices Crop Management

CROP MANAGEMENT PRACTICES

DSIR/TPDU/TIF 056 and E-28-GO-GOI Research Scheme on "Documentation and Preservation of Agricultural Traditional Knowledge through Farmer Participatory Approach"

S. No.	Documented ITK Practices	Rational
1.	**Mixed Cropping (Onion+ Sorghum+ Black Gram)** Onion, Sorghum and Black gram are raised in a same piece of land as mixed crop. In this, onion is the main crop; Sorghum is the border crop, which is raised along the sides of the bunds. In between onion and sorghum, black gram is raised. Black gram is a leguminous crop which fixes atmospheric nitrogen into the soil. It increases the soil fertility level and gives additional nutrients to the associated crops. Sorghum grows upto the height of 5-6 feet. It prevents heavy wind, especially northern wind which carries lot of disease causing micro organisms. So main crop (Onion) will be safeguarded	Based on experience • Mixed cropping helps in better land utilization. Cultivation of crops in varying height also utilized the sunlight effectively.

	from diseases. It is used as cattle feed and it acts as live fencing to avoid cattle grazing. Sorghum stalks are used for preparation of "Pattarai"-*in situ* onion storage structure. Aadi (July-August) is the suitable season for raising these crops. Seed rate for onion is 3-4 kg/ac; it yields about 250-300kg/ac.	
2.	**Mithi ulunthu (Rice fallow Pulses)** 10 days before harvesting the paddy, black gram seeds are sown in field (when field is in waxy condition) while paddy harvesting the black gram seeds struck in the stalks will be well stamped and pressed in the field. This induces the germination of those seeds. This will effectively utilize the fertilizer, water and land. Except rat control mechanism, all other technologies are still under use.	Based on experience • This technology saves the time and moisture.
3.	**Kalappu Payir (Ragi, lab lab, Caster and Red gram)** Mixed cropping is a common practice followed by the farmers of Thenkanikottai Taluk in Krishnagiri district of Tamil Nadu. Crops grown in mixed cropping system are ragi, lablab, castor, red gram. Ragi is a main crop rests are sown in one row, as mixed up with the spacing of 1m. Seeds used for mixed cropping are traditional/ local variety. In which red gram (Mara thuvarai) is a perennial crop. After the harvest of ragi, rests of the crops are allowed to grow in the same field itself. Mara thuvarai will give yield upto 3 years. In this, lablab and red gram is leguminous crop which fixes the atmospheric nitrogen in to the soil. It increases the soil fertility level. In pastures, a mixture of grass and legume provides balanced forage for animals and at the same time soil fertility is enhanced	Based on experience • This system helps to fulfill the household needs.
4.	**Moodu Payir (Guava + *Stylosanthus hamata* & Jack + *Stylosanthus hamata*)** Normally the orchard method of crop cultivation often faces weed menace and it requires intercultural operations frequently. It will reduce the net income of farmer and renders him bankruptcy. It converts the farming unproductive and if it is dry land, the damage is too high. To prevent those melodies and ensure sustainable farming they have been cultivating *Stylosanthus hamata* along with guava and jack fruit trees.	Based on scientific reason • The legume forage tolerates drought condition.

Salient features

- *Stylosanthus hamata* is a legume forage crop.
- It is highly palatable for goat and sheep.
- It often suppresses the weed growth in dry land as it is spreading in nature.
- The roots are deeper and prevent the emergence of all type of weeds.
- It requires less sunlight and grows well under fruit trees.
- It has deeper roots to conserve soil moisture and suitable for dry land.
- It is one of the components in mixed farming.
- Duration of the crop is 180-200 days.
- It increases milk yield in goat.

5. **Kathiriyil Ottu kattuthal (Indigenous grafting technique for Vegetable)**

Brinjal is a widely used crop for edible purpose and has high market value compared to "sundai". Based on the experience, both crops are belonging to same genus and come to a conclusion to generate brinjal tree by crossing with "sundai"

The technique adopted is epicotyl grafting. The rootstock and scion are "sundai" and brinjal respectively. The stem of required young tender brinjal variety is to be collected and grafted with "sundai' tree and tied with a cloth. The field is to be irrigated regularly. After a period of 15 days the "sundai" tree bears brinjal branches and produce brinjal flowers.

Advantages:

- This will increase yield 300% than the individual brinjal plant.
- As "sundai" is disease free in nature, the brinjal generated from "sundai" plant is also disease free.
- It requires less water and can survive in black, red and saline soils.
- It requires less fertilizer compared to brinjal plant.
- Intercropping is possible.

Based on scientific reason

- *Solanam tarvum* is wild sps belongs to the family of solanaceae. It is having genetic nature of resistance to pest and diseases especially nematodes. It also has the tolerant to abiotic stress like high temperature and alkalinity. Fruit contain high amount of solasidine alkaloids content. It will provide immune to the most of the pest and disease.

6. **Oodu payir (Pine Apple + Rubber)**

Contour is an imaginary line connecting slopes of different elevation. The system followed for pineapple and rubber is row intercropping, two crops are grown simultaneously in 2 distinct rows.

Based on scientific reason

- This practice controls the soil erosion.

Method

At first, rubber seedlings are planted in June-July with pits of 1m^3 and spacing of 3x2/5x5m. The slopy areas should be first cleaned and contours are made. Rubber starts yielding only after 7 years. During the first 7 years, pineapple gives yield every year. Pine apple is raised in rows first, with 90 cm spacing between two rows or trenches and row to row spacing is 60cm with 30cm between plants. The season of planting is July-September. Duration of the crop is 18-24 month. Yield ranges from 45-50 tonnes/ha.

Advantages:

Abundant sunshine hours permit pineapple to grow well throughout the year. Farmers get additional income; more over the land is effectively used. It also minimizes and controls soil erosion.

7. **Kalappu Payir (Black gram + Tapioca + Paddy)**

Black gram is a pulse crop with duration of 90days, which is sown after 10days of tapioca planting. This combination is raised in main field whereas in 'saalu'-irrigaton channels, paddy is cultivated. Paddy is with duration of 115days and so it is taken up sowing at last. Black gram is harvested by just cutting the terminal portion of the plant.

Benefits:

Legume effect- legumes fixes atmospheric nitrogen. The succeeding crop paddy gets benefited by N fixation of legumes. It has been observed that 15-20%N can be saved in the N budget of the succeeding crop. Carry over effect of nutrients/manures to the succeeding crop. Nutrient recycling-crop residue of previous crop viz., paddy straw, and stubbles can be incorporated in the soil and can be used for succeeding crop as organic manure. Nutrient pumping from deeper layer. This is based on the principle that uptake pattern of nutrients will depend upon the rooting pattern of individual crops.

Based on experience
- A raised bed concept is used. Irrigated dry crops are planted in the beds and rice in the channel.

8. **Oodu Payir (Coconut + Banana)**

Banana is the intercrop planted in between the rows of main or base crop coconut with a view to obtain some extra intercrop yield without sacrificing the main or base crop yield. Coconut is planted in pits of 1m^3 size, spacing of 8x8m between rows and 6m between plants. Coconut yields 80 nuts/tree/year. Banana should be

Based on scientific reason
- Better utilization of inter space and utilized the nutrients and moisture effectively.

planted only after first five years of coconut. Because, the coconut comes to certain height after 5 years only. Banana is raised in January-February or November-December, with duration of 12-15 months. The yield range of banana is 40-60 t/ha/year.

Advantages

- Higher returns from a unit area
- Optimum solar radiation provided for banana
- Nutrient recycling is provided for coconut by residues of banana.

9. **Rubber+ Banana**

Rubber and banana are raised together in lowlands of kanyakumari district in petchiparai and kuzhithurai blocks. Rubber is planted in June-July in pits of $1m^3$ size. Adopted spacing for rubber plantations is 5x5 m. Banana is planted in January-February or during November-December. The spacing followed for banana is 2x2m with duration of 12-15 months. Rubber starts to yield only after 7 years. So, banana is planted till 6th year. Banana yields 40-60 tonnes/ha/year. Here, competition for solar radiation and soil factors are less influenced. Because, rubber needs less solar radiation and lesser nutrients than banana, till it starts yielding. After 6th year, rubber alone is grown and gives yield from 7th year onwards.

Based on scientific reason
- For higher income and better utilization and nutrients and moisture.

10. **Black gram + Amaranthus**

Black gram is an important pulse crop sown in June-July and harvested in September-October after 90 days. Amaranthus is also a short duration crop raised together with black gram. It usually comes well in drained loamy soils. Sowing of amaranthus can be done throughout the year. Irrigation channels / saalu are left as such without any crop. Black gram is risen near to 'saalu' and amaranthus was raised at the inner side of the field and harvested once in 3-4 months. A single crop can be continued to remain till 5 years and more. Yield ranges from 5 tones/ha/year.

Based on scientific reason
- Well drained loamy soils are best suited for vegetable crop particularly amaranthus.

11. **Banana + chillies**

Chilli is the intercrop planted in between the rows of main or base crop banana with a view to obtain some extra intercrop yield without sacrificing the main or base crop yield. Banana is planted in January-February

Based on scientific reason
- Chilles are planted in the dry periods of january-february to avoid ill effects of excess water.

or during November-December. The spacing followed for banana is 2x2 m with duration of 12-15 months. Banana yields 40-60 tonnes/ha/year. In the irrigation channels/saalu water gets stagnated. To overcome this, chilli is sown as a border cum inter crop at sides of irrigation channels. Chillies are usually sown in January-February, June -July. The yield ranges from 2-3 tonnes/ ha.

12. **Border cropping (Sorghum + Kuthiraivali + Rice)**
It is the traditional red rice variety growing in Paramakudi taluk of Ramnad district. The rice grain weight of kallurandaikaar is slightly more than kuliparichaan/varappukodainchan varieties. It is the drought resistant and tall variety. The outer skin is slightly dark. The inner grain is bold and round. The seed rate of Kalurundaikar variety of an acre is 45 kg for direct sowing. This variety withstands hunger for several hours than fine rice. The yield is 1500-1800 kg per acre. Paddy straw yield is approximately 3 tons per acre. In some field rice + kudiraivali (in border areas) system is followed. Kudiraivali yield is 600 kg per acre and it fetches higher price in market. (Rs.6 per kg.).
In some fields 3 types of sorghum are grown as border crop. They are red, white and black. These sorghum varieties differ in colour. They are grown for cattle feed purpose. The traditional sorghum varieties grow very tall in nature.

Based on scientific reason
• Best practice for dry regions avoids risk on crop failure due to vagaries of monsoon.

13. **Betel vine + Jaffna Moringa:**
The traditional "betel vine" variety is grown along with "Velimurungai" which gives anchorage to betel vines. It is a suitable crop for coastal sandy soils. In the border areas banana trees are grown and it gives higher crop intensity leading to additional income. There are 150 banana trees surrounding around the border area of the 2 acres land and most of them are border crops in betelvine field. The field is irrigated by 3 wells and the water is available at 10 feet depth.
The traditional 40 years old Jaffna Moringa has been grown in Thangachimadam areas and the cutting was brought from Jaffna, Sri lanka during the cargo service between India and Sri lanka. The 40 years old tree gives 1500 moringa sticks daily between March-June. They apply only organic manures such as goat manure to the tree. It is not usually affected by pests and diseases.

Based on scientific reason
• Utilized the available soil moisture in effective manner and it provide support the in betel vine.

The length of individual fruit is upto 12 feet and it fetches Rs.3-5/- based on demand. The rainy season of October, November and December is the off-season for moringa tree.

14. **Wild Brinjal + Snake gourd (Kottai Pudalai)**
Wild brinjal is traditionally cultivated in Devoor village at Salem district. Cultivation starts during the month of August and January and the duration of this crop is 6-7 months. This variety is mostly preferred in Erode market and has good market value while compared to other varieties. This variety possesses high pest and disease resistant, so that the farmers used to prefer to cultivate this variety. Cow dung, goat manure and poultry waste are used as manures. After first harvest the plants are left with 1 foot height from the ground level, so that other harvest can be done. Normally the yield obtain is upto 30-40 bags in 3 months. The rate fixed in the market is Rs.15 per kg. So that, the farmer fetches net profit of upto Rs. 30,000- 50,000. Inter cropping is also possible in this crop, at the time of harvest "kottai Pudalai' seeds are sown. The pudalai plant uses the wild brinjal plant as a support for its growth. From the intercrop upto Rs.20, 000 additional income is gained by the farmers.

Based on scientific reason

- Wild type has insect resistant.
- Brinjal plant with tall stature provides support to the snake guard.
- Location is specific.

15. **Betel vine + Agathi (vellaikodi) Piper betel**
It is followed in kunnipalayam village at Namakkal district. Well drained fertile clay loams are suitable. It does not tolerate saline and alkaline conditions. Betel vines require a cool shade with considerable humidity and regular supply of moisture in the soil. The vines are propagated by terminal stem cutting or sets about 30-45cm long. Sets obtains from the top portions of the vines are easy to root and hence best for planting. Provide drainage trenches of 0.5m width, 0.5m depth in between two adjoining beds. Plant the seeds of the live supports *Sesbania grandiflora* (agathi) in long rows. About 300 banana suckers are planted at the edges of the beds which are used, for tying the vines on the live support and for packing the betel leaf. When the agathi support grows to a height of about 4cm they are topped. The crop is planted in two rows in beds of 180 cm width on agathi plants with spacing 45 cm between plants in the row.

Yield

30 to 35 lakhs leaves/ac/year.

Based on scientific reason

- This method is normally followed practice in all areas.
- Cheap source of support to betel vine.
- Agathi leaves used as a cattle feed.

16. **Amla + Ramnad Jasmine**
District: Namakkal
Varities: Amla NA-7, jasmine Ramnad jasmine,
Spacing: 20x20ft
Season: Amla june-july, jasmine june- july
Amla is a subtropical plant and prefers dry climate. It can be grown in variable soil condition. It is a hardy plant and tolerates salinity and alkalinity condition. Well drained fertile soil is suitable for jasmine. Irrigate the plants initially for establishment. FYM 10kg /plant, and fertilizers 200:500:200g NPK /plants to be given immediately after pruning. Ramnad jasmine is cultivated as intercrop with Amla (Na-7). The jasmine plants are purchased from nursery, one plant cost is Rs.1.50/-. In one acre, 400 plants are planted.
Neem oil spraying is done for controlling pest and disease. Income from amla trees is Rs.50, 000 to 60,000/-. Income from jasmine flower is Rs. 5,000-6,000/- per year. Total investment for these two crops is Rs.8,000 to 10,000/- and net profit is Rs. 50,000/-.

Based on scientific reason
• Better utilization of space and land.

17. **Cotton +Red gram +Castor**
It is followed in Kalathiyur at Vellore district. In a cotton field, red gram and castor crops are cultivated as an intercrop, during the month of July. Cotton seeds are sown at the same time. Red gram seeds are sown in line 12 feet spacing of 12 feet interval. Castor seeds are sown as a border crop around the cotton field. Seed rate required for red gram is 2 kg/ac and ½ kg /ac for castor seeds. Cotton is harvested with in a period of 5 months. Red gram is harvested in the 6 months. Duration of the castor crop is 10 month. Yield obtained from red gram is 400 kg/ac. The cost of 1 kg red gram is Rs. 50/-. Yield from castor seeds 120-150 kg/ac.

Based on scientific reason
• Different plant height uses the sunlight and soil nutrients effectively.

18. **Ground nut + red gram**
In ground nut field, farmers normally grow red gram as an intercrop. It will give extra income to the farmers and it will also be used for household purpose. In one acre of land, 2 kg of red gram seeds are sown. Ground nut is harvested on the120th day. Red gram is harvested in180th day. The yield from red gram 400-450 kg/ac. The price for One kg of red gram is Rs 50/-. This system is traditionally followed in northern district of Tamil Nadu.

Based on scientific reason
- This method is commonly followed practice
- It is a rain fed system of cultivation.
- Avoid risk due to drought conditions.

19.	**Mixed cropping (Chillies + Castor + Agathi)** Chillies, castor and Agathi are raised in same piece of land as mixed crop. In this, chillies are the main crop. Castor is the border crop, which is raised along the side of the bunds. In between chillies and castor, Agathi is raised. Agathi is leguminous tree fodder which is a protein supplement to small ruminants and also used as green leaf manure. Agathi growth is limited upto 10 feet height for easy management. The castor crop is used for castor oil extraction and the castor leaves are used as green fodder to live stock. **Yield**: 2-3t/ha of dry pods,10-15t /ha green chillies	Based on experience • Castor crop acts as a trap crop. • Agathi leaves used for cattle feed. • In this practice, pest incidence is less.
20.	**Groundnut + Black gram** Intercropping of black gram in groundnut is practiced in Villupuram District. Black gram was sowing at an interval of2-3 m in between ground nut. It is an additional income before harvest of the groundnut. **Grain yield:** 100 Kg/Acre as an intercrop **Colour of the grain**: black	Based on experience • It is common practice in irrigated areas. • This practice generates higher income to farmers.
21.	**Groundnut + Amaranthus** Groundnut cum amaranths inter cropping is practiced in Erode district. Amaranths are harvested with in one month. Ground nut harvested at end of three months period. Amaranths crops will give additional in come to needy farmers by selling amaranths green leaves daily in the market. **Yield** 25 days after sowing for mulai keerai 4Tonnes/acre 40 days after sowing for Thandukeerai 7Tonnes/acre	Based on experience • Amaranthus crop grows depends upon the soil type. • It gives additional income to growers.
22.	**Kasevu Neer Kuttai (Percolation Pond)** In the farming system, different types of irrigation method are followed. Since it is a coastal village, sandy soil allows easy excavation of farm pond. Thus with a width of 20 feet (approximately) and depth of 10-15 feet a farm pond is excavated, where water oozes out even at 10 feet. For effective utilization such farm ponds are dug in centre of the field. Since this coastal garden land has no other source of irrigation, these farm ponds are the only possible source. This kind of farm pond can be seen in almost every farm in the coastal villages especially Vettaikaraniruppu, Vellapallam, Nalluvethapathy, Vizhuthamavadi, Periyakuthagai, Sempodai of Nagapattinam district.	Based on scientific reason • Soil type is best suitable for excavation of pond. • It also provides irrigation facility.

23. **Snake gourd sowing**

Pit size of 1.5 feet depth and 2 feet width is dug in the pit ,a bunch of fresh Pungam (*Pongamia pinnata*) leaves are incorporated ,this will acts as green manure as well as preventing termites and ant damage. Due to the incorporation of pungam, the plant will emerge quickly and grow sturdy. After that, pungam is covered with thin layer of soil, 2.5 kg well decomposed FYM is applied over it and then it is covered with 5cm of soil. The layer of soil and FYM should be mixed well and pit is rilled at centre, in which 3 seeds are sown. Sharp edge of the seed is dibbled inside the pit and then covered with soil.

Finally the entire pit is covered with leaf litters in order to avoid excess evaporation which in turn avoids damage due to rodents. Immediately after sowing, the pit should be watered. The seedlings will emerge out 7 days after sowing. After 3 weeks, the pandal is erected using Palmyra fibre/coconut coir pith rope/ Karung kodi- as a twiner.

Karung kodi is a twine found in all coastal villages. Matures wine is used for preparation of pandal. Instead of other ropes, this is used, due to its hardy nature, after drying.

From 50-60 days after sowing, it starts bearing. Naturally snake gourd grows in an irregular manner which is least preferred in the market. In order to avoid this economic loss, coconut pith is tied at the posterior end of the fruit. There are two types of local snake gourd varieties that are prevailing. One is lengthy snake gourd (Neetu pudal), which grows up to a length of 6 feet and it weighs about 2 kg. Second is short and stout snake gourd (Pani pudal) which weighs about ½ kg-1kg and it grows upto 1-1.5 feet length. Suitable season for sowing snake gourd is July-August. Existence of the local variety is for 75 years.

Based on scientific reason

- It is a local method
- Additional organics provides conducive seed bed for better germination of seeds.

24. **Paddy straw bunches**

Paddy straw bunch is kept in one corner of the nursery to avoid soil erosion and up rooting of plants.

While irrigating vegetable nursery, care has to be taken so that soil is not eroded, this may up root the seedlings. To avoid such damage, a small pit is dug in the corner of the plot and filled with sufficient paddy straw. Thus pouring water upon it seedlings are watered but do not damage it.

Based on scientific reason

- Paddy straw bundles are smoothened the flow of water hence, it arrest the soil erosion.

25.	**Muthalai Poondu/Agayathamarai -** ***Pistia stratiotes*** Agayathamarai is a common weed found in all water bodies. In the month of January, whole plants are removed from the water bodies and allowed for drying. The fully dried plant materials can be incorporated to the saline and sodic soil for reclamation. It can be used as green leaf manure, in wet land condition. Before puddling, whole green plants are harvested from the water bodies and used as green leaf manure. It has to be puddled well in the wet land and allowed for decomposition. Otherwise it will emerge as weed in wet land field. After incorporation of Muthalai poondu, discoloration of water i.e. red colour indicates that the salt is dissolved and removed from soil. No specific dosage in agayathamarai incorporation. Dosage depends on the quantity of availability of the weed.	Based on scientific reason • Addition any green biomass will increase the organic matter content, produces acid drip decomposition and also reduces the salt content.
26.	**Pitcher pot irrigation** Pitcher pot irrigation is a method of drip irrigation in which mud pot is used for irrigation. Capacity of pot is 10 litre which costs about Rs10/. Pitcher pot irrigation is a suitable method of irrigation for the area where water availability is less. Mostly it is used to irrigate the tree crops like Mango and Sapota. One foot away from tree's trunk, a pit with size of 1 feet depth and 2 feet width is dug. Small size hole is made at the bottom of the pot. Pot is buried in the pit, in such a way that it's rim is slightly above ground level. Water is poured into the buried pot and the pot is closed with mud lid, one pot of water is enough to irrigate a tree for about 4-5 days. In this method, water droplet is applied directly to the root zone of the crop. Water wastage will be avoided and water loss due to evaporation will be minimized. So, water can be effectively saved through this method of irrigation	Based on scientific reason • It is very effective method in drought prone regions. • In this method water wastage will be avoided and also water can be saved.
27.	**Stubble mulching** Tillage with leaves to the surface of the soil, cloddy should be mulched with crop residues for an effective moisture and soil conservation. Properly mulched soil surface protects soil particles from rain beating, washing or drift. Thus erosion does not take place. The traditional methods of tillage developed for soil (Water) moisture conservation. Locally available materials such as straw and stalks could be used for mulching. Soil loss and runoff from cropped area can be	Based on scientific reason • Mulches act as barrier and reduce the soil erosion. • Adds organic matter and also improves the infiltration rate.

reduced. Mulches improves infiltration rate, increasing termite and biological activity.
Mulches have been used to reduce evaporation, increase infiltration, prevent the soil from blowing and washing away, control weeds, improve soil structure and to increase crops yields.
Mulching is a traditional agronomic practice, in which crop waste and leaf litter are used to cover the soil. Mulching is done under the tree ie, around the tree trunk. Through this method soil moisture is conserved effectively, evaporation loss is also minimized and checks the weed growth.

28. **Kodai Uzhavu (Summer Ploughing)**
Plough the field during summer and two feet height, earthen bund is made around field which checks run off of the rain water.
Advantages of summer ploughing
It increases water capacity of the soil.
It destroys weeds and pupal case of the pests.

Based on scientific reason
- It conserves soil moisture and reduces the pest incidence.
- Water requirement for field preparation of succeeding crop is reduced.

29. **Kidai Adaithal (Sheep penning)**
During non cropped season or after harvest, sheep penning is done with 100 sheep's in the field for about 4-5 days. Every day morning the place of penning can be changed. So that the entire field will be covered. At the time of penning, sheep manure enriches the soil fertility. Wages for sheep penning is Rs 80/night.
Existence: .75 years

Based on scientific reason
- In this practice, the field is enriched by organic matter.
- Sheep urine having the valuable nutrients, it also enriched the soil.

30. **Padal izhuthal/ Mull izhuthal (Leveling)**
After sowing, proper leveling is highly required for germination. Hence, leveling is done by farmers using wooden plank of size 5 feet length and 1 feet width. The size of width varies from place to place. The wooden plank when pulled by farmer covers the seeds with soil. A rope is tied on the 2 sides of the plank and for increased efficiency on leveling. Stone is kept on the plank. One member can level one acre a day. In the place of wooden planks, some farmers are using Prosopis (*Prosopis julifera*) branches for leveling and this is called as padal izhuthal or mull izhuthal. Even spade is used for leveling but it is labour intensive.
Leveling gives even /proper covering of seed, which facilitates germination

Based on scientific reason
- The following advantages are incurred by this method.
- Covers the soil.
- Soil compaction.
- Uniform germination.
- Avoid damage due to birds.

	Practice	Rationality
31.	**Bench terracing** On steeply sloping and undulated land, intensive farming is possible only with bench terracing. It is usually practiced on slopes. Bench terracing consists of principally transforming relatively steep land into a series of level strips or platform across the slope of the land. It reduces the slope length and consequently erosion. The field is made into a series of benches by excavating the soil from upper part of the terrace and filling the lower part. A good soil depth is required to avoid exposure of unproductive soil during leveling.	Based on scientific reason • Reduce as the soil erosion. • Better utilization of slope land.
32.	**Puzhuthikal nursery (Dry land paddy Nursery)** Entire nursery for an acre is divided into 10 plots. 18 hours pre soaked seeds are sown in that nursery. In this method, seeds are not kept in the dark room. After sowing, nursery is covered with crop waste. This is called as "Mootam Poduthal". Five days after sowing, seeds will emerge out.	Based on scientific reason • Suitable for less water availability. • Seedlings withstand in water scarcity areas.
33.	**Contour Trenching (Watershed management practices)** Watershed is a natural hydrologic land unit. It is an area of internal drainage above a common point of outlet. Its size and shape are determined by surface topography. It is bound on all sides by a divide or ridge line. **Advantages** • Control the runoff which in turn influences the rate of soil less, moisture availability to plants, sedimentation, ground water recharge, water quality etc. • Useful to manage rain water falling within the unit for various purposes including agriculture without damage to the natural resources like soil The size and number of trench varies based on the spacing of the crop. Trenching reduces the velocity of runoff. The water collected in the trenches or pits percolated through the soil and improve soil moisture conditions. Number of these structures helps to improve the recharge of ground water and reduces the runoff.	Based on scientific reason • It is a soil moisture conservation practice. • It is suitable for dry lands.
34.	**Paanai Neer Pasanam (Pot irrigation)** Locally available waste plastic bottles are used to irrigate the Sapota and Mango crops. A plastic bottle has the capacity of 3 litres of water is buried into the soil. . Fifteen days once, the water is filled in to the	Based on experience • It is very effective method because use of available little water for crop cultivation.

bottle. At the bottom side of the bottle a small hole is made. The water is coming from that hole only. Due to this method, the water wastage is restricted and it reduces the evaporation loss also. The plants are getting the water uniformly throughout the day.

35. **Eatram (Water lifting device)**

Using the traditional water lifting device is called as Eatram. This is used to lift the water from well. Eatram panai made up of iron, its capacity is 10 litres. Two iron rings are fixed on two rims of the eatram panai in which the bamboo pole is tied with the help of rope.

Operating mechanism

A member can keep two legs on two tamarind poles and hold the mun kazhi by using hands. While pulling the mun kazhi simultaneously eatram panai will take the water from well and eatram panai with water will be uplifted and discharge the water automatically due to the weight of the rock stones on the vasam.

Most of the farmers in that hilly track are using eatram for irrigation still now. This type of traditional irrigation method can be used for Rice, Sorghum, Horse gram and Gingelly.

Based on scientific reason

- It is a local traditional and energy saving method.

36. **Varappu Poosuthal (Bund Plastering)**

Raising the bund height, strengthening the bund and plastering the bunds are the common traditional practices followed by wet land farmers.

- Raising the height of the bund prevents water over from rice field.
- Strengthening the bund and plastering with available clay in the rice field will check the water wastage due to seepage and leakage.

Based on scientific reason

- Avoids seepage of water and also reduces the weeds.

37. **Pasunthal Uram (Green leaf manuring)**

It is the application to land of green leaf, gathered from shrubs and trees growing in waste lands. Green leaf is gathered from all sources by farmers during the rice planting season. Different kinds of shrubs growing on tank bunds, waste lands, field bunds etc., are used. In addition, lopping from trees on the road side and forest area also gathered for use as green leaf manures. Common shrubs and trees useful for this purpose are, Neem, Teak, Pungam and Erukku leaves.

Based on scientific reason

The following advantages are incurred by this method.

- Addition of organic manure.
- Availability of nutrients.
- Improvement of soil structure.

Advantages of green leaf manuring

- Leguminous green leaf manure crops fix atmospheric nitrogen which enriches the land.
- When the green manure crop is ploughed in the soil, it decomposes easily, without leaving much of residue in the soil.
- It improves the structure of the soil. It acts as a binding material in light soils and increases their water holding capacity.
- Crumb formation is induced in clayey soils, leading to an improvement in the absorptive capacity of the soil and percolation.
- The green manure added to the soil promotes the activity of soil micro- organisms, carbon dioxide is liberated, the chemical reactions in the soil are speeded up and more plant nutrient is made available.
- Organic acids produced during the decomposition of the green manures help to release phosphate and make them available for the use of the next crop.

Incorporation of chopped roots, stubble, green-stalks, tree leaves (neem, teak, pungam) etc., it increases the organic matter and bacterial population which in their turn to improve the soil structure.

38. **Natural fruit ripening structure**

Normally, carbide stone is used in artificial chamber for fruit ripening. This is used for mango, sapota and jack. But it fetches poor market price and make the fruits into bitter in taste. To overcome these problems Mr. Abraham has developed a natural fruit ripening structure by using a local carpenter.

Description of the structure

The size of the structure is 1x1x1 m. It is made by using clay soil and covered by a wooden shield. The entire structure is smeared with cattle dung. The mini stove is attached with a whole tube so that to supply a smoke by firing farm waste. The capacity of the container is 15 kg.

By burning wood ash and natural waste the smoke is supplied into the hole. Thus, the smoke curing technique ripens the fruits.

Advantages

- The duration for fruit ripening is 2-3 hours.
- This method does not ripen the tender fruits and prevent poor quality, low price in market.

Based on scientific reason

- By this method, triggering the ethylene production and gets early repining of the fruits.

- The fruits ripened by this technique are sweet in taste and free from foul odour normally occurred during inorganic ripening method.
- The cost of preparation of the structure is less than Rs.200/-. But the stony method of fruit ripening requires Rs.200/- for single attempt.

It is a permanent structure and is not affected by termites since is smeared by cow dung.

39. **Covering Pineapple with its slips**

Pineapple is a tropical and subtropical fruit found in humid hill slopes of kanyakumari district especially, Paraliar, Thadikarankonam. Crops grow about 1-1½ years and harvested in 12th month onwards. Slight fruit colour at base of the fruit indicates maturity. The matured fruits create fragrant smell which invites crows and other birds to taste it off. To avoid this, the fruits are covered with the slips around them with a rope. It also prevents the breakage of overweighed matured fruits from the plant. The operation is carried out after 1-1½months of fruit setting.

Benefits:

- Prevents the major economic loss.
- Quality of the fruit is maintained.

Based on experience

- Slips are acted as wind breaker as well as birds scarer.

40. **Kanchi Kambu (Propping Pole for banana)**

Propping is a process by which the kanichu kambu –a pole is kept at an angle of 45^0 to banana trees. This is mainly to prevent the crop from lodging on heavy rains, winds also the over weight of the bunch. The pole is also used to make holes in the fields.

Merits:

- Crop yield maintained
- Plant is made safer from natural disasters
- Highly effective to control lodging

Simple to follow in farmer's own field.

Based on experience

- Avoids lodging.

41. **Saalu (Irrigation Chennels)**

In general irrigation is done through channel and water reaches the plant roots directly. In kanyakumari district, large trench like structures called 'saalu' are created. It is mainly followed for banana and paddy. Saalu is created between the rows with a breadth of 1m and depth of 1.5m. The main reason for this structure is soils of kanyakumari has proportionate amount of clay

Based on experience

- It is trench method.
- It is used wherever the soil region is deep.
- It is useful method in higher water availability areas.

No.	Practice	Rationale
	in paddy and to some extent in banana fields. To mitigate this problem and avoid water stagnation in the root zone of plants with regular supply of soil moisture 'saalu' is made. **Advantages:** • Number of irrigations minimized. • Optimum amount of water is effectively used for irrigation. Border crops can also be raised on the edges of saalu.	
42.	**Chev vaazhai- Koodai podal (Bunch covering** Dried coconut leaf are collected and made as coverage i.e., basket like structures to protect the scorching sunlight falling on the fruits. This is mainly practiced in chevaazhai and nendran varieties. Banana is a crop known for value addition and better price in the market. For this, the cover should be made and tied on the fruits 15-20days before harvest. This results is uniform coloring and ripening through out the bunches. **Advantages:** • Easily made with the farm available dried coconut leaves. • Simple to follow. • No chemical is involved. • Better price in the market.	Based on experience • This is one of the oldest technologies. Things this method, quality of the produce is enriched and exportable quality will be maintained. Apart quality it also reduce the pest and disease incidence.
43.	**Pakku Mattai Nar (Arecanut outer husk)** Dried outer husk of arecanut are collected and finely made into small pieces. In home gardens and nurseries, these dried outer husks are spread on the mud pots kept for ornamental plants and flowers. Irrigate the pots, the arecanut husk absorbs moisture and retains with in it. So, the arecanut husk makes plants avail of moisture for longer time than normal one. This result in minimizing irrigation, effective utility of waste products, eco-friendly and no making cost is incurred.	Based on experience • Arecanut husk is easily decomposable in nature that decomposed one is used as soil mixture and as mulch. This is used as organic mulch and mixture for growing plants.
44.	**Rubber Moodu Payir (Rubber' cover cropping)** ***Pueraria phaseoloides*** Cover crops are grown primarily to cover the soil, reduce the loss of moisture and erosion by wind and water. Pueraria provides 8 t/ha of Nitrogen. They are sown after 6 months of rubber establishment. Flowering is noticed yearly once. After seed setting, fall on the ground and leaves cover the full field and automatically get decomposed.	Based on scientific reason • Cover cropping with rubber reduces the weed population and also soil moisture evaporation.

45. **Mazhai Neer Semippu (Rain water harvesting)**
In rubber plantations, pits of 1m^3 size are dug seedlings are planted. The spacing between the plants is 3x2 / 5x5m. Planting is usually done in June- July. In between two rows of planting pits of 6'x 2'x 2' are dug, this is mainly to store the rainwater harvested in plantations. When heavy rains occur, water is stored in it and used at the time of scarcity. It protects the crop from drought during non rainy days. The dried leafs also fall into the pit and gets decomposed over a period. They are collected and applied as manures.

Based on scientific reason
- It is very useful method in water scarcity areas and also increasing the ground water level.

46. **Brinjal raised in coconut ring basin**
Coconut is grown in larger tracts of Eetamozhi, thengai patnam and Marthandam of kanyakumari district. Red sandy loam, laterite and alluvial soils are suitable to raise coconut. Planting season is June-July and December-January. Normally spacing of 25'x25' between rows and 20' between plants is adopted. Irrigation is done through ring basin method. Basins of 1m length are made around the coconut trees. In it, sow the seeds of brinjal at its periphery. Spacing between brinjal plants is 30 cm. The seeds germinate in 2-3days. Once the seedlings are emerged, arrange the coconut husk in every brinjal seedlings and form an inner periphery with it. Once the coconut is irrigated, the coconut husk absorbs moisture and retains with it. The brinjal gets moisture from the coconut husks. Brinjal starts giving yield in 3months. Effective land utilization, minimize water loss, benefits the farmer with additional income.

Based on scientific reason
- Effective land utilization and irrigation water.
- Water saving method.
- It generates additional income.

47. **Paddy Manual Threshing**
Harvested paddy bunches are threshed manually i.e stricking on the stones. Two members are needed for manual threshing (one female labour and one male labour).
Female labour is helping the male labour for taking harvested paddy bunches. Pidikayaru is used to hold the bunches for threshing
Coverage/ efficiency – 33 cents/day
Wages: Rs. 90 / day.
Advantages
- Germination percentage will be more
- No damage or less damage to grains
- Less number of broken grains.

Based on experience
- Less damage to grains.
- It is very common practice followed in all regions.

48.	**Thokai Kattuthal (Shellter belt)** Banana leaves are tied around the betel vine field up to 5-6 feet height. This method is useful to control the high wind velocity and temperature. It will give cool atmosphere to the betel vine field. More wind and temperature can be affecting the betel vine leaf. It helps maintaining the moisture condition inside the field. It is low cost method followed in Namakkal district.	Based on scientific reason • Maintenance of relative humidity in the field.
49.	**Marakilai Panthal (Tree branch panthal)** It is followed in Vadavathure village at Namakkal district. Vathanarayanan and oonchai tree branches are used as a pandhal. It is easily available in field and is easy to stag. There is no need of labour to stag the tree branches. After the crop period, it is easy to dislocate from the field. Then we can go for further cultivation, it is useful for the growth of a variety of vegetable crops namely lab lab and bitter gourds. This pandhal used by many crops as it is economic.	Based on experience • Any locally available materials are used. • It is economic method. • It is very easy method for further cultivation.
50.	**Banana trees tying with ropes / Plastics tapes** A plastic rope is tied 2/3 from the bottom of the banana tree to prevent the lodging and bunch damage, While high flow of wind during the rainy season. The end of plastic ropes from 4 banana trees is tied with one feet height wooden plug which is firmly attached with earth. The cost of this indigenous technique is Rs. 4 per tree which includes the labour charges. The duration of this plastic ropes will exists for three year, which is after harvest of banana this plastic ropes can be used in subsequent seasons.	Based on experience • Any local materials are used for tying. • It is useful to prevent to lodging.

6

Rationality on Indigenous Technical Practices Crop Protection Measures

CROP PROTECTION MEASURES

DSIR/TPDU/TIF 056 and E-28-GO-GOI Research Scheme on "Documentation and Preservation of Agricultural Traditional Knowledge through Farmer Participatory Approach"

S. No.	Documented ITK Practices	Rational
1.	**Cow urine + Neem leaf extract** **Ingredients** • Neem – 10 kg • Cow urine – 8-10 litres **Preparation** Neem leaves are pounded well by using ural and ullakai and extract is taken. This extract is mixed with 10 litre of cow urine. Neem leaves bunch is taken for applying the extract. **Dosage**: 10 lit / ac **Time of application**: 30-45 (DAT) **Existence:** > 150 years	Based on scientific reason • Azadirachtin content of neem leaves in known scientifically pest control.

General profile of the ITK holder
Th.Mohan,Thiruvalluvarpuram, Kaliyapettai, Uthramerur Tk, Kanchipuram Dt.

2. **Poochi Viratti (Pest Repellent)**
Eruku-3Kg, Pungam-3kg,Umatham-3kg, Notchi-3kg , Kadal palai-3kg, Neem Seed-1/2kg , Pungam Seed-1/2kg, Neem Cake-5kg, Cowurine-10lit, Water -30lit.
Eruku, Pungam, Umatham, Notchi, Kadalpalai leaves are pounded well. Neem seeds, Pungam seeds are pounded separately. Fifty litre capacity mud pot is buried in the ground, because soil will give favourable microclimate for fermentation. Pounded materials are put into the pot. With this neem cake 5kg, cow urine10 litre and water 30 litres, are added. After that the mouth of the mud pot is tied with gunny bag or waste cloth. It is kept as such for about 10-15 days. Every day morning, it has to be stirred well to ensure well and quick fermentation. After 10-15 days fermented solution is filtered and ready for spraying.
Dosage: 5 lit/ac (500 ml/10 lit water).
Additional information: as stated by Mr. Marimuthu, Poochivirati spraying will control the pests and also promote the plant growth. Cow urine has nitrogenous compound which will promote the plant growth.

Based on scientific reason
- Pungam, Notchi, and Neem leaves/ seeds all already known for their insecticidal properties.
- Any protect after 10-15 days of fermentation will have foul odour which will deteriorating the insect.
- Extraneous application of nitrogenous compound in any form will stimulate plant growth and may invite more pests.

3. **Coconut Eriophid mite kattupaduthuthal**
"Nethili karuvaadu" dried fish and made as powder, mixed with salt in the ratio of 1:1 and applied as such or spray (250gm/litre).

Based on scientific reason
- Dried fish meal and salt may aberate microscopic mites and prevent their damage.

4. **Illupai-Seeds and oil cake**
Illupai (*Madhuca latifolia*) belonging to Sapotaceae family. Nut of Illupai has various uses. It is documented in Vellapallam village at Keevalur taluk of Nagai district .Each illupai fruit has 4 brown coloured seeds with hard seed coat. Seeds are elliptical in shape with pointed edges.
Health aspects
The oil extracted from the nuts reduces heat and cools the human body. For the extraction of oil, the outer skin is peeled off and the mesocarp is removed from each seed, by using ural and ullakai. This mesocarp is sun dried for 4-5 days. Now the meso carp becomes ready for oil extraction by using oil expeller.

Based on scientific reason
- Illupai oil/ cake are also known for insecticidal property. But large scale availability is questionable

Cooking
Oil extracted from illupai nuts is good for cooking/ culinary purpose, because of its cholesterol controlling property.

Agriculture
Oil cake obtained from illupai nuts is also advantageous. Illupai oil cake is applied at the rate of 50 kg/ac. Before transplanting or at the time of transplanting to the main field will effectively control the pest and disease incidence in the main field. Even though this is an effective method, due to the non-availability of illupai oil cake, it rarely exists only in few areas.

Apart from this, it is also believed that illupai trees have the capacity to absorb rain bearing cloud. So it is said that wherever illupai trees are found, rainfall is received and it also absorbs carbon-di-oxide.

5. **Pest control in Rice field**

Neem Cake Application
Five kg Neem cake is taken in gunny bag and kept in the main channel of the rice field. During irrigation, water passes through Neem cake and takes essence from Neem cake which will control the root borne diseases and pests. It can be done after transplanting.

Existence: > 25 years

b. Cow Dung Smoking
Smoking with the help of dried cow dung in one end of the field will reduce the pests in field.

Cow Urine Tonic
Cow urine (10 lit) is taken in a mud pot or any container and boiled well until 10 litres cow urine comes to 5 litres i.e. half of the quantity. With this, neem oil 2 litres and cow urine 5 litre are added. Again the above said contents are boiled well until 12 lit solutions come to 6 litres. After that the boiled solution is cooled.

Dosage: 1 lit/ac (100ml/10 litre water).

Cost: cost incurred for the preparation is Rs. 50/ac.

Based on scientific reason

- It is sort of pestisidal method. Instead of using foliar spray of NSKE, Neem cake is allowed to dissolve in irrigation water. Azadirachtin rich water in non available in a root zone of each and every plant. Plant is able to absorb azadirachtin rich water through roots. (Systematic action) thus it may kill root bone pest and disease leaf feeders and sap feeders as well.
- Smoking of dried cow dung will reduce the pests

6. **Kunabum- plant growth promoter**

Preparation
Meat (1 kg), Beef (1 kg) and Pork (1 kg) are boiled well. After boiling, it has to be cooled and cow urine (10 litres) has to be added into it. These contents are kept for fermentation for about 10 days. It has to be stirred well daily. After 10 days the solution is filtered.

Based on scientific reason

- Practically this method cannot be used. Even at villages large scale availability unsold/ waste flash is not possible. In other word is highly

	Dosage: 3.5 litre /acre. (50 ml/10 litre water) **Additional information** Meat, Beef, Pork and Cow urine are having rich nutrients. So this kunabum is used as plant growth promoter. Unsold /waste flesh can be used for the preparation of the kunabum.	nutritive plant material is available at cheaper rate at large quantity. Any material after boiling or before boiling under goes fermentation thus will be more microbial fauna (like panchakaviya) in the fermented product that will be nutritious.
7.	**Ginger + Garlic + Chillies + Cow urine** Ginger (1/2 kg), Garlic (1 kg) and Green chillies (1kg) are pounded well and made to a paste. This paste and Cow urine (10 litres) are put into the buried mud pot. Mouth of the mud pot is covered with waste cloth and the contents are allowed for fermentation for about 10 days. To obtain well fermented solution, every day morning it has to be stirred well by using wooden stick. After 10 days fermented solution is filtered and used for spraying. Spraying can be done in weekly intervals. Cost incurred for the preparation of the solution is Rs.50-100/ac. The solution controls 40-50% of the cotton boll worms. **Dosage:** 3.5 lit/ac (0.5 lit/ 10 lit water). **Note:** Fresh cow urine should be avoided.	Based on scientific reason • Pungent taste of spray after fermentation of ginger, garlic and green chillies may not like by larval stage. Same condition may deteriorate the adult settling behaviour. Cost of 3 G's to be worked out for adoptability.
8.	**Rice bran + Neem oil** **Symptoms of damage** Nymphs and adults congregate at the plant above the water level and suck the sap from the tiller. The affected plant dries up and gives a scorched appearance called "Hopper burn". Circular patches of drying and lodging of matured plant are the typical symptoms caused by the pest. It is vector of grassy stunt, ragged stunt and wilted stunt diseases. **Procedure** Rice bran or husk (5-10 kg), Neem oil (1 lit) are mixed well and sprinkled over the rice crop at early morning i.e. 5-6 a.m to control Brown Plant Hopper damage. Cost involved for the preparation is around Rs.50/ac. **Existence of the practice:** More than 50 years.	Based on scientific reason • Rice bran+Neem oil admixture after application may be settled on the water the role of rice bran under the situation is not known. But Neem oil through its azatirachtin content deterioting property and systemic activity will not allowed BPH to move from one tiller to another tiller.

9. **Pig Manure Smoking**
Pig manure, dried neem and notchi leaves are used to make smoke. This is being followed to repel the pests and disease incidence.

Based on scientific reason
- Again smoke due to pig manure, Neem leaves and Notchi leaves will deter away the pest.

10. **Graveyard Ash Sprinkling**
Method I
Graveyard ash, cow urine (10 litres) and turmeric powder (½ kg) is mixed and the solution is sprayed over the vegetable crop.
Existence: > 125 years
Method II
In early morning ash powder is sprinkled in bhendi and brinjal field. It reduces the pest incidence.

Based on scientific reason
- Has very limited practical usage, due to non-availability of any ash and cow urine
- Ash will not dissolve in cow urine / water. They may deposit on leaves/ fruits as such and aberate the

11. **Payir Valarchi Ookki (Plant growth promoters)**
Method I
Cow urine (5 litres), Neem oil (5 litres) and Nayuruvi (1/2 kg) are taken in mud pot and the contents are allowed fermentation for about 10 days. The fermented solution is transferred into another mud pot and small hole is made on the bottom of the pot. This pot is kept in the entrance of the irrigation channel, so that crop can get the solution for its growth along with irrigation water.
ii) Fresh cow dung (10 kg) is taken in gunny bag and kept in the entrance of the irrigation channel or cow dung is mixed with irrigation water as such.
These indigenous practices can be followed during 30-40 days after sowing, when crop is in the main field.
Method II
100 g Sheep manure is mixed with one litre of water and this solution is sprayed in the early morning.

Based on scientific reason
- Fermented neem oil solution is allowed thro' irrigation channel and is reached to individual plants (Site specific organic manuring).
- Again organic manuring in a site specific

12. **Paravai Thanki (Bird Perch)**
- It is used to control insect in rice field.
- Square shaped polythene sheets are used.
- Coconut branches are kept in the field randomly used as paravai thanki.
- 'T' shaped stick is used for this purpose.
- Total numbers required: 15 Numbers/ acre.
- Expenditure: Rs.150 /ac.

Based on scientific reason
- Bird perches with T shaped stick can be used to invite predatory birds to pick up rats and larvae.
- Use of polythene sheets or coconut branches in different practice to avoid certain birds which will damage the crops. Polythene sheets give shining light and coconut leaves will produce sound, which will scare away the birds.

13. **Eli Viratti (Rat Control)**

- Palmyra frond is tied on the stick and kept in the rice field. It will create sound due to air movement, which will scare the rat away.
- Total numbers required: 15 Numbers/ acre.
- Expenditure: Rs.150 /ac.

Based on experience

- Initially rats can be scared. But sound sharpness will be developed at rats. In general, rats are little clever, they can develop shyness to feed baits and even chemicals.

14. **Eli Kattuppaduthuthal (Rat Control)**

- Mud pot is kept upon 3 pegs in which some popped rice are placed. A mud lid is kept beneath with some popped rice. 4 sets of pot are placed in 4 corners of the field. First rat will come to take the popped rice in mud lid, after that it will go to take the popped rice in mud pot, at that time pot is fell on the lid. The rat is trapped and killed.
- Pots are filled with water and buried in the bunds of the field. When rat enter into the field, it will be trapped and killed.
- By leaving space (nearly 1 ft) in all four sides of field, rat damage will be reduced.
- Cow dung balls are immersed in kerosene and kept in 8-10 places of the field randomly. Odour comes from kerosene and cow dung will control the rat damage.

Based on experience

- Smell from popped rice may attract rats.
- The different smell omitted from kerosene + cow dung may not like by the rats and get deteriorated.

15. **Neekal Poduthal (Rat control in Rice Field)**

During panicle initiation and milky stage, rat damage in rice field is one of the major problem. In order to avoid this problem, Neekal poduthal is one of the traditional practices followed in Kanchipuram District. Neekal poduthal is nothing but giving one feet space in 4-5 places of the field randomly when the crop is at milky stage in the rice. So that rice field will get more light penetration which prevents rat entry and movement. It also provides chance to owl to catch the rat in the rice field during night time.

Existence of the storage structure: > 100 years

Based on experience

- Providing more interfaces between plants and rows leads to vigorous growth of plants.

16. **Paravai Viratti (Bird scaring)**

Among all the growth stages, economic part development is the most important and crucial period. During such period, external damages especially by birds are more

Based on experience

- Sounds normally deter away the birds. Whistling sound due to tapes and common and more intensive. Due to such damages farmers face loss in yield. Normally, to avoid such loss happening due to birds, various local bird scaring techniques are being followed by the farmers. Following are some of such techniques which are being followed by farmers in Trichy district. These techniques are highly popular among small and marginal farmers, because it involves low / no cost expenses.

Tying Palmyra fronds

Dried palmyra fronds are tied to the poles which makes sound when wind blows. In an acre nearly 5-10 such fronds are tied to a pole and kept sparsely. Thus, by producing sound, birds like sparrows and parrots are scared and the crop is saved from its damage.

Tying Unused Recordable Tapes

Unused Recordable Tapes are tied across the field in a cris-cross manner and it looks like a net spread over the field. So birds perceive that a net is bring spread to trap them and avoids to land on that field. When light falls on it the polished sides, reflects the light scares the sparrows, parrots and other birds.

Tying polythene sheet

Polythene sheets/papers are tied to a pole as like that of dried palmyra fronds and kept in the field. When wind comes in contact with it, sound is produced and this will scare the birds.

Beating drums and using crackers

Sometimes to scare the birds, sound is produced by beating drums or by beating a plate with a stick. Crackers are also fired and thus sound is produced. The disadvantages of this method is, it needs some persons to perform, by firing crackers environmental pollution happens, which is not an encouraging activity.

All the above techniques, uses a simple logic i.e., by producing sound in a way or other, the birds are made to scare and thus the crop is protected from damage. Tying polythene sheet is followed both in garden land and wet land crops (Rice). Remaining other three techniques are followed in garden land only. (Ragi, Cumbu, Sorghum and Maize). The farmers of Perahambi and Pachamalai (Trichy district) are practicing these techniques.

shining appearance may deter away the birds.

- Beating drums and using crackers costly affairs to deter away the birds.

17.	**Manjal and Mor Vithai Nerthi (Seed treatment with turmeric and butter milk)** 1 kg of turmeric is taken and it is soaked in 2 litres of skimmed butter or Neeragaram. It should be pounded well and the extract is used for seed treatment. This can be done in two days after harvesting of the rice grains. Treated seeds are dried in shade for 8-10 hours. This will control the storage pest moth and enhance the germination percentage. Treated seeds are stored in "Kottai"- storage structure made out of paddy straw (now it is not in use). Cow dung is smeared over the kottai to avoid moth damage.	Based on scientific reason • Seed treatment with turmeric powder aberate the approaching the insects. Turmeric smell sometime deters away the storage pests. • Cow dung smearing is not allowed the pests to enter into "kottai".
18.	**Mann pzhlu uram (Vermi compost)** It is practiced at Bogasandiram village of Thenkanikottai taluk in Krishnagiri District. He has been practicing vermi compost for past two years. Raw materials used : FYM, Coconut sheath and Banana sheath Vermi tank size : 2.50x2.25x0.65 Vermi wash tank size: 0.65x0.60x0.50m Verm Name : *Perionex escavatas* The raw materials are changed once in two months. He collected the vermi tonic from vermi wash tank at monthly once. Vermi tonic acts as a growth regulator. The 15 litre of vermin tonic is diluted with 60 litre of water and sprayed in to the field. It reduces the pests and disease incidence. The vermi tonic applied field totally prevented from animal grazing. He sold the verms and vermi compost. Until he sold the 30 kg of verms around the cost of Rs.5000 and he earned the money from selling the vermi compost also.	Based on scientific reason • Scientifically it is proved that vermicompost is good organic manure. Emerging worms from vermicompost enrich the soil condition and nutritive status. • Vermi wash / tonic is already known for their plant growth promoting activity.
19.	**Lanther Pori (Light trap)** Light trap is used to trap the pest in the rice field. Now a day's bulb is used as light source, previously Aricane or lanther light is used as light trap. Lanther is kept on the stool which is in one corner of the field. Below this structure, water and kerosene mixture is kept in a pan. Pests are attracted by light and fall in the pan. Once the pest fall into the pan, wings get damaged due to kerosene. So it can not fly again. Light trap can be used between 6.00-9.00 p.m.	Based on scientific reason • Most of the insects are phototrophic and attracted to light sources during night hours. Attracted insects, if they fallen pan which is entering kerosenated water, insect will be killed due to kerosene.

20. **Poochi Pori (Pest Trap)**

It is an organic method of plant protection i.e. pest trap in a low cost manner. This organic method of plant protection measure is used to trap the pests particularly the Rhinocerous beetles and root grub.

Description

A mud pot is filled in with 0.5 litre castor oil and 4 litres of water. It is mixed well and buried up to pot's neck inside the field. This solution creates an odour which attracts the Rhinocerous beetles and root grubs. Thus the incidence of damage to the leaves and economic parts are diverted into this trap. If sunlight is more, the odour is increased which attracts more insects into it. Nearly 40 pots have been buried in his field @ 1 pot per slope i.e. he has made 40 contours with very less slope in which each contours has one pot buried with the solution.

About 2000 beetles are being trapped in a day through a single pot. These beetles are in turn used as organic manure to field. Since the wings are wetted with oil, it sticks to the solution and due to weight; beetles are unable to fly out of the pot. Thus beetles are trapped and such beetles are used as organic manure by incorporating it near /around the trees.

Cost effectiveness

Cost of pot is Rs. 10-12. Caster costs Rs.10-12/kg. So totally Rs. 20-22/ is the only expenditure to be incurred in making this pest trap. So compared to other ways and means of controlling pest, it is much cheaper as well as easier.

Advantages

This method is highly advantages as well as highly useful. By cost factor, it is much cheaper, by made of operation and maintenance; it is highly easier and comfortable to handle. Moreover, it controls the pest and makes it available as manure to crops. This method turns the negative factor, so called pest into a positive factor called organic manure to crop. Thus it is highly advantageous.

Disadvantage

Though it is highly advantageous, the only disadvantage is that, it fails to trap all pests.

Other details

Initially, he tried with curd and castor but it was damaged by dogs due to the preparation and taste. So he avoided instead of curd and using water.

Based on scientific reason

- Castor oil will not dissolved in water. Smell from castor oil alone may attract rhinocerous beetles and root grubs.Attraction to a level of 50 beetles /trap is a good attempt.

21. **Mooligai Poochi Viratti (Herbal Pest Repellent)**

a. "Oomathai" (*Datura metal*) leaves – 2 kg.
b. "Erukku" (*Calotrophis gigantea*) leaves – 2 kg.
c. "Peenaarisangu" (*Clerodentron inerme*) leaves – 2 kg.
d. "Thumbai" (*Leucas aspera*) leaves – 2 kg.
e. Water or cow urine – 10 litres.

Method of preparation

The aforesaid leaves are to be ground well by using grinder and soaked in an earthen or plastic vessel. This has to be allowed for fermentation by keeping the vessel as such for 15 days. Then, the solution is to be filtered and added with water for spraying. The recommended ratio is 1:10. It is used to control all types of pest in fruit trees and especially fruit borer.

Advantages

- The cost of preparation of pesticide is Rs.150/ per ha. But the chemical pesticide cost is Rs.500-800 per ha.
- It is eco-friendly and will not create any detrimental effect on human body while applying and ensures sustainable farming.

Based on scientific reason

- It is like panchakaviya. Fermented liquid, when sprayed on fruits, the foul odour and unpalatable taste may lead to avoiding of fruit borer larvae from biting and adults from egg laying on fruits.

7

Rationality on Indigenous Technical Practices Post Harvest Methods

POST HARVEST METHODS

DSIR/TPDU/TIF 056 and E-28-GO-GOI Research Scheme on "Documentation and Preservation of Agricultural Traditional Knowledge through Farmer Participatory Approach"

S. No.	Documented ITK Practices	Rational
1.	**Vallai Nar Oorithal (Banana Sheath Fibre Extraction)** Well matured banana trees are selected. The pseudo stem of the trees are removed and allowed for optimum drying. After drying, with the help of scrapper-Vaazhayinaa kathi the sheath is removed. Now, the pseudo stems are converted as raw banana sheath fibre. It is repeated to get better quality fibre. The fibre is used to make toys, fancy articles, bags. Fetches better price in the market and even sometimes the products are exported. They are easy to use and highly economical.	Based on scientific reason • This technology is being followed Trichy and Kanyakumari district as a homestead venture by self help group peoples.

2.	**Pattai Oorithal (Dried Cinnamon barks and leaves)** Cinnamon is a spice crop which is used in soap industries. The barks are falling off from the tree even after maturity. Similarly the leaves fall off from the tree are also collected in a gunny bag. They are weighed and sold for better price in market.	Based on scientific reason • The bark making from the cinnamon plant is called quills.
3.	**Panam Kalkandu (Palm candy extraction)** Palm candy is an alternative to palm gur-"karuppukaktti". It is used by the people of kanyakumari district in their tea, coffee making. This is extracted from Neerah-"padhaneer". **Procedure** Fresh neerah is collected from the palm tree in early morning. Water should not be added to it. The collected neerah is brought to the palm candy manufacturing centre. It should be sweeter. The ph and brix value are maintained to above 9 and 14 respectively. Super phosphate is added to neerah it serves as filter and made it clear solution. Super phosphate helped to remove the lime present in it. Again, neerah is filtered using a filter cloth. The filtered product is again given a uniform heat up to 108^{0}C. The heated neerah is poured to the crystallizer of 30 liter capacity. Sand is kept around the crystallizer to avoid movement in it. Sometimes, paddy straw and bricks are also used. Crystallizer is kept in the same position for 41 days. After 41 days, the palm candy can be taken out from the crystallizer frames. Immediately after removing out from crystallizer, the palm candy is washed with cold water and kept for sun drying. Threads on them are removed. Based on their size, the grades are separated but the quality remained the same. Finally, they are packed in polythene bags for sale. From a crystallizer 12-14 kg of palm candy is extracted in a single time.	Based on scientific reason • Commercial practice is being followed in Thoothukudi, Kanyakumari and thirunelveli district.
4.	**Panai olai petti thayarithal (Palm leave basket making)** Palm leave basket and hat making process is the community based profession (of "kuravars", "Sakkiliyars" and "Paraiyars") in Uchipuli region of Ramnad district. Mat weaving is done by the rest of communities (Nadars and thevars). They buy palm leaves from adjacent areas. Palm leaf box, hat and mat weaving are done by "Sathuvakathi" especially made for palm leaf products. The edged portion of "Sathuvakathi" is	Based on experience • This technique is followed in Kanyakumari district.

used for tearing of palm leaves and back point is used for punching of palm leaves while making palm leaf products. It is specially made curved portion on top. The palm leaf petti ranges from Rs.20, 30, 50, 60 based on size and colouring given to palm leaf petti. The plastic wire is also added to give good support to the palm leaf petti. The matured palm leaves are bought for 10 paise and "kuruthu" bought for Rs.3-5/-. The customers are allowed to take only the palm leaves. The palm leave mat is made from matured leaves and the younger palm leaf (kuruthu) is used for "petti", "Kottan" and hat making .The palm leaf mats are used in ships, drying of grains and in packing of fish ice boxes. The younger palm leaves known as "vidali" are used as brushes .The palm "Ekki, and palm"Thumbu" are exported to foreign countries from Trichy and Tuticorin areas. As rain fed farming is not profitable one, these people are interested in palm leaf product business. The bigger kuruthu palm leaves are used for "narpetti" making and smaller kuruthu palm leaves are used for "fine mat" weaving. These fine mats have the longevity of several months.

5. **Panai Olai Pai Pinnuthal (Palm leaf mat weaving)**
Palm leaf mats are produced by matured palm leaves and palm leaf "kattu" was bought from nearby places (one palm leaf kattu = 40 leaves). Some time palm leaves are alone bought leaving mattai and "eekki". There are three types of mats are made viz., ordinary mat, nice mat, and kurutholai mat. The longevity of ordinary mat is just one day (Rs.2-3) and nice mat is kept for 2 months (Rs.6/). The kurutholai mat has longevity of more than 2 months (Rs.15) and is used for drying of paddy and keeping small children inside the mat. Ordinary mats numbering 6to 7can be made by single person in a day. Where as only one kurutholai mat can be made in a day. Small type of "kottans" and Archani thattus are made by palm leaves. The jaggeries are kept in "kottans" for safe transport.

Based on experience
- Commercial practice is being followed in Thoothukudi, Kanyakumari and thirunelveli district

6. **Munthiri Pathapaduthuthal (Cashew nut processing)**
Cashew processing
Cashew is one of the important cash crop which are grown in larger area of Perambalur District of Tamilnadu. From seed selection to packing several

Based on experience
- Commercial practiced in Kanyakumari district with local cashew nut. Raw cashew also imported and

indigenous practices are involved, they are detailed below;

Seed selection

Nuts are dried for about 2-3 days. One kg of cow dung is mixed in three litre water and made it as solution, which is locally termed as 'Saanipaal.' Dried nuts are put into the solution, some of them floated and rest of them sunk in the solution. Floaters are discarded, because these are immature or ill filled nuts. Rest of the well filled nuts is selected for sowing.

Processing

Fresh nuts separated from apples are sun dried for about 2-3 days. (Less or more sun drying yields increased number of broken nuts) after that, nuts are ready for processing, for manual breaking of the whole nuts, plate like square shaped rock stone and small cylindrical stone are used. While breaking the whole nut is hit vertically and horizontally. So that the nuts is separated easily from the outer shell. Small iron needle is used for separation. During this operation, exudates is oozing out from the outer shell, which will create wounds on fingers, in order to avoid this wood ash is smeared over the fingers. Mostly women are employed for this cashew processing operation. One woman can process 21kg of nuts in a day. One kg whole cashew nut yields about 270 gram processed nuts. Wages for this processing operation is Rs 2.30/kg.

Only disadvantage is that, more broken nuts are obtained from manual breaking compared to machine breaking.

Peeling

After braking, nuts with light orange colour peels are obtained. Peels are removed by using small knife, specially made for this purpose, which costs about Rs 35. One woman can peel up to 7 kg nuts/day. Wages for this is Rs 4/kg.

then processed cashews are re-exported.

7. **Puli Pathapatuthuthal (Tamarind processing)**

The main concept behind tamarind processing is purification. i.e., Tamarind is deseeded and dried by which the pulp alone is separated and stored. By doing so, its shelf life increases and moreover as days passes the taste increases in such processed tamarind.

The method in purification (processing) of tamarind differs from area to area.

Based on experience

- It is practiced in many of the rural parts of Tamil Nadu and Anthra pradesh

Method 1

Tmt. Rasathi, residing in Siruganoor, Manachanalur taluk of Trichy district processes Tamarind using stone mortar and wooden pestle. This method of Tamarind processing is almost followed by most of the women in Trichy district particularly in Manachanallur taluk.

In this method, a stone mortar, both of circle as well as rectangular shape is used. Dried tamarind fruits are put into this mortar and pounded by using a wooden pestle (ulakkai). So that the entire content is deseeded. To avoid the pulp sticking the pestle, either castor oil or gingelly oil is smeared at the edges of pestle

Efficiency

Nearly 30 kg of tamarind fruits can be processed at a time by 4 women members. After pounding, tamarind pulp and seeds are removed separately by hand. The advantage of the method is the bulk quantity that can be pounded simultaneously.

Method 2

The dried tamarind from Arival manai (a kitchen gadget with curved iron blade fixed to a base and used for the cutting vegetable) or by using suthiyal (hammer). This method is used only at household level i.e., for processing minimum quantity of tamarind that is used for household consumption. This is most common practice followed in almost all eco systems.

8

Rationality on Indigenous Technical Practices Seed Storage Structures

SEED STORAGE STRUCTURES

Seed Storage Structure

Kuluppai / Kulumai (Mud Pot)

Th. Gopalakrishnan, 52 years old is residing in A.Edayapatti village in Manapparai taluk of Trichy district. He is educated up to 10th standard and has agriculture as his primary occupation. He possesses a post harvest storage structure called 'kuluppai/ kulumai'.

It is a storage structure used to store grains, which is suited to all ecosystems. This structure is invented by ancestors and maintained for more than 100 years until now.

This entire structure has a construction process similar to that of constructing a house. Stone pillars (Nearly 10 feet) are buried upto 6 feet into the ground in circular manner. The distance between each pillar is about 3-5 feet. Thus a basement is made and upon these pillars, wooden planks (Mostly of Neem – *Azardiracta indica*) are laid down. These planks are plastered by using clay. A platform is raised on the base about 1-1.5 feet height, in a similar circular fashion. Above this platform a circular bin (storage structure) is erected with 6-7 feet height, 2.5– 3 feet width. Nearly at 6 feet height of this bin, a small window is provided with a lock that serves as an entrance into the bin. The side walls are made up of bricks (Dried stalks of *Dodoneae angustifolia* – virali are also used) and both inner and outer walls were plastered with clay.

In the total 6 feet height nearly bottom 3-4 feet of the inner portion is compartmentalized into four equal compartments using wooden planks.

The top portion is closed with wooden planks, (generally neem) and plastered by using clay. Above of this, a hut like thatching is made using either sorghum or maize stalks. This thatched roof is pillared by 6-10 neem pillars (up to 5-6 feet), which are raised upon the platform margin.

Existence

Its existence is found to be more than 100 years and it is a core structure mostly erected by farmers in and around Trichy district. It is believed that the structure has a spiritual power i.e. the holders state that, members from other than their family cannot get into the structure and can't even sit/sleep during night hours. Moreover, even the female members of their own family are avoiding stepping into the platform.

Cost Effectiveness

The entire structure is more or less similar to a hut in which a four member's family can dwell. So this structure, comparatively involves some investment on pillars, wooden plank, bricks etc.

Feasibility and Relative Advantage

Though there is a spiritual belief on the structure, it is feasible to construct and maintain the same in other location too. The advantages can justify it.

The Advantages are

- Size- a small hut sized structure enables to store more quantity of grains.
- Compartmental structure provides space for storing various grains simultaneously.
- At a time 100 bags of grains can be stored.
- Multipurpose- since it is constructed on pillars, where the bottom portion underneath the platform can be used to herd goats. Thus it is used both for grain storage as well as shed for small animals.

Moreover, the platform (veranda) rose around the circular structure provides enough space for the family members to use it for other purposes (like drying, staking any materials etc.)

Condition for Adoption and Limitation

This structure needs a separate place to be allotted for its construction. Since it is an out of house storage structure, cost is one of the limiting factor (due to its size). But the benefits out of this structure will definitely mask these limitations.

Though the structure is made with enough space, the entry window is a limitation i.e., the window provided is very small, that requires care. The trained person's only can enters in to this window.

Thaniya Kuthir (Grain Storage Structure)

Type I

Th.kuppamuthu, 49 years old is residing in kambalathupatty in Manaparai taluk of Trichy District. He is educated up to 10th standard and agriculture is his primary occupation. He possesses a post-harvest storage structure called **'Thaniya kuthir'.**

It is a storage structure used to store cereals in almost all ecosystems. This was invented by ancestors and carried by present generation. Nearly 5 bags of cereals can be easily stored in a safe manner within the house.

This structure is spindle shaped and circular frames are used to construct it. These frames are, made up of clay and rice bran mixture. Based upon the quantity to be stored, the size of the structure and number of frames to be prepared are determined. To attain spindle shape, each frame is reduced from the size of the centre frame. Finally, the position of the frame is fixed with dots (clay balls) i.e. bottom most frames are fixed with one dot and the second with two dots and so on. Finally to top opening is based with a mud made lid. This helps to ensure balance and order of frames while positioning it. These frames are generally made by potters and bought by farmers.

These frames are positioned and plastered using clay and cow dung mixture. For outlet, a hole is made in the bottom most frames, which is plugged with paddy straw and plastered by using clay, similarly the lid is also plastered.

Existence

It exists for more than 100 years, but now faces an extinction phase due to the arrival of modern iron storage bins. Even though iron bins are introduced, due to its limitations there are lots of scopes for reintroducing such mud made storage structure in the rural houses.

It is completely made out of raw materials that are easily available. Its cost of making (including labour charge for potter) will come around Rs.200/-.

The structure is neither too small nor too big, so size is compatible for having it inside the house. The maintenance is very simple as well as damage is very meager which are the added advantages of this storage structure.

Since it is a mud made structure, it can be placed only in the rural houses which are made out of mud/concrete, but not in any other type of houses. This structure is mainly used to store grains that are to be used for seed purpose. So it doesn't fit into urban centres.

Other Details

Whatever to the number of frames, the topmost frames is encased with a symbol of growing moon (Pirai Nila) . They believe that, as like the moon grows, their grains wealth also grows, ensuring their food security.

Type II

It is a storage structure used to store cereals in almost all ecosystems. This was invented by ancestors and carried by present generation. This structure is cylindrical shaped and circular frames are used to construct it. These frames are, made up of clay and rice bran mixture. Based upon the quantity to be stored, the size of the structure and number of frames to be prepared are determined. The structure is provided with a small hole at the bottom which serves as outlet and bigger hole at the top which serves as inlet. The bottom hole is used for taking out the grain. Out let is covered by using coconut shell and plastered with clay whereas inlet is covered with a mud made lid or wood. The capacity of the structure ranges from 250-300 kg. These frames are generally made by potters and bought by farmers. These frames are positioned and plastered using clay and cow dung mixture.

Existence

It exists for more than 200 years, but now faces an extinction phase due to the arrival of modern iron storage bins. Even though iron bins are introduced, due to its limitations there are lots of scopes for reintroducing such mud made storage structure in the rural houses.

It is completely made out of raw materials that are easily available. Its cost of making (including labour charge for potter) will come around Rs.200/-.

The structure is neither too small nor too big, so size is compatible for having it inside the house. The maintenance is very simple as well as damage is very meager which are the added advantages of this storage structure.

Type III

Purpose

This drum is traditionally used for storing paddy, cumbu, ragi etc. This method is in existence for about 20 years in kakarunkalpatti village at Salem district.

Making

This storage drum is made up of iron steel plate. The cylindrical shaped drum is having a flattened bottom and the top. The height of the drum is 7 feet and the diameter is 3 feet. The top portion of the drum have the inlet, through this grains are filled. The bottom portion of the drum is having one small opening for withdrawal of grains whenever it is needed.

Cost

The cost of the drum is Rs.1000 in 10 years ago; nowadays it is Rs.3000.

Special Feature

1. It will be purely protected from ants, rat and termites.
2. The capacity of the drum will be 550 kg.
3. Thombai koodu (Grain Storage Structure)

Type I

Th. Senguthuvan, 45 years old is residing in Vellapallam village in keevalur taluk of Nagapattinam district. He is educated up to 10th standard and agriculture is his primary occupation. He has a storage structure called **Thombai koodu'** which is used to store grains. This storage structure is suited to all ecosystems but possessed mostly in coastal areas only.

The purpose of this structure is to retain the quantity of the seeds to be used either for seed or consumption purpose.

This structure is mostly spindle shaped with an average height of 4-5 feet. It is made out of bamboo palm splits, paddy straw, rock pillars and few poles. Either bamboo splits or palm frond splits are intertwined in a criss-cross fashion and a spindle shaped basket with a wide opening at the top is made. An opening of small size is made at any one side at the bottom outlet. This basket is placed upon a platform made out of poles which are raised on four small stone pillars (approximately 1 ft).then the basket is plastered by using clay and cow dung mixture. This is left to dry for one day. By plastering, it is ensured that there is no opening for pest entry.

After drying, the outlet is plugged with straw and the grains are filled into it and covered by using paddy straw. Finally the entire structure is enclosed either by paddy straw or palm fronds. Now, the entire structure is tied by using a rope made out of palm fronds. (i.e. palm fronds are split into many then and long fibre are made into rope).It is nearly stored upto 60 kgs.

Existence

The practice of having **'Thombai koodu'** has its origin beyond more than a century, as stated by him. This type of storage structure when compared to other structures followed in other location (eco system) is different i.e. it is made without completely depending either mud or wooden planks. This kind of storage structure is much prominent in coastal ecosystem (especially Nagapattinam), because of non-availability of clay that is needed for making storage structure.

The storage structures are made out of materials that are locally available like palm fronds, paddy straw etc. so it doesn't incur more expenditure for constructing this structure. The cost of making one such storage bin comes around Rs.50/- depending upon the size. Generally, the bin is bought from carpenter through orders. Thereafter, the outer plastering and covering with paddy straw/palm fronds are done by the holder.

The whole structure can be easily shifted to another location whenever needed. Moreover, it doesn't need more space to erect. Therefore space is not a constraint. Similarly, the maintenance is very less, but need to do twice every year (like plastering, cleaning etc.).Thus every handling, less maintenance, cost effectiveness is the relative advantages of this storage structure that makes it feasible to erect.Since it is an out of house structure, there is no specific condition for adopting it.

Limitations

First, when compared to other storage structure, capacity of holding is less. Secondly, when proper plastering is not done there is a chance of termite damage. Similarly when the outlet is not plugged properly, loss / damage of grain may occur in both quantity as well as quality wise.

Type II

The purpose of this structure is to retain the quantity of the seeds to be used either for seed or consumption purpose. This structure is mostly spindle shaped with an average height of 4-5 feet. It is made out of bamboo palm splits, and then the basket is plastered by using clay and cow dung mixture. This is left to dry for one day. By plastering, it is ensured that there is no opening for pest entry.

After drying, the outlet is plugged with straw and the grains are filled into it and covered by using paddy straw. Finally the entire structure is enclosed either by paddy straw or palm fronds.

Existence

The practice of having **'Thombai'** has its origin beyond more than a century, as stated by him. It is made without completely depending either mud or wooden planks. Cylindrical structure made by using bamboo splits and plastering with cow dung. It doesn't incur more expenditure for constructing this structure. The cost of making one such storage bin comes around Rs.75/- depending upon the size. Thereafter, the outlet and inlet are covered with paddy straw bunches and plastered with cow dung and clay mixture.

The whole structure can be easily shifted to another location whenever needed. Moreover, it doesn't need more space to erect. Therefore space is not a constraint. Similarly, the maintenance is very less, but need to do twice every year (like plastering, cleaning etc.).Thus handling, less maintenance, cost effectiveness is the relative advantages of this storage structure that makes it feasible to erect.Since it is an out of house structure, there is no specific condition for adopting it.

Limitations

- When compared to other storage structure, capacity of holding is less.
- Similarly when the outlet is not plugged properly, loss/damage of grain may occur in both quantity as well as quality wise.

Type III

Thombai is an out door storage structure found in the villages of pachamalai. It is a circular storage structure made out of clay and bamboo splits. Wooden poles are arranged in a criss cross manner on four stone and smeared with clay. Hence it looks like a floor or thinnai. Five feet lengthy bamboo splits are erected on the floor or thinnai in a circular manner. Inner and outer side of the bamboo splits are coated with clay for about 5cm. inner side of the thombai is divided into two rooms. Roof is made on the top portion of the thombai. Cereals, pulses can be stored in thombai with out pest attack and rat damage.

Existence: More than 75 Years.

Name: Mrs.Revathi Address: Periyapakalam, Pachamalai, Trichy district.

Ragi Kuzhi-(Grain Storage Pit)

Th.M.S.Sivanna, 60 years old and 40 years experienced in agriculture is residing in Mallan kuzhi village of Sathy taluk of Erode district. His primary occupation is agriculture and the predominant ecosystem is garden land ecosystem, where all pulses are grown along with minor millets like Ragi, Cholam. As a post harvest storage practice a storage structure called Ragi pottu (Ragi crop waste) is used. This post harvest storage structure is typical serving as a multipurpose structure.

Description and Uses

A big (about 1 feet height) area is exhausted below ground level either in circular or square shaped, with a small mouth that provides as inlet. Inlet is dimensioned such that one person can enter into it. Stones are used to line the inner wall so that, soil is not eroded in the inner side. Harvested produce are stored inside the structure which ensures the quality of produce for longer period. The capacity of this structure is about 4 qtl. While storing cholam, the gunny bags are plastered by using mud and this type plastered and stored cholam is used only for consumption and not for cultivation. Banana ripening can also be done in this structure. A bunch of banana is kept inside the structure flames are raised using Sambirani. By doing this, within 3 days banana gets ripened.

Cost Effectiveness

Except excavation cost, there is no additional investment in this structure. This type of structure is entirely different because, it is not erected /constructed completely but excavated and a room like structure is made.

Advantages

No maintenance cost, is the main advantage similarly, stored produce is not affected by any pest and diseases which is the desired advantages of any storage structure.

Disadvantage

This cannot be followed in all ecosystem and all types of produce cannot be stored in it.

Existence of the ITK: more than 100years.

Pakka Kothi (Seed Storage Structure)

One such storage structure is Pakka kothi which is nothing but the modified 'kadage' –old traditional storage structure made out of bamboo splits. A bamboo made barrel shaped container or storage structure is called 'kadge'

Description

Pakka kothi is made out of brick and sand. It is a concrete structure which differs in shape but usually in square /rectangle shape. Walls are raised in four sides with an inlet on the top and an outlet at the bottom. Before fitting in the grains the inner wall is lined with plastic sheet to avoid damage to grains due to moisture. It capacity is 12 qtl.

Advantages

Prior to this structure, gunny bag storage was used (during extinction of kadaga) where rat damage was more. But it has been avoided by this structure moreover the holding capacity can be modified as per the need while construction. Grain quality deterioration also prevented due to plastic sheet lining.

Disadvantages: It occupies space with in the house.

Cost Effectiveness

It requires some expenditure to be spent on concrete work. But while calculating the benefits and advantages, the cost of Rs.360/ for making one such as structure seems to be an investment not expenditure.

General Profile of the ITK Holder

Name and address : M.Pomman, S/o Mathan, Malasajanoor,Burgur (post), Bhavani (TK), Erode district. Age: 35 years. Educational status : 5th standard.

Existence of the ITK: more than 85 years.

Thirugai/ Endhiram Processors /Hand Hullers

The stone made quern is called 'thirugu' or 'Endhiram'. The circular stones with a diameter of 30-35 cm and 15cm width are placed one upon the other. The upper stone plate is provided with a hole at the centre that serves as an inlet, through which the grain is powdered and poured in. Similarly, a wooden handle is inserted at the margin of the circumference of the upper stone. This handle is used as a grip while grinding.

The lower stone has a peg inserted at the centre of it, upon which the upper stone is placed. When the grains are poured into the centre hole and grounded. The grains get struck in between the stone plates and breaks into halves due to centrifugal movement, the split halves of grains fall out and get collected around the 'Thirugai'.

Advantages of Using the Hullers/Processors

- Cereals (rice, sorghum), minor millets Cumbu, ragi, samai, kuthiraivalli, thenai, varagu), Pulses (red gram, black gram, green gram and cowpea) are processed by using these traditional processors.
- While processing through these processors nutrients like minerals and vitamins are conserved as such, whereas in machine processing some of the nutrients are removed. Because, while operating the machine lot of heat is produced, that will affect the nutrient status of the grain.

Seru / Kalanjium Storage Structure

Name and Associated Crop with Ecosystem

Seru / kalanjium are a storage structure to store cereals, mostly found in garden land ecosystem.

Description

Seru / kalanjium is a storage structure which can be erected as both in house as well as out house structure. Generally, this structure is built as a part of house, by just making a partition across the opposite walls. Such type is considered as in house structure whereas in out of house structure, entirely a new house like construction is made, fixed with doors and compartments within the structure.

Stone pillars of one feet height are laid as basement; above which wooden planks are arranged to form a platform. This is plastered with clay so that there is no gap in between the planks. Walls made up of either brick or simply mud is raised along the margins of the platform. The height and size of the structure varies depending upon the needs and requirement. Irrespective of size and height, the width of the wall remains nearly 1.5 - 2 feet. After raising the walls a room like structure is formed which is either divided into two or three compartments. Each compartment has an outlet (mostly circular with a diameter of 15cm approximately) plugged with paddy straw after filling the structure with grains. The upper portion is left uncovered which serves as inlet. But like that of thatched roof made for houses. This entire structure is roofed.

There is no provision/staircase to reach the upper opening (inlet).But by stepping on doors and windows, one can climb to the inlet and other person should help him by lifting the grain filled through gunny bags or vessels.

This entire structure is built attached to the house and a common roof is erected both for the house and the storage structure.

In the case of out house structure construction process remains the same, but both the portions are compartmentalized and a common entrance is provided in-between those two portions. Since this structure is out of house to ensure its safety, a door is fixed with a lock.

Existence

As stated by the holder, the origin of this kind of structure goes beyond a century and is mainly followed in the garden land ecosystem.

Cost Effectiveness

In the first type of Seru the cost of construction is less when compared to the second type (i.e. out of house model).The reason behind is in the first model, the storage structure forms a part of the house whereas in the second model, a separate house like structure is built. So it requires, additional investment (Rs. 1000 approximately)

Feasibility and Relative Advantage

In the first model, safety is secured. Since it is within the house. In the second model, the capacity is two times more when compared to the other model i.e. each compartment has a capacity of holding 20 bags of grains (1200kg) obviously the holding capacity of second model increases.

Whenever required, based on the quantity, both outlet as well as inlet can be used for taking stored grains. I.e. if bulk quantity of grain is needed, then the inlet is used as outlet.

Constraints/Limitations

Though this structure is advantageous it fails to mitigate from rat damage, which leads to grain loss. Since, the inlet is left uncovered, pest and rat invasion is not checked and so this become a big disadvantage.

Similarly, climbing to the inlet is a risky task which can be performed only by trained persons. Also, when compared to other storage structures like (Kuthir,Thombai)this requires more space.

General Profile ITK Holder

Name & Age: Thiru.Natarajan, 62 YearsAddress: C.R.Palayam, Manachanallur Taluk,

Trichy District. Occupation: Agriculture-Farming, Farming experience: 45 years.

Saal /Urai (Traditional Storage Container)

Mud pot or mud jaadi used to store the grains. It is locally called as saal. Quantity of the saal may vary; it depends on the quantity of harvested grain availability while storing the grains in the saal. Ash, neem and notchi leaves are kept to control storage pest like moth. Inlet of the saal is covered by using cow dung.

Existence of the storage structure: > 100 years

Ambaram (Paddy Grain Storage Structure)

Five feet lengthy bamboo poles are place in the ground. With in the space available between the four poles (space will vary according to the paddy grains to be stored). Paddy straw is spreaded over the ground; above that gunny bags are spreaded. Over the gunny bags paddy grains are heaped. (After harvest, the paddy grains are dried in the sun light for about 7 days and winnowed before storage). Around the four poles paddy straw rope which is locally termed as 'piri' is used to cover the heaped grains. Over that polythene sheets / Tharpai is used to cover the entire ambaram from rain water. Upto 6 months grains can be stored.

Expenditure: Rs 200-300 / Ambaram

Existence: > 125 years

General Profile of the ITK Holder

Thiru:karuppaiah, Sukkankollai, Chengalpet tk, Kanchipuram Dt. Age:60, Experience in farming: 40years, land holding:3ac.

*Onion Pattarai – (*In Situ *Onion Storage Structure)*

Name with Associated Crop and Ecosystem

Onion storage structure – "Pattarai" (local name) practiced in garden land.

Origin: It seems that this practice is being followed for long past by their ancestors.

Purpose: It is used to store onion i.e. for in-situ storage.

Description

A leveled place (9'x3') at any corner of the field is selected. Four stones are placed, so that it serves as a base pillar. Over this base pillar, either bamboo or poles of any other tree is laid down. Harvested onion are uniformly spread over the poles and covered with coconut fronds (paddy straw is also used) to protect it from sunlight. This sort of storage is practiced only to store seed onions. Nearly 75 years old technology invented and followed by group. It is still existing and used by almost all onion growers.

Feasibility and Advantage

Since the material needed are locally available and the method of construction is very simple, it is much feasible as well as advantages in economic aspect i.e. *in-situ* storage and no transportation cost. Thus because of its cost effectiveness it is considered relatively advantageous.

Extent of Adoption

Due to its relative advantages, almost all of the onion growers adopt this storage practice.

General Profile of ITK Holder

Th. Raman.Address: Perahambi (P.o), Manachanallur (T.K), Trichy District.

Different Types of Bamboo Baskets

Since the farming is done in hilly area, the system and implements as well as the storage structures are somewhat different from that of plains. Among those structure various container made out of bamboo is typical and traditional to this locality.

a. Vidhai Petti (Seed box)

Bamboo is splitted into many small splits of 1 cm wide and length of 40-45cm. using these splits a container like structure is made i.e. splits of 2 cm wide is kept one upon the other in a star shaped fashion. This forms as a frame/skeleton to give a shape to the box. Then the small splits are in a criss-cross fashion by inserting in between the frame splits.

Thus a basket is made to a height of 15-20 cm with a diameter of 20-25cm. It is plastered with dung and dried in sunlight. Since it is used for sowing purpose, size is made such that it is neither too big nor too small such are filled in and broadcasted in the field.

b. Periya Koodai (Big Size Basket)

It is similar in material and making of seed box. But the only difference is its size. It varies from 1 feet height with 1-1 ½ diameter to 1 feet height with 2 feet diameter.

c. Moonkil Thattu (Bamboo drying plate)

Materials and making are similar to seed box and periya koodai. But the size differs i.e. height is very less up to 10-15 cm only, but diameter is more when compared to other baskets i.e. up to 3-4 diameter. It is used to dry the seeds.

d. Moonkil muram (Bamboo winnower)

Materials and making principle are similar to seed box, but the shape differs. The outer i.e. top most split is the base frame for this winnower, it serves as a rim on which the entire structure is shaped. In this winnower, one side is raised up to 20-25 cm and it is gradually sloped on both the sides and finally finished with a flat end. It is used to collect the harvested produce from the field.

Paddy Seed Storage Method

To prevent the attack of pests in stored paddy grains, a plant ‘kanjanamkorai’, is kept along with the grains. It emanates a pungent smell which drives away the pests. Sometimes, neem leaves are placed over stored grains to keep the grains pest free.

Soaked in water for 3 days, preferably in river water. The poisonous substances in the fruits get removed off. In the same way, again the fruits are soaked in water, boiled and washed. Finally added shriveled coconut pieces to it and kept for storage.

The above twelve practices are given to rational analysis and the practices are rational based on scientific reason and also experience. The following points are given below.

- Due to air tight nature, the moisture exchange with seeds and external environment is less. This improves storage environment and avoid pest attack.
- The moisture content of the seeds may be low for safe storage based on their experience.

DSIR/TPDU/TIF 056 and E-28-GO-GOI Research Scheme on "Documentation and Preservation of Agricultural Traditional Knowledge through Farmer Participatory Approach"

S. No.	Documented ITK Practices	Rational
13.	**Vaalai pazham Mootampoduthal (Banana Ripening Method)** **Gereral profile of the ITK holder** Thiru.Vellaiyan, Perahambi,Manachanallur(TK),Trichy (DT),Age: 45 years. **Preservation** Well matured banana bunch is harvested, which is cut into several bunches. The bunches are put into water for a week. It will increase the shelf life and freshness of the banana. Mostly Monthan variety is preserved by this method. This technique prevents the ripening for banana, until used for cooking / culinary purpose.	Based on experience • This technology is now converted into zero energy cool chambers. By which self life of the produce will be increased.
14.	**Mambalam Palukkavaithal (Ripening method in Mangoes)** For quick ripening of mangoes, normally sarakondrai flowers are used in coastal areas of nagapattinam. As stated by Th.Senguttuvan, residing in vellapalam in keevolur taluk of Nagapattinam district. The practice of ripening mangoes using Sarakondrai (*Cassia fistula*) flowers is alone in almost all mango growers farm and it exists for more than 60-70 years. A pit is dug in any corner of the field and the dimensions vary based on the quantity of mangoes to be ripened. The leaves of sarakondarai are put into it. Such that, it forms a bedding layer. Above this layer, the mangoes are arranged and covered with sarakondrai flowers. Finally a thin layer of soil is spread over it, so that it is completely covered. It is left undisturbed for one week after which it ripens. By doing so, the keeping quality of mango is high and the taste as well as aroma is retained. If the number of mangoes to be ripened is less, instead of pits, mud pots can be used. The practice of doing such ripening method exists for more than 90 years.	Based on experience • The sarakondrai plant has to accumulate more heat in the plant system. When the leaves and flowers are using for bed purpose, the heat will be released and this heat will be triggers the ethylene production by which mangoes get early ripening.
15.	**Maravalli kilangu Padhapatutahal (Tapioca preservation)** **Method I** In January month, the matured tapioca is dried under sun for a day. After this boil the tapioca till the raw	Based on experience • Another method of dehydration process with olden technology. Once moisture is fully removed

	taste removes and gets completely cooked. Leave after sometime and again boiled. This is to maintain stableness and again sun dried. Uluppai poochi which creates holes on the tubers are kept away during this storage. **II. Tapioca storage- vathal:** The freshly harvested tapioca are cut in to pieces and boiled in water. After boiling they are stored in the house, if required it can also be fried and stored over years. **III. Tapioca storage: Unakkan kizhangu:** freshly **harvested** tapioca is sun dried The sundried tapioca are taken in a gunny bag and tied. the tapioca in the gunny bag is kept for storage even up to 3 years	the produce gets long storability. • It is one of the oldest by products and value addition practices. • Gunny bags having holes, and through this aeration will be maintained by which storability life is lengthened.
16.	**Pirandai Storage Edible stemmed vine (Pirandai Processing)** The pirandai inside it is white in colour allowed to get soaked for a day, the quality or nature of the fruit is changed. Again, they are taken out and cut in to small pieces and put in water for 3 days preferably in river water. The poisonous substances in the fruits get removed off. In the same way, again the fruits are soaked in water/boiled, washed. Finally added shriveled coconut pieces to it and kept for storage.	Based on experience • This is one of the oldest technologies for the removal of poison alkaloids from the plant parts.
17.	**Snake Gourd Seed Processing** **Storage** Fresh Snake gourd seeds are extracted from the well matured fruit and mixed with fresh cow dung. This cow dung treated seeds will be placed on the mud wall of the house, which absorbs excess moisture. Cow dung which acts as antifungal and pest repellent agent, prevents pest and disease attack on the seeds until the next season comes. Seeds can be stored up to one year. **Sowing** Pit size of 1.5 feet depth and 2 feet width is dug in the pit ,a bunch of fresh Pungam (*Pongamia pinnata*) leaves are incorporated ,this will acts as green manure as well as preventing termites and ant damage. Due to the incorporation of pungam, the plant will emerge quickly and grow sturdy. After that, pungam is covered with thin layer of soil, 2.5kg well decomposed FYM is applied over it and then it is covered with 5cm of soil. The layer of soil and FYM should be mixed well and pit is rilled at centre, in which 3 seeds are sown. Sharp edge	Based on experience • It is one of the oldest technology for seed extraction and keeping for viability. From this only now panchakavya is formulated. • Application of more organic sources maintaining best physical and chemical chambers of the soil. By which growth of the crop is increased. • This technology is mainly used to provide conducive climate for plant growth as soil organic mulch. • Panipudal is preferred for single family for usage their daily life. Long type is best for big families.

of the seed is dibbled inside the pit and then covered with soil.

Finally the entire pit is covered with leaf litters in order to avoid excess evaporation which in turn avoids damage due to rodents. Immediately after sowing, the pit should be watered. The seedlings will emerge out 7 Days after sowing. After 3 weeks, the pandal is erected using Palmyra fibre/coconut coir pith rope/ Karung kodi- a twiner. Karung kodi-is a twine found in all coastal villages. Matured vine is used for preparation of pandal. Instead of other ropes, this is used, due to its hardy nature, after drying.]

From 50-60 days after sowing, it starts bearing. This is least preferred in the market. Coconut pith is tied at the posterior end of the fruit. There are two types of local snake gourd varieties are prevailing. One is lengthy snake gourd(Neetu pudal),which grows up to a length of 6 feet and it weighs about 2kg.Second is short and stout snake gourd (Pani pudal) which weighs about ½ kg-1kg and it grows up to 1-1.5 feet length. Suitable season for sowing snake gourd is July-August. Existence of the local variety is for 75 years.

18. **Ulunthu vithai nerthi (Black gram seed storage method)**

- Black gram seeds to be stored are presoaked in water for one day.
- Red earth is made into slurry and to this slurry presoaked blackgram seeds are soaked for 2-3 hour.
- Then it is sun dried for 2-3 days.
- While sun drying the slurry enclosing black gram is hardened and form a thick shield like enclosure around the seeds.
- Seeds treated by this method can be kept until next season without deteriorating the quality as well as the quantity of the seeds.
- This is a no cost technology that has the advantages of improving shelf life, preventing the seeds from storage pest incidences like bruchids.

Pre storage processing practices of pulses varies widely depending upon the type of pulse. Black gram is treated with coconut oil so that all the seeds are coated with oily surface. On an average 50 ml of coconut oil is enough to treat 1kg of black gram. Thus treated black gram seeds are stored in a container made up of mud or

Based on experience

- Red earth slurry pelleting prevents the egg laying of bruchids and seed coat.

plastic. These black grams can be stored as such or they may be split into two halves of using traditional stone splitters called 'Thirugai' splitted pulses can be sieved by using iron sieves. This type of iron sieve depends on the grain to be sieved and it varies from grain to grain. For e.g. big sized hole is used for sieving red gram, medium sized hole for black gram and small sized hole for green gram.

19. **Ellu vithai nerthi (Gingelly seed preservation)**

Gingelly is mostly affected by storage beetles that drastically reduce its quality and shelf life. So to mitigate from such damage, Tmt. Rajalakhsmi follows many methods that involves very low or no cost. They are found to be eco friendly, and the process is alone without deteriorating the quality of gingelly seeds.

a) **Using vasambu:** Just by placing Vasambu (*Acorous calamus*) on stored gingelly will safeguard it from pest damage. I.e. vasambu has a property of repelling the pest through which the seeds are safeguarded. This is a low cost technology that costs only about Rs.10/-.but the disadvantage of this method is that, only a little quantity of seeds can be processed and stored.

b) **Using dried chillies:** In this method, a handful of dried chillies are spread over the surface of gingelly. The concept of repellation is used here to control storage pest. This is also a no cost technology.

c) **Using ash:** The concept of mixing is similar but the content is different. Here ¼ kg of ash is used to treat 1 kg of seeds. Seeds are mixed well with ash and stored separately. This is a low and no cost technology that has various benefits in such pre-storage.

d) **Without treatment:** Neem and pungam leaves are known to have pest repellent property. So invariably neem and pungam leaves are just spread over the grains that prevents it from pest attack. This technique is very much eco friendly and so it doesn't deteriorate the quality of seeds. Similarly even nochi leaves are used. It may be spreaded into the grain or the twigs of notchi may be inserted between the gunny bags filled with grains.

Based on scientific reason

- Essential oil has insecticidal effects inhibit feeding activity that is repulse and sterilize pest. Vasambu is powdered if the efficacy is for short period only.
- Mortality pest injured articulatious desiccation and obturating of respiration apertures.
- The active ingredients have an insecticide and repulsive effect which inhibits feeding activities.

9

Rationality on Indigenous Technical Practices Agricultural Tools

AGRICULTURAL TOOLS

DSIR/TPDU/TIF 056 and E-28-GO-GOI Research Scheme on "Documentation and Preservation of Agricultural Traditional Knowledge through Farmer Participatory Approach"

S. No.	Documented ITK Practices	Rational
1.	**Katti udaikum karuvi (Clod breaking tool)** - Sathy taluk of Erode district. Iron rods of 10 cm length is sharpened and made flat like a chisel on both ends. These rods are inserted into the roller such that one end projects out. Rods are inserted in an alternative row arrangement. At the centre of the cylindrical roller and / long wooden pole is inserted in the centre. While bullock draws the tool, roller is rolled and the iron rods break the clods which are smashed by the roller.	Based on scientific reason This is constructed by farmer's experience. But the tool obey the principles of simple machines i.e. simple machines are tools that makes work easier and they have few or no moving parts. These machines use energy to do work with lever principle.

2.	**Kaluvai (Payaru ootum karuvi)-Leveller** Kaluvai is used in Mallan kuzhi village of Sathy taluk at Erode district. It is a wooden tool of about 5-6 feet length; it is used to level the ragi field. A long wooden pole is fitted with a handle in the middle, facing upwards and nearly 10-15 pegs are fitted facing downwards. It is made out of any wood but kaluvai is preferable. **Uses-**After sowing Ragi it is used to level the field. This facilitates uniform spread of seeds and easy sprouting of seeds. **Cost-** It costs about Rs.200 for making a single kaluvai	Based on experience This is constructed by farmer's experience. But the tool obey the principles of simple machines i.e. simple machines are tools that makes work easier and they have few or no moving parts. These machines use energy to do work with inclined plane principle.
3.	**The wooden Leveling Board** **Purpose** It is used for leveling the paddy field and also used for spreading and sweeping the grains while sun drying. **Operation** When the operator stand over the board, it stinks soil lightly in to the loose mud, excess soil is collected and released to the undulated place. Since it is field operation it is attached with cattle but in drying floor it will operate manually. **Cost** -Rs.200/-	Based on experience This is constructed by farmer's experience. But the tool obey the principles of simple machines i.e. simple machines are tools that makes work easier and they have few or no moving parts. These machines use energy to do work with inclined plane principle.
4.	**Spade** **a. Manvetti** -Kanyakumari district. The lengthier handle allows the farmers to work longer time without getting tired. **b. Pal manvetti** This is to dig the highly compacted soil and break the clods, remove off the bushes and shrubs. **c. Mammaty** Square shaped iron blade is inserted in the 3-4 feet lengthy handle. It gives less strain to back bone. It is used in coastal sandy soil of Keevalur taluk of Nagapattinam district. Cost of the tool is Rs.150/-	Based on experience This is constructed by farmer's experience. But the tool obey the principles of simple machines i.e. simple machines are tools that makes work easier and they have few or no moving parts. These machines use energy to do work with lever principle.
5.	**Koru Kalappai (seed drill)** It is used for sowing and weeding operation. Major components are Mozhi (Handle used to hold the plough / koru), Thindu (wooden plant (4 feet length) in which mozhi and iron pegs are inserted),Munsattam (connecting pole), Erukaal(wooden pole, attached to the Nukathadi), Kuzhai(wooden tube made out of bamboo shoot through	Based on experience This is constructed by farmer's experience. It is an improved machine. An improved machine manages several operations. This tool manages mass, space and

	which seeds are sown) and Iron pegs are used for ploughing. In Kolathacherry village of Kanchipuram Dt.) **Coverage:** 2 ac/ day.	time i.e. it delivers certain quantity of seed, at certain spacing at a determined time.
6.	**Nattu kalappai (Country Plough)** Wooden plough is comfortable for straight and easy plough; soils are pulverized up to 20- 25 cm depth. It covers the area of 2 ac / day. **Cost** The cost of the plough is Rs.300/-	Based on experience This is constructed by farmer's experience. It is a combination of two (or) three simple machine. These machines are use energy to do work with wedge, lever and incline plane principles.
7.	**Bose plough or Melur plough** It is a wooden plough which takes a mould board cum share instead of the usual small iron share. The mould board share is slightly tilted. It cuts a new furrow slice. The slice is twisted slightly and pushed to side. There is a complicated fitting or adjustments. The draught is low and less than that of mould board plough and country plough	Based on experience This is constructed by farmer's experience. It is a combination of two (or) three simple machine. These machines are use energy to do work with wedge, lever and incline plane principles.
8.	**Kalai Kothi (Small Hand hoe)** It is made up of iron and weighs about ¼ kg. It is used to uproot the weed completely. Less weight compared to ordinary hand hoe. No occupational health hazards to farm women. One woman can cover 1 cent / hour and 6 cents /day. **Reason for intervention** Previously used hand hoe weighs more than the modified new one. So, the farm women get tired up quickly. Previously one removes the grass and other monocot weeds at top level only. But modified hand hoe removes the shrubs completely.	Based on experience This is constructed by farmer's experience. But the tool obey the principles of simple machines i.e. simple machines are tools that makes work easier and they have few or no moving parts. These machines are use energy to do work with wedge and inclined plane principle.
9.	**Koorgai (Thinning tool)** It is a wooden tool of about 5-6 feet length; it is used to thinning the ragi field. A long wooden pole is fitted with a handle in the middle, facing upwards and nearly 4-5 pegs are fitted facing downwards. **Uses** It is mainly used for thinning ragi crop and also used in sowing when the soil is wet condition.	Based on experience This is constructed by farmer's experience. But the tool obey the principles of simple machines i.e. simple machines are tools that makes work easier and they have few or no moving parts. These machines are use energy to do work with wedge and inclined plane principle.

10.	**Bagunda (Weeding tool)** The entire structure looks like a plough. More or less this is also similar to the koorkai but the difference is, the wooden pole in which the pegs are fitted is slightly bigger in size i.e. the diameter is up to 15 cm. Two pegs are fitted at the bottom which is fitted with a flat iron strip. It is fitted such that, two pegs are connected at the both end of the strip/blade and the mid portion of the strip/blade alone gets buried into the soil while operation. **Use:** It is used for weeding in groundnut field as well as for earthing up activity	Based on experience This is constructed by farmer's experience. But the tool obey the principles of simple machines i.e. simple machines are tools that makes work easier and they have few or no moving parts. These machines use energy to do work with lever principle.
11.	**Arganakunda (Weeding tool)** This is similar set up of bagunda, but instead of one such wooden pole, a pair of set up (i.e. wooden pole fitted with two pegs and an iron strip/blade) is fitted and operation handled with two hands. **Uses:** it is used for thinning operation, in many crops like gingelly, groundnut, especially in ragi crop. While bullock draws the implement, person who operates it should press well so that the thinning is performed effectively. **Advantages** Implements can be made out of locally available forest trees. Less preparation cost. Easy for operation. co friendly and involves less cost i.e. in mechanized farming diesel and petrol usage leads to air pollution and needs additional cost. Less maintenance cost.	Based on experience This is constructed by farmer's experience. But the tool obey the principles of simple machines i.e. simple machines are tools that makes work easier and they have few or no moving parts. These machines use energy to do work with lever principle.
12.	**Payaru kunduvay-Weeder** It is a tool used for thinning and weeding operation in the field at the earliest stage of the crop i.e. 11 days after sowing. Six number of one feet iron pegs are inserted in the wooden pole of about 2 feet length facing downwards. Number of pegs will vary according to the size of the wooden pole, capacity of the bullock and stage of the crop. The iron pegs are gradually broadened towards the bottom. The bottom of the peg is sharpened for efficient and effective weeding and thinning operation. 'T' shaped wooden handle of about 3 feet length is attached at the centre of the pole which faces upwards. Two bamboo poles are attached as shaft. It covers 1.5-2 acres / day. Note: Shape of the peg can vary. Sharp edged pegs can be used. It cost is about Rs.5000/-.	Based on experience This is constructed by farmer's experience. But the tool obey the principles of simple machines i.e. simple machines are tools that makes work easier and they have few or no moving parts. These machines use energy to do work with lever principle.

13.	**Mara Polluku** It is also used for thinning and weeding operation. But it is used at the later stage of the crop i.e. 45 days after sowing. It is similar to that of Payaru kunduvay. But dimension will vary .Wooden pole of about 3.5 feet, pegs are made up of wood about 1 feet length. Number of pegs attached are 12. Architectural works are done in the wooden pole. It costs about Rs.200 (approximately). After finishing the one stripe of the entire length of the field the thinned and weeded plants are collected and disposed.	Based on experience This is constructed by farmer's experience. But the tool obey the principles of simple machines i.e. simple machines are tools that makes work easier and they have few or no moving parts. These machines use energy to do work with lever principle.
14.	**Uzhavaram (Weed removing tool)** It is an indigenous tool used to remove the weeds (grasses, broad leaved weeds). It has two parts one is wooden hand second is iron blade. It's' shape is cylindrical and the length is 1.5-2.0 feet. One end of the handle is sharpened. 'V' shaped iron blade which has the 30 cm lengthy cylindrical provision is attached to the sharpened end of the handle. Total weight of the uzhavaram is 1.5 kg (approximately. Continuous use of the tool may reduce efficiency of the iron blade.	Based on experience This is constructed by farmer's experience. But the tool obey the principles of simple machines i.e. simple machines are tools that makes work easier and they have few or no moving parts. These machines use energy to do work with wedge principle.
15.	**Burme Saturn (Dry land weeder cum mulcher)** It is similar to that of cultivator tynes. The wedge shaped angle designed at 60 degree and it has designed with a hole and knob. The plough end is similar to that of chisel plough to remove deep rooted weeds. It is made up of galvanized iron. The entire length of plough is 30 - 40 cm and the curved portion length is 15-18 cm. Similary, 10 tynes have been designed and fitted into the cultivator. If it is operated by bullock, 3-5 tynes are sufficient. The depth of the plough should be 5-10 cm only. **Operation mechanism** The 10 tynes should be fitted instead of cultivator tynes and operated for burial of weeds. By using tractor lever mechanism the depth should be maintained at 5 cm since the plough is to be operated in the paddy field one week after sowing. Thus, the initial stage weeds are uprooted and mulched as manure to the emerging paddy seedlings. It can not be operated in water stagnated field. **Salient features** 1. The cost of plough (10 tynes) is Rs.1500/-	Based on experience This is constructed by farmer's experience. But the tool obey the principles of simple machines i.e. simple machines are tools that makes work easier and they have few or no moving parts. These machines use energy to do work with wedge and lever principle.

2. Coverage is 2 hrs / acre. Thus, the cost of coverage per acre is Rs.300. But the manual weeding costs Rs.1000/- (20 w @ Rs.50/-).
3. One time weeding is sufficient for the crop. But, the manual method needs 3-4 times.
4. Bullock drawn method needs 3-5 tynes and the coverage is 4 hr/acre. Here also the depth can be adjusted accordingly.
5. As this instrument is compact in size it can be operated in the corners areas also.
6. It reduces the manure cost as it mulches the weed

16. **Mamaty kalaikothi (Spade Weeder)**
It is used in Kolathancherry village of Kanchipuram district. It is existence in more than 10 decades. One feet wooden handle is inserted in iron blade whose length and width is 10 cm and 10 cm respectively. Iron blade looks like a square shaped one. It is a type of hand hoe used for weeding in the widely spaced crops like ground nut.

Based on experience
This is constructed by farmer's experience. But the tool obey the principles of simple machines i.e. simple machines are tools that makes work easier and they have few or no moving parts. These machines use energy to do work with wedge and lever principle.

17. **Oosi kalaikothi (Needle type Weeder)**
It is used in Kolathancherry village of Kanchipuram district. In Oosi kalaikothi one feet wooden handle is inserted in iron blade whose length and width is 15 cm and 5-7 cm respectively. It is a type of hand hoe used for weeding in the closely spaced crops like ragi, vegetables etc.

Based on experience
This is constructed by farmer's experience. But the tool obey the principles of simple machines i.e. simple machines are tools that makes work easier and they have few or no moving parts. These machines use energy to do work with wedge and lever principle.

18. **Kothu kalai kattu (Pick axe)**
It is used in Kolathancherry village of Kanchipuram district. Pick axe is half feet lengthy iron blade with hole at top in which the wooden handle is inserted. Wooden pole serves as handle. It is a tool used to dig the land left while ploughing i.e. corner of the fields. It is also used to remove the interference of the root to the field and used to loosen the soil in hardy surface.
Cost: Rs.200/-

Based on experience
This is constructed by farmer's experience. But the tool obey the principles of simple machines i.e. simple machines are tools that makes work easier and they have few or no moving parts. These machines use energy to do work with lever principle.

19. **Palluvai maram (Ragi weeder)**
Used for weeding in ragi field.
Parts
- Mozhi – Wooden handle
- Thindu – wooden plank – 3-4 feet length, in which mozhi is inserted on the upper portion, iron pegs are inserted on the lower portion.
- Pal: 10-12 iron pegs. Length – 1- 1 ½ feet
- Coverage : 2 ac/day
- Wages – Rs.200/ac

Based on experience
This is constructed by farmer's experience. But the tool obey the principles of simple machines i.e. simple machines are tools that makes work easier and they have few or no moving parts. These machines use energy to do work with lever principle.

20. **Kondi ('C' Shaped Hook)**
Purpose
It is used for pulling the thorn push after cutting the plant. It will help the farmers from thorn injury.
Making
This kondi is intervention by the farmer himself. It is made up of bamboo stick and iron hook. A question mark shaped hook is attached with bamboo stick.
Cost
It cost is Rs.15/-.

Based on experience
This is constructed by farmer's experience. But the tool obey the principles of simple machines i.e. simple machines are tools that makes work easier and they have few or no moving parts. These machines use energy to do work with lever principle.

21. **Kala karuvi (Indigenous Dry land Cycle Weeder)**
In Kothali village of Thoothukudi District a farmer has developed dry land weeder by modified bicycle wheel. It is easy to operate with the help of long handle and the bottom side is fixed with iron plate which is used to remove the dry land weeds.
Working Principle
The operation of the implement is just by walking and application of gentle pressure by hands.
Uses
It is much useful in crops which are planted in ridges and furrows. Cost-Rs.750/-
Cost
It costs about Rs. 750 /-.

Based on experience
This is constructed by farmer's experience. But the tool obey the principles of simple machines i.e. simple machines are tools that makes work easier and they have few or no moving parts. These machines transfer energy to do work with wheel and axle principle.

22. **Vallam (Iron measuring device)**
Purpose
It is specially used for measuring the grains (cereals, Pulses and oil seeds etc.). The capacity of this equipment is 4kg.

Based on experience
This is constructed by farmer's experience. But the device obeys the principles of simple machines i.e.

	Making This equipment is made up of iron and it is round in shape. The height of the equipment is 23 cm, the diameter is 20 cm. **Cost** The cost of the measuring device is Rs. 200/-.	simple machines are tool that makes work easier and they have few or no moving parts. These device transfer energy to do work with lever principle.
23.	**Padi (Brass measuring device)** **Purpose** It is an measuring equipments used for measuring grains like Cereals, Millets, Pulses and oil seeds **Making-** It is made up of brass. It has role shaped flattened bottom and round shaped top portion. The height of the equipment is 25cm and the diameter is 8cm. It is golden in color. The capacity of the equipment is 1kg. **Cost-** Cost of this equipment is Rs. 150 /-.	Based on experience This is constructed by farmer's experience. But the device obeys the principles of simple machines i.e. simple machines are tool that makes work easier and they have few or no moving parts. These device transfer energy to do work with lever principle.
24.	**Thorati (Hook)** **Purpose** It is used to plucking the tree branches (younger branches) and harvesting tamarind, drumstick, etc. **Making** It is made up of bamboo stick and iron hook. The iron hook having 2cm thickness with the height of 1ft. the top of the hook is blended in down ward position and the bottom is attached with bamboo stick. **Cost** This cost of the hook is Rs.15/-.	Based on experience This is constructed by farmer's experience. But the tool obey the principles of simple machines i.e. simple machines are tools that makes work easier and they have few or no moving parts. These machines transfer energy to do work with lever principle.
25.	**Arivalmanai (Cutting knife)** It is used in Thenkanikottai taluk of Krishnagiri district. It is a cutting tool / knife used to the cut the fodder for cattle feeds. Earlier it was used for cut the mulberry leaves for silkworm rearing. Onc fcct iron knife is attached at one end of the wooden plank. Length of the wooden plank is 3 feet and width is 0.5 feet.	Based on experience This is constructed by farmer's experience. But the tool obey the principles of simple machines i.e. simple machines are tools that makes work easier and they have few or no moving parts. These tool transfer energy to do work with lever principle.

26.	**Aakathi (Harvesting knife)** It is used in Marthandam of Kanyakumari district. The knife is an iron made with curve at it top and bottom to get expanded 'U' shape. Knife is sharpened at one end and other side blunt. The knife is mostly used for harvesting operations, in coconut, pineapple.	Based on experience This is constructed by farmer's experience. But the tool obey the principles of simple machines i.e. simple machines are tools that makes work easier and they have few or no moving parts. These tool transfer energy to do work with lever principle.
27.	**Manjal Vettum Karuvi-1 (Turmeric Harvester)** It looks like a hand hoe. The handle is made up of tree pole, the length of the handle is 60 cm and 30 cm metal blade is attached with the end of the pole. **Cost-** The worth (Cost) of this harvester is Rs.80/-. **Manja kothi-2 (Turmeric Harvester)** It looks like a wood (Pick Axe) cutter. The handle is made up of tree pole. The bottom of the harvester is 8 cm thick and the top of the portion is sharp ended. **Cost** -Rs.150/-	Based on experience This is constructed by farmer's experience. But the tool obey the principles of simple machines i.e. simple machines are tools that makes work easier and they have few or no moving parts. These tool transfer energy to do work with wedge and lever principle
28.	**Korai Aruva (Korai harvester)** **Purpose** It is specially used for harvesting korai. **Making** It is made up of very thin iron blade and wood. The straight iron blade has a breath of about 6cms and length is 35cms one side of the serrated edges. The handle is made up of neem or teak wood. The wooden handle is machine made. **Cost** It is available in local market and its cost is Rs.150/-.	Based on experience This is constructed by farmer's experience. But the tool obey the principles of simple machines i.e. simple machines are tools that makes work easier and they have few or no moving parts. These tool transfer energy to do work with wedge principle.
29.	**Iruppu Nagam (Betel vine leaf harvester)** It is used in Kunnipalayam village of Namakkal district. **Purpose-** It is used to harvest the betel vine leaf. In a single cutting some fifty vine leaves can be cut depending upon the age and capacity of the users. **Making-** It is made up of iron. A ring shaped tool is put to the thumb finger. The edge of the tool look like an	Based on experience This is constructed by farmer's experience. But the tool obey the principles of simple machines i.e. simple machines are tools that makes work easier and they have few or no moving parts.

	inverted blade the ring is of 2cms width. The length between the iron ring and shaped of the inverted blade is 5cms. **Cost-** The cost of the tool is Rs.25/-	These tool transfer energy to do work with wedge and lever principle
30.	**Kodikkal Kathi (Sesbania Pruner)** **Purpose** It is used to remove the side branches of the Agathi plants. **Making** It is a small sized knife with a 'C' shaped iron curve attached to a wooden handle. It is made up of wood and iron. The length of the curved iron piece is 8cm and the wooden part is 7cms. The wooden handle has several ring shaped curves giving grip to the users' hand. **Cost** The cost of the Sesbania pruner is Rs.100/-.	Based on experience This is constructed by farmer's experience. But the tool obey the principles of simple machines i.e. simple machines are tools that makes work easier and they have few or no moving parts. These tool transfer energy to do work with wedge principle.
31.	**Pachakkal Eani (Bamboo ladder)** **Purpose-** The ladder is used to harvest the betel vine leaves and remove the side branches of the Agathi. **Making-** It is made up of bamboo poles. Two 10ft length bamboo poles are attached with vertical serving as steps in between them. This is called a ladder. Another two poles of same size are tied at the top of the ladder in the shape of an inverted "V" these poles help to fix the ladder in the betel vine field. **Cost** The cost of this bamboo ladder is Rs.250/-.	Based on experience This is constructed by farmer's experience. But the tool obey the principles of simple machines i.e. simple machines are tools that makes work easier and they have few or no moving parts. These tool transfer energy to do work with incline plane principle.
32.	**Vanku arrival (Sickle)** It is used in keevalur taluk of Nagapattinam district. The Length of the tool is 1-1.5 ft. in the top 'C' shaped portion is sharpened for cutting rest of the portions are having blended end to safeguard hands during cutting operation. It is used to cut thorny bushes and used to clean coconut trees.	Based on experience This is constructed by farmer's experience. But the tool obey the principles of simple machines i.e. simple machines are tools that makes work easier and they have few or no moving parts. These tool transfer energy to do work with wedge principle.
33.	**Arrival (Cutting Tool)** It is used in Thenkanikottai taluk of Krishnagiri district. It is a 'C' shaped traditional cutting knife made out of iron. The length of the iron knife is 15 cm. Handle	Based on experience This is constructed by farmer's experience. But the tool obey the principles of

is made up of wood and special provision is made in the handle at the lower end to give grip to the hand. Knife is inserted in the upper end of the handle and covered with brass ring. This knife is used for weeding and harvesting.

simple machines i.e. simple machines are tools that makes work easier and they have few or no moving parts. These tool transfer energy to do work with wedge principle.

34. **Vaashayinaa Kathi (Scrapper)**
It is used in Marthandam of Kanyakumari district. It is a rectangular shaped one, made of iron and has a wooden handle. The length and breadth of the scrapper is 10 cm and 6 cm respectively. The handle is to hold on the knife and remove the sheath. Wet banana sheath preferably un-diseased one is taken and kept for extraction of fibre. The scrapper is kept on it and scrolled towards the bottom, the sheath is removed and the fibre is extracted. Extracted fibre is used for making toys and other fancy articles.

Based on experience
This is constructed by farmer's experience. But the tool obey the principles of simple machines i.e. simple machines are tools that makes work easier and they have few or no moving parts. These tool transfer energy to do work with incline plane principle.

35. **Sathuvakathi (Palm leaf making knife)**
It is used in Thuthivalasai village of Ramanadhapuram district. It is a knife used only for Palm leave mat weaving and palm leaf basket making. It is specially designed by black smith as per the instruction of palm leaf mat weaver and the cost of the knife is Rs.20/- only. There are two portions in the knife. The curved frontier portion is used to tear the palm leaf mat and the back long portion is used to punch and weave the mat/basket. As it is tiny and long it is very easy to handle while weaving. The weight of the knife is 200-250 g.

Based on experience
This is constructed by farmer's experience. But the tool obey the principles of simple machines i.e. simple machines are tools that makes work easier and they have few or no moving parts. These tool transfer energy to do work with lever and wedge principle.

36. **a.Groundnut Grading Sieves**
It is made up of iron plate and it is round in shape. Size of the sieve is 2.5ft length and 1.5ft width. Purpose of the sieve is grading the good quality seeds
Cost of this tool is Rs.150/-

b. Mallattai Salladai (Ground nut sieve)
Purpose-It is used for grading the ground nut seeds. The good quality seeds are stag over the sieving equipment. The immature and damaged seeds are fall down from the equipment.
Making- This sieve is made up of iron. The round shaped sieve having the diameter of 45cm. Bottom of the sieve

Based on experience
This is used by farmer's experience. It is an improved tool. Grading of groundnut kernels is based on the thickness / length (Diameter of the grain). This tool transfer energy to oscillate the sieve with lever principle.

have vertical thin holes up to 60nos. the length of the hole is 5cm and the width is 1cm.

Cost

The cost of the sieve Rs.50/-.

37. **Ragi Grading Sieves**

It is made up of iron plate with wooden stick and the size of the sieve is 3x3 ft Square in shape. The diameter of the mesh hole is 0.4 cm (Round). It is used to Cleaning the ragi Grains (To remove the dust and other unwanted materials)

Cost of the tool is Rs.300/-.

Based on experience

This is used by farmer's experience. It is an improved tool. Grading of ragi is based on the thickness / length (Diameter of the grain). This tool transfer energy to oscillate the sieve with lever principle.

38. **Grains and Pulses Grading Sieves**

It is made up of iron plate. Size of the sieve is 1.5 ft and the diameter of the sieve is 0.4 cm. It is used to cleaning and grading the good quality seeds from immature seeds

Cost of the tool is Rs.150/-.

Based on experience

This is used by farmer's experience. It is an improved tool. Grading of grains and pulses is based on the thickness / length (Diameter of the grain). This tool transfer energy to oscillate the sieve with lever principle.

39. **Thuvarai Salladai (Red gram sieve)**

Purpose

It is used to clean the broken red gram Dhall.

Making

This sieve is made up of iron. The shape of the sieve is round. The diameter of the sieve is 45cm the sieve has vertical thick holes up to 33 nos. the length of the bole 7cm and the width is 1.5cm.

Cost

The cost of the sieve is Rs.50/-. This equipment is being used by the farmer for the past 80 years in Vettuvanam village of Vellore district.

Based on experience

This is used by farmer's experience. It is an improved tool. Grading of pulse is based on the thickness / length (Diameter of the pulse). This tool transfer energy to oscillate the sieve with lever principle.

40. **Arisi Salladai (Rice sieve)**

Purpose

It is used to clean the rice from stone dust and other impurity seeds.

Making

This sieve is made up of iron. The shape of the sieve is round and the diameter of the hole is 0.4cm.

Based on experience

This is used by farmer's experience. It is an improved tool. Grading of grains and pulses is based on the thickness / length (Diameter of the grain).

	Cost The cost of the sieve is Rs50/-. This equipment is being used by the farmer for the past 80 years in Vettuvanam village of Vellore district.	This tool transfer energy to oscillate the sieve with lever principle.
41.	**Anil kanni (Squirrel Traps)** **Purpose** These traps are used for trapping squirrels in coconut trees. **Manufactured** It is made up of iron wire which is used for power looms. Waste iron wire are collected and used for making traps (Kanni). **Mechanism** If the squirrels runs on the coconut trees it gets trapped in the traps are tied around the coconut trees. **Special feature** It is no cost-effective and easy to prepare.	Based on experience This is used by farmer's experience. It is a device used to trap the squirrel.
42.	**Peruchali Kanni (Rodent traps)** **Purpose** It is used to trap the rodent in the field and shape guard the plants from damage **Manufactured** It is made up of iron wire which is used for power looms. Two iron wires are tied on a 15 cm stick which acts as a trap. It is prepared by farmer itself. **Mechanism** This is placed near a rat burrow or on the rat path. If rat comes in /goes out of the tunnel it gets caught in the trap. **Cost** No needs to spend money for preparation.	Based on experience This is used by farmer's experience. It is a device used to trap the rodent.
43.	**Eli koondu (Rat traps)** **Manufacturing** It is made up of iron wire. It is 45 cm long and 20 cm wide and the weight is 1.5 kg. **Mechanism** The trap can be placed where the rats problem is predominantly achieve both home and field. In order to trap the rat food materials like Groundnut, Tomato, Vada are kept inside the trap. **Specialty** In this trap is 5-10 rats will be caught at a time.	Based on experience This is used by farmer's experience. It is a device used to trap the rats.

Killing
After caught the rats, rats are collected in gunny bags in the outlet of the trap. Then it will be killed by beating it the ground or dipped in water.
Cost
The cost of the trap is Rs.150/-

44. **Vilangu Veratti (Animal scaring device)**
The principle behind is avoiding damage due to pigs and birds by producing sound
Description
A drum/ locally available oil tin is made like a bell, by at opening the bottom portion and hanging a peg inside. This tin is then suspended by using two poles (where in between the poles tin with peg is tied).
The peg is tied with a long string and held by the operator, who is provided with a shed like structure made out of bamboo splits and coconut fronds. This structure is erected upon four wooden pillars to a height of 3-4 feet.
Cost effectiveness
It is low cost technology, where all the needed materials are available locally, for which no expense is to be spared.

Based on experience
This is used by farmer's experience. It is a device used to scaring the animals.

45. **Thenkkai Mattai Uri (Coconut Dehusker)**
It is an iron made having a long rod as its main portion. To it a sharp edged "Uri" is attached with a screw connecting them. A handle is present at the side to remove the outer husk i.e. to get coconut. At bottom is a straight platform like structure is present which is to place the 'Uri' stable in a place. Cost of the implement is Rs.75/ efficiency is 200-250 coconuts / hr can be dehusked / removed from its shell. Easily operated by an single person.

Based on experience
This is developed by farmer's experience. But the tool obey the principles of simple machines i.e. simple machines are tools that makes work easier and they have few or no moving parts. This tool transfer energy to do work with lever principle.

46. **Padhaneer kuduvai (Neerah jar)**
It is purely made out of palm and coconut leaf and fibre. Inside, the jar is coated with red earth to maintain chillness. Neerah collected from the Palmyra tree is directly poured in it and kept for use. Its capacity is 2.5 litres, has a handle at the top to carry from one place to other. No cost is incurred in making the jar. It is naturally made one, no chemical is added. Neerah kept in this jar remains its quality and taste

Based on experience
This is developed by farmer's experience. It is a container used to carry the neerah.

47.	**Panai Vellam Kopparai (Palm Candy Kopparai)** The vessel is made up of iron. The top portion of the vessel is wide and it narrows down to the bottom. The capacity of the equipment is 54 liters. **Cost** The cost of the equipment is Rs.350/- **Man panai Aachu** Palm treacle is poured into the pot to get the pit condition. **Mara Aachu** It is made up of wooden piece. The length of the wooden piece is 1 meter and the diameter of the pit has15 cm; the depth is 10cm. In this pit only treacle is poured. Longevity of the tool is 3-4 years. **Special feature** It is used for longer time and un- broken.	Based on experience This is developed by farmer's experience. It is a heat exchanger used to prepare the palm candy.
48.	**Traditional Cane Jaggery making Tools** **a. Kopparai** It is used for cane jaggery preparation. **b.KarumpuVettum Karuvi** It looks like axe. The flat shaped iron blade is attached with wooden pole. The length of the blade is 15 cm, width is 12 cm. **c. Matti Pattai** It is used to remove the dust from the sugarcane juice during the time of jaggery preparation. **d. Kuruppi** It is used at the time of cane treacle cooling, to avoid sticking. **e. Muttukattai** It is used for broke the clot. It is used to get fine powdered jeggery. **f. Pagu Aathi** It is used to stir the cane juice and help to avoid sticking of treacle.	Based on experience The principles behind this tools are, a. Kopparai- Lever principle b. Karumbu vettum karuvi- Wedge principle c. Mattipattai- Lever principle d. Kruppi- Lever principle e. Muttukattai- Lever principle f. Pagu Aathi- Lever principle These is developed by farmer's experience
49.	**Mathu (Curd Mathu)** Mathu is used for prepare butter milk preparation. **Manufacturing** It is made up of tree wood and bamboo sticks. A wooden stick with a hemispherical bottom having a grooves collect the butter. In that bamboo sticks streaks are made to fix the rope because the rope is used for rolling	Based on experience This is developed by farmer's experience. But the tool obey the principles of simple machines i.e. simple machines are tools that makes work easier and they

	the curd mathu. The length of the bamboo sticks is 5 feet and the width of the groove is 15 cm. **Cost** The cost of the curd mathu is Rs.50/-	have few or no moving parts. It is a tool used to prepare butter milk with agitator principle.
50.	**Ariakal (Ragi grinder)** **Purpose** It is used in V. Mettuppalayam village at Salem district. Aria kal is traditionally used for grinding grains and to break the pulse grains. Mainly it is used for grinding ragi, red gram, black gram, green gram, cow pea and lab lab etc. The pulses are soaked in water and then sun dried. So that it will be in easy for breaking. Some times pulse seeds are fried and then broken, and it gives good flavor and helps easy broken. **Special feature** Dhal obtained through this type of grinding process possessed good taste.	Based on experience This is developed by farmer's experience. But the tool obey the principles of simple machines i.e. simple machines are tools that makes work easier and they have few or no moving parts. It is used to grind the pulses involving the grinding mechanism
51.	**Mara ural and ualkkai (Wooden Pestle and Mortar)** Used to pounding the grains and pulses. **Manufacturing** It is made up of teak wood. Shape of the tool is cylindrical with slight curve in both the side. The height of the pestle is 3 feet. Centre of the top portion have one feet shallow depth with the diameter is 15 cm. The capacity of the pestle is 2 kg. It is used to pounding the grains like paddy, combu, wheat, thenai and samai etc. For pounding mortar is used, it has the 6 feet long and 7 feet wide. Fixed with an iron ring both side of the motor. It helps to remove the outer skin of the grains. **Special Feature** Hand pounded grains are minimizing the grains broken. It has good taste compare to mill pounding. **Cost** -The cost of the tool is Rs.150/-	Based on experience This is developed by farmer's experience. But the tool obey the principles of simple machines i.e. simple machines are tools that makes work easier and they have few or no moving parts. It is used to pound the grains.
52.	**Traditional Water Lifting device** **a. Koonai** It is used to carry water from well. It is made up of iron plates and the mouth is tied with leather tube for uniform flow of water. By using long ropes and with the help of bullocks it carries around 60 litres of water at a time. Because of the invention of diesel and electric motor pump sets this koonai is just the thing in museum. Now a day in rural areas the so called koonai is used for Poultry shed or kozhi koodu.	Based on experience This is developed by farmer's experience. But the tool obey the principles of simple machines i.e. simple machines are tools that makes work easier and they have few or no moving parts. It is used to lift the water from the well.

	b. Kavalai (Urulai) Kavalai is made up of wood (Neem, Teak,) shape of the roller is round. The diameter of the roller can be 30cm the centre of the roller holes are provided it's moulded with iron blade. It is used to lift the water from the well. The cost of the roller is Rs150/-.	
53.	**Theevanapetti (Indigenous feed box for small ruminants)** Iron trunk boxes which are previously used for keeping clothes and valuables now a day used as the feed box for small ruminants. Goats and Sheep will not eat the grasses which are mixed with the soil, dung and urine. So this iron plate box is very much useful for the safe feeding of small ruminants. This practice has been followed in Kothali village of Ottapidaram Block of Thooththukudi district.	Based on experience This is developed by farmer's experience. It is used to feeding the animals without wastage.
54.	**Thali (Animal watering pot)** This structure is build with red soil and bricks. The height of the structure is 4 ft. The diameter of the pot can be 60cm. It cost about Rs.80-100/-. It is traditionally used in Vellore district of Tamil Nadu.	Based on experience This is developed by farmer's experience. It is used to feeding the animal in liquid form.
55.	**Manja kopparai (Indigenous turmeric boiling stove)** The size of this stove is five feet length and the three feet breath .The depth is 6feet from the top of the earth. The harvest turmeric leaves are used as the fuel material. The fuel material are inserted in one way (3 Feet breath area) and out let smoke in other side of the breath area. Indigenous boiling drums having the capacity of 90 Kg each, can be kept on the Indigenous turmeric boiling stove for a period of 45minutes.this is the cost effective stove locally made in the field itself. They are using the harvested dried turmeric leaves as the fuel which is having no cost. To set up an Indigenous turmeric boiling stove 2 laborers are needed and the total cost of this set up is Rs. 500 only.	Based on experience This is developed by farmer's experience. It is a heat exchanger used to prepare the cane jaggery.

10

Rationality on Indigenous Technical Practices Medicinal Plants and Ethnoveterinary Practices

MEDICINAL PLANTS

DSIR/TPDU/TIF 056 and E-28-GO-GOI Research Scheme on "Documentation and Preservation of Agricultural Traditional Knowledge through Farmer Participatory Approach"

S. No.	Documented ITK Practices	Rational
1.	**Vishnu kiranthi -*Evolvulus alsnoides*** A fistful of Vishnu kiranthi leaves, 25 gram pepper and 25 gram seeragam are boiled well. It should be boiled until one litre of water comes to half litre. It can be taken twice a day. It can be mixed with milk also. This medicine cures the Tuberculosis, cold and fever.	Based on scientific reason • Plant contains alkaloid and it stimulate resistant against diseases in the body.

2.	**Purusalachanam** A fistful of Purusalachanam root is taken and cleaned. To this little quantity of seeragam is added and ground well. Paste is made out of it and it is mixed with 100 ml of water. This medicine should be given twice a day for 2-3 days to women who is having uterus problem. It controls menstrual problems.	Based on experience • The root contain alkaloid and it leads to promote and rectifying the hormonal imbalance in the body
3.	**Ariyanangai** A fistful of Nilakunni root, a fistful of Ariyanangai leaves and little quantity of pepper are ground well and paste is made out of it. This paste is given twice a day to person who is affected by snake bite.Care to be taken while administering the medicine.Food items to be avoided upto three days (sweet, salt and sour items). Up to three months the person should not take fish, dried fish, chicken, lab lab. These treatments were learnt from their ancestors.	Based on experience • It is layman and rural medicine.
4.	**Unni mul thazhai- Wild sage -*Lantana camera*** It is a thorny bushy plant whose leaf extract is used to heal the cut wounds and other wounds	Based on experience • It also used as anti-insecticidal properties.
5.	**Karu umathai -Indian Thorn Apple -*Datura fastuosa*** Flower of the karu umathai is used to cure suffocation due to asthma. Inhaling of smoke comes out from dried flower will give relief from asthma.	Based on experience • Same may be used for cows and cattles in the rural areas.
6.	**Notchi - Five leave chaste tree -*Vitex negunda*** Handfuls of Notchi leaves are taken and ½ lit of water is added to it. Notchi decoction has to be prepared using hot coal in the absence of burning. This should be done in previous day night .Next day morning; the filtered decoction will be given to the person suffering from fever and cold. It should be taken, before intake of food and to be continued for 3-5 days. It should be given twice a day.	Based on experience • It is very old medicine. Same tree, there is also three leaves type also having medicinal value.
7.	**Mudakatran- Balloon Vine - *Cardiospermum halicacabum*** a) Mudakatran decoction is used to cure joint pains and constipation. Handful of Mudakatran leaves are taken and ½ lit of water is added. The contents are boiled and	Based on experience • The same leaves can be added at the time of making rasam. Leaves will used to make chutney

	filtrate is given to persons affected by joint pain and constipation. This can be taken in twice a week. b) Handful of Mudakaran, 4 bulbils of garlic, 4 dried chillies, required quantity of dried ginger (*Gingiber officinale*), Asafoetida, Tamarind and salt are taken and paste is prepared out of these. This paste is also used to cure joint pains. This can be taken with normal food, twice a day.	
8.	**Perumarunthu -Indian Birth Wood -*Aristolochia indica*** Handful of Peru marunthu leaves are taken and it has to be fed to the off feed animal with dub grass (*Cynodon dactylon*).This should be done twice a day for 3 consecutive days.	Based on experience • The same leaves will also used to external application for mammals to control insect damage in hairs.
9.	**Aanai nerunji- Bathala Mooligai -*Pedalium murex*** Handful of Aanai nerunji leaves are taken and soaked into ½ lit of butter milk for 10-15 minutes. The contents are smashed well and the filtrate is fed to the animal affected by diarrhoea. This should be given in twice a day.	Based on experience • The seeds also contain alkaloid it stimulate the vigorous.
10.	**Sangu -*Azima tetracantha*** A fistful of sangu leaves, 2 numbers of betel vine leaves and required amount of salt are ground well. The extract is added to the eye of dull and drowsy animal. It should be administered till the animal come to normal condition.	Based on experience • This plant is urban plant and not need any care for cultivation.
11.	**Adathoda (*Adathoda vasica*) Malabar nut** A fistful of Adathoda leaves are soaked in half litre of water. This should be boiled well and the decoction is to be given twice a day, for one month. This medicinal extract will give best relief for Asthma.	Based on experience • Also used for relief of cold.
12.	**Veli paruthi -*Fergularia daemia*** Handful of fresh veli paruthi leaves are fed to the animal thrice a day. This will give relief for indigestion.	Based on experience • The same will be utilized for rectification of gas trouble in man.
13.	**Thulasi (*Ocimum sanctum*)** 10-15 fresh Thulasi leaves are given to person suffering from cold. This is recommended twice a day for a week	Based on experience • Better to use from child age to get resistance.

14.	**Nayi vellai -wild mustard -*Cleome viscosa*** A fistful of Nayi vellai leaves, 4 bulbils of garlic, 2 nos. of dried chillies, required amount of Asafoetida and salt are macerated well and extract is taken. Two drops of extract is added to the eye and ear of the dull and drowsy animal. This should be administered once a day upto three days.	Based on experience • It is occurred in the road side areas and tank fed areas.
15.	**Nanchu murichan -Emetic swallow wort- *Tylophora indica*** A fistful of Nanchu murichan leaves, 5 nos of pepper and 3 nos of betel vine are taken and made into paste which is then fed to the animal affected by poisonous bite. This should be continued for a week, thrice a day. The same paste recommended to human beings.	Based on experience • It is vine like crop and grows very fast.
16.	**Papaya -*Carica papaya* and Guava-*Psidium gujava*** Continuous consumption of papaya and Guava will cure the stomach ulcer.	Based on experience • Proteolytic enzyme and protein content in respective fruit will take care the human food canal system.
17.	**Garlic-*Allium sativum*** Five to six bulbils of garlic should be consumed and should drink enough water. After 1 hour this will cure indigestion.	Based on experience • Diallyl disulphats is chemical responsible for curing the problem in human body.
18.	**Aamanakku / Castor –*Ricinus communus*** Mustard, chopped onion, kuppai meni leaves are fried in castor oil. This medicine is given usually to the women after delivery for a period of 3-5 days. It prevents puss formation. Applying castor oil over stomach and massaging will relieve the pain.	Based on experience • Castor oil is used for curing diseases because it's having many medicinal properties.
19.	**Banana pseudo stem** A fistful of sprouted horse gram and cooked banana pseudo stem consumption will cure kidney stone.	Based on experience • Because contain fibres.
20.	**Arugampul -Bermuda grass- *Cynodan dactylon*** 100 ml of Arugampul (*Cynodan dactylon*) consumption will serve as first aid against cold, fever and poisonous bite.	Based on experience • Same may be given for stone problem.

21. **kuppai meni -Indian Acalypa -*Acalypa indica***
It is used to cure scabies and internal wounds after delivery (in human beings).Handfuls leaves are taken and water is added to make it into paste. This paste is smeared over the affected portions. This should be continued until scabies heels.

Feasibility and Relative advantages

Since all the ingredients are locally available. It is very easy to prepare the medicine. Similarly, the method of preparation is also very simple which doesn't involve cumbersome process. Herbal cure is relatively advantages over other types of medicines due to its nil side effects..

Based on experience

- Crops are available through out year and most occurred in waste places and manurial pit places.

22. **Sudukadu mallikai- Periwinkle -*Catharanthus roseus***
It is often linked with the graveyard as it thrives there without much care. The flowers and leaves of this herbaceous plant would never find its place in a garland, bouquet or in a maid's hair as the odour emitted by the plant is so bad that it always brings out a negative response in any individual. Later, this plant has gained great popularity of its medicinal use. The plant may have a few branches and grows to about 50 cm high. The glossy leaves are oblong in shape. The flowers are normally white or a purplish- pink in colour. The fruit is somewhat cylindrical with a length of about 2-3 cm. It is seen growing in some gardens, seashores and sandy areas. The whole plant is used for medicinal purpose.

Medicinal use

- Said to be anticarcinogenic or an agent which works against cancer.
- Used for diabetes.
- Used for diarrhoea.
- Used as a vermifuge or a substance this expels worms from the body.
- Good remedy for toothache and works well as a purgative.
- The decoction of the flower is given for treating tuberculosis.
- The decoction of the plant is used as a wash for mouth ulcers and tired eyes to clear the vision.

Caution: women should not use this plant during menstruations.

Based on scientific reason
Roots, bark, stem and leaves contain vinblastine, vinbastine alkoloids. It contains anti- cancerial activities.

23. **Mahuli keerai -Wool flower -*Celosia argentea***

The wild cock's comb grows in waste land as an annual, smooth, erect, branched herb, reaching upto 1 metre high. The leaves are linear to ovate lanceolate. The flowers are silver-white, with a tinge of pink in erect or drooping spikes. It is rarely cultivated plant. The tender flowers and leaves are eaten in tropical countries as a vegetable.

Medicinal Use

- Used in cases of dysentry, diarrhoea, acute abdominal pains and inflamed stomach.
- Heals wounds, sores, ulcers, skin eruptions and relieves itching.
- Used in the treatment of tuberculosis.
- Used to cure the ulcer of the mouth and for tired eyes to clear vision.

How to use

1. As an infusion, the whole plant is used for cases of dysentry, diarrhoea, acute abdominal pain and inflamed stomach.
2. For external use, the pounded leaves, flowers and seeds are applied over wounds, sores and skin eruptions.

Dose: 12 to 20 grams of plant material for 1 litre of water is taken in spoonful doses.

Based on scientific reason

- In rural parts of Tamil Nadu and especially in dry tracts, the tender leaves are used as regular green vegetables.

24. **Tottalvadi- Sensitive plant -*Mimosa pudica***

Sensitive plant grows in the low lying land and land tropics. It is a common diffused spreading prickly, an herb which grows everywhere in southern India. The stems are branched, with bristly hairs. The leaves are small leaflets on a stalk, and when touched they fold together. They are sensitive to the touch. The flowers are numberous, long-stalked, pink, solitary globose heads. The pods are small, curved and numerous. The whole plant is used for medicinal purpose.

Medicinal use

- Used for liver disease, constipation, dysentry and kidney problems.
- Used for mouth and throat inflammations.
- Used for cancerous ulcers, skin diseases and glandular swellings.
- Used in scorpion sting.
- Used for urinary complaints and as a purgative.
- Used for treatment of mumps.

Based on scientific reason

- It also used for external piles.

How to usee infusion of the leaves is used for liver disease, constipation, dysentry and kidney problems.

1. The decoction of the root takes care of the urinary complaints. The infusion of the whole plant is used for mouth and throat inflammations.
2. The seeds are used for sore throat and hoarseness.
3. Externally used, the juice of the plant makes a good lotion to be applied over cancerous ulcers and skin diseases.
4. The crushed leaves made into paste are applied topically on glandular swellings.
5. Use cold compress on swollen glands for 30 minutes at least 3 times a day. Apply crushed past of leaves after each cold compress session.
6. The crushed leaf and stem is applied over the scorpion sting to reduce the pain and swelling.
7. **Dose**

   10 to 15 grams, for 1 litre of water. Use 3 to 4 cups a day.

Caution

In large quantity the root is toxic.

25. **Thotta sinigi- Touch-me-not-** ***Mimosa pudica***

Habit : Herb

Parts used : Root, Leaves and Seeds

Uses : Minor skin wounds, Diarrhoea, Piles, Diabetes

Based on scientific reason

- It is used in siddha medicine.

26. **Andhimalligai - Four O' clock plant -*Mirabilis jalapa***

The four O' clock plant is a perennial, herbaceous, bushy plant which grows about 1 meter high. The leaves are narrow, pointed, and ovate. The flowers are variously coloured, showy and sometimes striped. The black seed is twice the size of a pepper corn, and the tubers, when fully grown, are more than 20 kilos in weight. It is grown in gardens for its shown flowers.

Medicinal use

- Promotes flow of urine.
- Used in any inflammatory viral disease of the skin.
- For the treatment of abnormal accumulation of liquid in the cellular tissue.
- Treatment of boils, inflamed and enlarged lymph-nodes, and for the accumulation of pus in a cavity.
- Heals wounds, bruises, and allergy of the skin caused by indigestion.

Based on scientific reason

- The same plant also having the properties of anti

- Used for earaches.
- Used to remove freckles on the skin.

How to use

The root tuber is used in any inflammatory viral disease of the skin, treatment of abnormal collection of liquid in cellular tissue, boils, and inflamed lymph-nodes. Use 20 grams of root tuber in 1 litre of water, by decoction and 4 to 5 cups are to be taken per day.

1. A hot poultice of the leaves is used to mature and resolve boils, inflamed lymph – nodes, and accumulation of pus in a cavity.
2. The juice of the fresh leaves is a healing gent for wounds, bruises, skin allergy, caused due to indigestion also used for earaches in children.
3. The flowers are applied topically over the skin in the case of any inflammatory virus disease and for freckles. The powdered seeds are also used for the same problem.

Parts used

The root tubers, leaves, flowers and seeds.

Dose

10 to 15 grams, for 1 litre of water. Use 3 to 4 cups a day.

27. **Marul - Bowstring hemp- *Sansevieria zeylanica***

The bow string hemp is an herbaceous perennial plant, used in the garden as border plants. The leaves are sword like, stiff and smooth. Some of the leaves have marble patterns, while others are gold edged and sometimes concave on the upper surface. The flowers are tubular, white to yellowish, in groups of 4 to 6, scattered along the stalk. Strong fibers are used from the leaves for making cloth, twine, and paper.

Medicinal use

- Gives tone and vitality to the body.
- Promotes the removal of mucous secretions from the bronchial tubes.
- Used as a purgative.

The decoction of the plant root is used as a tonic, purgative and as an expectorant

Based on scientific reason

- The plant is grown to prevent snake entry to the house

28. **Marul -Bow string hemp- (*Sansevieria zeylanica*)**

The succulent portion of marul is fried; the liquid that comes from it should be allowed to cool. And the liquid is dropped into the ear. Dosage: One time /day.

Based on scientific reason

- The plant is grown to prevent snake entry to the

29. **Pachaiyalari - Yellow oleander - *Thevetia peruviana***
The yellow oleander is an erect smooth branched shrub, with milky latex. The leaves are linear and shiny. The calyx is green and pointed. The yellow corolla is bell shaped. The fruit is green, smooth and rounded. It is an ornamental plant which is cultivated in gardens.

Medicinal use

- Used as an insecticide.
- Used to reduce eliminate fevers.
- Mild laxative.

How to use

1. The decoction of the seeds can be used as an insecticide.
2. The decoction of the bark of the plant, in moderate doses, is used to reduce or eliminate fever.
3. The leaves act as a mild laxative.

Parts used: Bark, seeds, and leaves.

Dose: For internal use only with medical supervision. Externally as needed.

Caution

Unless in extreme need, this plant should be used under medical supervision for internal use. The plant is poisonous.

Based on scientific reason

- In rural, seeds with pod making killing poison for human.

30. **Thaivelai**
A handful of leaves are taken, cleaned, ground and juice is extracted for about 20 ml. 10 ml of gingelly oil added to it. This mixture is warmed and given to the person who is affected by cold and fever. A fistful of thaivelai leaves are taken, cleaned and juice is extracted. 4-5 drops of juice is poured into ears of affected person. With in 10 minutes pain due to scorpion bite can be cured.

Dosage: Three times /day.

Based on scientific reason

- Used in siddha medicine.

31. **Omavalli – Karpuravalli - *Coleus aromaticus***
A fistful of Omavalli leaves are boiled with ½ litre of water. It has to be boiled until ½ litre comes to ¼ litre of water. This treatment can be continued for about 5-7 days.

Dosage: ¼ litre / day.

Uses: Juice mixed with sugar is given to children in asthma, chronic cough and fever.

Based on scientific reason

- Leaves juice also contain properties of anti-insecticidal activities especially for mosqutoes.

32.	**Vallarai- Indian Peeny Wort -*Centella asiatica*** **Preparation** A fistful of green vallarai leaves are collected, cleaned and taken as such internally. The regular intake will increase the memory capacity.	Based on scientific reason • It contains vallarine alkaloids.
33.	**Sathisaranai – *Trianthema decandra*** A fistful of sathisaranai root, dried chillies 2 nos, a little amount of salt are taken and ground well. Finally paste is made out of these contents. It gives relief from gas trouble. ***Dosage:*** Twice a day.	Based on scientific reason • It will be eaten as green vegetables.
34.	**Manathakgali – *Solanum nigrum*** A fistful of manathakali leaves, a fistful of murungai leaves, 2-3 numbers of onion, 5 garlic bulbils, 1 tomato, a little quantity of seeragam and goat's intestine are fried. These substances are ground well and paste is made out of it. This should be continued up to 3 days for curing uterus problem.	Based on scientific reason • It contains solanin alkaloids and it will take care
35.	**Karunthulasi- *Ocimum basilicum*** Regular consumption of this leaves (a fistful) controls diabetes and blood pressure. ***Dosage:*** Twice a day.	Based on scientific reason • Regular usage in siddha medicine.
36.	**Thaneervittan kizhangu- Wild asparagus – *Asparagus recemosus*** A fistful of tuber is cleaned, add required amount of water and ground well. This has been filtered by using cotton cloth. The filtrate is given to lactating women for about 4 days. In order to increase the milk secretion for lactating women. Flowering Season is May to December	Based on scientific reason • It contains aspargin alkaloids. The juice will take care about joint pain.
37.	**Palperuki – *Euphobia heterophylla*** **Uses:** The filtrate of the leaves is used to wean infants from their mothers and it's protecting new born babies from germs.	Based on scientific reason • Used in siddha medicine.
38.	**Keelanelli – *Phyllanthus amarus*** Erect, annual herb, leaves closely arranged, flowers yellowish green in colour, pendent from the axils of leaves.	Based on scientific reason • Leaves contain phyllanthin alkaloids.

Uses

- The drug is highly reputed as a single drug remedy in the treatment of jaundice in traditional medicine. The whole plant is medicinal.
- It is used in chronic dysentery, cough, indigestion, diabetes, urinary tract diseases and skin eruptions.
- Whole plant pounded, is used as a poultice for ulcers, sores, swellings and skin affections. In practice, it is mainly used for the treatment of jaundice.

Ingredients

Keela nelli-A handful, Cow milk- 100 ml

Medicinal Preparation

A handful of keelanelli leaves are taken, cleaned and ground well. Required amount of water is added to it and made it colloid. This has to be filtered and the extract/juice is obtained out of it. This juice is mixed with 100 ml cow milk. This treatment can be continued up to 3 days for about thrice a day.

39. **Seenthil (*Tinospora cardifolia*)**

Family: Menispermaceae

Ingredients

- Egg- 1
- Seenthil leaves

Preparation

A fistful of seenthil leaves are cleaned and fried well. With this egg (without egg yoke) is added and again fried well. This fried mixture is given to the person affected by viraivatham. This can be administered for about 3 days, thrice a day.

Additional Care

While administering the medicine, the person affected by jaundice should not take salt, sour and spicy items in food.

Based on scientific reason

- Secretion of mother milk. It contains tinosporin, columbin, berberin, giloin and starch.

40. **Periyaannangai -*Andrographis alata***

Name of the disease: "Snake bite in human"

Conditions

Within 30 minutes period the medicine is to be administered into the body of person. Before this, the type of snake bite is to be identified by seeing the intensity of wound.

Ingredients

1. "Periyaanangai" (*Andrographis alata*) - a whole plant with root.

Based on scientific reason

- The plant contains bitter glucosides, andrographolidc, andrographin are responsible for medicinal properties.

2. "Syrianangai" (*Andrographis paniculata*)
   - a whole plant with root.
3. "Muluvanangai"
   - a whole plant with root.
4. "Vellarugu" root (*Enicostema auxillare*)
   - a single plant root
5. "Arugampul" root (*Cynodon dactylon*)
   - a single plant root.
6. "Kurundhampulluruvi" (*Dendrophtoe falcata*)
   - a single plant root.
7. "Neykaaya" root (*Cucumis trigonus*)
   - a single plant root.
8. "Ponmustai" root (*Sida acuta*)
   - a single plant root.
9. "Vellaipoodachedi" leaves (*Sterculia urens*)
   - 20-50 g.
10. "Semmanampullurivi" leaves (Indian mistletoe)
   - 20-50 g.
11. "Karunkattukodi" leaves (*Coculus hirsutus*)
   - 20-50 g.
12. "Keeripachilai" leaves (*Ophiorhiza mongoose*)
   - 20-50 g.
13. "Paapaaranangai" leaves
   - 20-50 g.
14. "Kalarchikaai" (*Cesalpinia bonduc*)
   - a whole plant
15. 'Kilichi" leaves
   - 20-50 g.
16. "Muthirikkai" root (*Ampelocissus latifolia)*
   - 20-50 g.
17. "Kothupaalai" leaves (*Wrightia tinctoria*)
   - 20-50 g.
18. "Aduthinnaapaalai"leaves (*Aristolachia bracteolata*)
   - 20 – 50g.
19. "Vekkanotchi" leaves (*Vitex sp.*)
   - 20-50 g.
20. "Usilai" flower (*Albizia amara*)
   - 1
21. "Madukaarai" root (*Randia dumetorum*)
   - 1 (2-3") length.

The aforesaid ingredients are to be taken, dried under sunlight for 2-3 days and ensure complete drying of ingredients. This is to be ground well either by grinder or "ammi" and sieved to prepare the fine powder. The resultant content is dull green colored.

Administration
The aforesaid prepared ingredient is to be administered orally. One teaspoonful of medicine is recommended immediately after snake bite and from the next day onwards during morning and evening hours for 7 days continuously. The time is between 6-7 a.m. in the morning and 3-4 p.m. in evening hours.

Conditions to be Followed

1. Chemical medicine should be strictly avoided. If it is taken a tumbler of coconut milk would be sufficient to weaken the medicinal effect of the content.
2. Salt is strictly avoided in the food during the course of treatment.
3. Non-vegetarian food is strictly avoided and rice 'kanji' (gruel) is recommended.
4. The intake of medicine in an empty stomach has better healing capacity.
5. If the symptom is severe the treatment period goes upto 15-18 days.

Relative Advantages

- It slowly cures the bite and doesn't create any side effects.
- The treatment fetches free of cost as Mr. Muthadaikan renders free service.
- The patients' remarks are in favourable manner and it explains the vigour of medicine.

41. **Sivappu Nayuruvi -Prickly Chaff Flower-** ***Achyranthes aspera***

Propagation : seeds
Parts used : whole plant
Flower : reddish in colour

Uses: demulcent, asthma, bronchitis, leprosy, skin diseases, renal and vesicle calculi, cardiac disorders, anaemia, general debility, conjunctivitis, cuts and wounds, dog bite, tooth ache in human.Retention of placenta, wounds or abrasions, eye ailments in animals.

Nayuruvi ***(Achyranthus aspera)*** **Prickly chaff flower**
Nayuruvi plant is uprooted from which root portion is cut and outer skin is peeled off. Peeled outer skin is shade dried for about 7-10 days. Ten gram of Neeli root's outer skin, a little quantity of Seeregam and 800ml of water are taken in new mud pot. The contents are boiled well until 800 ml comes to 200ml.Extract is filtered and the decoction is given for the poisonous bite.
Dosage: 10ml / time / day

Based on scientific reason

- Leaves, roots and seeds contains tryterpenoid, saponins and potassium are responsible for medicinal values

42. **Pirandai- Edible stemmed vine-*Cissus quadrangularis***
Propagation : stem
Parts used : stem
Uses
dyspepsia, general weakness and bone fracture in human. Swelling, wounds or abrasions, intestinal worms, anorexia in animals.

Based on scientific reason
- *Cissus triangularis* sp also available and it is highly medicinal valued species. From this chutney also make and take regularly to increase the tongue glands.

43. **Sadhavari kizhangu -*Asparagus racemosus***
Parts used : root and tubers
Uses : increases milking, cures stroke

Based on scientific reason
- Normally occurs in sup-tropical zones small mountains and forest areas.

44. **Oridhazh Thamarai- *Ionidium suffruticosum***
Family : Violaceae
Habit : herb
Propagation : whole plant
Uses: affections of urinary organs, to improve memory and vitality, a Remedy in tuberculosis, asthma, fever, leprosy, eye diseases.

Based on scientific reason
- Another Viagra plant in medicinal industries.

45. **Otrai Moolihai**
Used to cure wounds and abrasions, 20 leaves taken and add 1ltr of water mixed to get leaf extract. It cures intestinal worms in human. It has a bitter taste.

Based on scientific reason
- Also used for reducing sugars in the body.

46. **Cinchona: (*Cinchona succirubra)***
Propagation : seeds, grafting
Parts used : quinine isolated from tree bark
Uses : malarial fever

Based on scientific reason
- Very old medicinal value plant and one of the aromatic perennial tree.

47. **Vasambu- *Sweet flag- (Acorus calamus)***
Botanical name :
Family : Araceae
Habit : Rhizomatous herb
Propagation : Rhizomes
Parts used : Rhizomes
Uses : Antipyretic, nervine tonic, sedative, epilepsy, mental disorders, skin disorders, skin diseases, general debility. Thikkuvaai, to avoid snakes entering in to house-solution sprayed.

Based on scientific reason
- Highly demand and very old medicinal plant for siddha medicine.

48.	**Kumari/ sotru kathalai** *(Aloe vera)* **Indian aloe** Habit : herb Propagation : suckers Parts used : leaf juice (sorus), roots Uses : leprosy, piles, mental disorders, skin diseases, liver ailments, jaundice, menstrual problem	Based on scientific reason • Also used for making cosmetics.
49.	**Arokya pachilai/ kuravan kanda mooligai** Use : found by khani people of Kanyakumari district, khanis look young even they are aged, controls appetite, sterility control, immediate energy source, improves immunity power,	Based on experience • The arogya is denoted as good health.
50.	**Thippili-** ***piper longum*** (long pepper) Propagation : nodal cuttings Parts used : fruit, root, stem Uses : bronchial infection of throat, respi ratory diseases, carminative, sedative, insomnia, epilepsy and abortifacient.	Based on scientific reason • It also known as king of spices. And old medicinal plants used for siddha medicine.
51.	**Athi*(Ficus racemosa)* : gular fig/cluster fig/fig** Parts used : bark, fruit, latex, root Use : local inflammations, polyurea, diabetes, darkening of face and around eyes, burns, nausea, mouth ulcer	Based on scientific reason • It is improving potency and blood circulation.
52.	**Vilvam: Bengal Quince, Bael Fruit-** ***Aegle marmelos*** Parts used : Leaves, Roots, Seeds, Bark Uses : Diarrhoea, Constipation, Worms, Skin diseases, High blood pressure, Joint pain	Based on scientific reason • Popularly cultivated for making pulps. The bark contains coumarin and marmin.
53.	**Sathakuppai- Dill -*Anethum graveolens*** Parts used : Leaves, Seeds Uses : Colic pain Vomiting, Hiccoughs	Based on scientific reason • Important seed spices.
54.	**BiramadanduThuththi- Prickly poppyllow-*Argemone mexicana*** Habit : Herb Parts used : Seeds Uses : Leprosy, Psoriasis, Rat bite	Based on scientific reason • Occurring in dry tracts and stream bank areas.

55. **Adu-Tinna-Palai- Worm-killer- *Aristolochia bracteolate***
Parts used : Roots, Leaves
Uses : Anti-inflammatory, intermittent fevers

Based on scientific reason
- It is also used for worm cutting in hairs. it contains aristolochic acids, volatile oil and tannins.

56. **Pudhina -Mint-*Mentha arvensi***
Parts used : Leaves, Flowering tops, stem extracts
Uses : Cough, Head ache, Fever, General weakness.

Based on scientific reason
- Piperin and piperinol are contains in the plant.

57. **Kattamamakku -Purging Nut, Physic Nut- *Jatropha curcas*:**
Parts used : Leaves, Seeds, Oil, Stem
Uses : Laxative, to increase secretion of milk, inflammations, Ear-ache

Based on scientific reason
- The plant parts contain jatrophin.

58. **82. Vetchie- Ixora, Sarced Ixora Jungle-Flame *Ixora coccinea***
Parts used : Roots, Leaves, Flowers
Uses : Head ache, Boils, Nausea, and Sedative in hiccough

Based on scientific reason
- It also acted as blood purifier.

59. **Kalyana Murunkai- Indian coral tree- *Erythrina variegata***
Parts used : Barks, leaves
Uses : Increase secretion of milk and Menstrual flow, Cure sterility in Women, Diabetes,Eye infections, worms, joint pains

Based on scientific reason
- The greens are used making chappathi.

60. **Korai Kizhangu- Nut Grass- *Cyprus rotundus***
Parts used : Bulbous Root
Uses : Diarrhoea, Indigestion, Deodorant, Dysentery. Fevers

Based on scientific reason
- The tuber used to increase vitality and blood purification

61. **Sirukurinjan- Gurmar- *Gymnema sylvestre***
Parts used : Leaves
Uses : Snake-bite, malarial fever, Eye diseases, Diuretic

Based on scientific reason
- Gymnemic acid which regulate the insulin metabolism

62. **Vellai Kundarikum, vellai Kungiliyam, Indian Copal Tree- *Vateria indica***
Parts used : Bark, Resin, oil
Uses : cure impure blood, Burns and Ulcers

Based on scientific reason
- It also acts as anti-fungal and anti- bacterial

63. **Venthayam Fenugreek- *Trigonella foenum***
Habit : Herb
Parts used : Seeds
Uses : Skin inflammation, Carminative, Galactagogue, Ulcers

Based on scientific reason
- Trigonellin is responsible for regulation of insulin metabolism.

64. **Siriyanangai -Creat -*Andrographis paniculata***
- Roots commonly used as a tonic and febrifuge.
- Plant is astringent, anodyne, tonic and alexipharmic, used in dysentery, cholera, diabetes, consumption, influenza, bronchitis, itches and piles.
- Decoction used for sluggishness of liver and in jaundice.
- Irulas use the parts of the plant for snake-bite and for curing jaundice.

Flowering Season: September to March

Based on scientific reason
- It contains bitter glucosides, andrographolids and andrographin are responsible for medicinal values.

65. **Siriyanangai (*Andrographis paniculata*)**
5-10 nos are given as such to persons who are affected by poisonous biter. Whereas in the case of animals juice is extracted from handful of leaves and given as internal feed. Five numbers of Siriyanangai leaves are given internally as such for about 2-3 days.
Dosage: 5 nos. / time / day.
Care to be taken while administering the medicine: while taking the above said medicines intake of non vegetarian items should be avoided.

Based on scientific reason
It contains bitter glucosides, andrographolids and andrographin are responsible for medicinal values.

66. **Aavuri -Wild indigo- *Tephrosia purpurea***
Leaf extract used in bronchitis, boils, pimples and bleeding piles. Roots and seeds insecticidal. Decoction of roots given in dyspepsia, diarrhea, rheumatism, asthma, and urinary disorders. Ointment prepared from the roots is used in elephantiasis. Decoction of pods used as a vermifuge and to stop vomiting. Seeds yield an oil said to be a specific against scabies, itch, eczema and other skin eruptions. Irulas use the leaves for curing skin diseases and roots for treating snake-bite.

Based on scientific reason
Used in siddha medicine.

67. **Manjal karisalankani *Wedelia chinensis***
Leaves are used for dyeing hair and for promoting hair growth. Roots yield a black dye. Leaves are a tonic and also used in cough, cephalagia, and alopecia. Decoction of herb used in menorrhagea. Ethanolic extract of herb inhibits the growth of carcinoma. Irulas use leaves for curing jaundice.

Based on scientific reason
Plant contains lactone, widelolactone, stigmasterol and tannin.

Flowering season: year round

Method I

Preparation

Karisalankanni leaves are taken, cleaned and dried well. Then make it into a fine powder by using Grinder. One teaspoonful powder is mixed with milk.

Dosage: One time/day. It is continues up to 10 days.

Method II

Ingredients

- Karisilankanni leaves - A handful
- Small onion (*Allium cepa)* – 5
- Tomato - One
- Chilly(*Capsicum frutescens*) – 3
- Oil – Required amount

Preparation

Fry the leaves of karisilankanni along with onion, tomato and chilly. And made it into a fine paste.

Dosage: In the morning, the paste is taken in empty stomach for 10 days.

Additional information about Karisilankanni

Leaves are used for dyeing hair and for promoting hair growth. Roots yield a black dye. Leaves are a tonic and also used in cough, cephalagia, and alopecia. Decoction of herb used in menorrhagea. Ethanolic extract of herb inhibits the growth of carcinoma. Irulas use leaves for curing jaundice.

Flowering season: year round.

68. **Navakonchi-** ***Rauvolfia tetraphylla***

Roots often used as a substitute and adulterant of those of *Rauvolfia serpentine* for nervous disorders; their reserpine content, however, is low. Plant may cause fatal poisoning. Extract of the herb mixed with castor oil is applied to skin aliments. Fruit yields a black dye. Irulas use roots as one of the snake-bite medicines.

Flowering Season: May to September

Based on scientific reason

- It contains serpentin and it is responsible for antidote to snake bite..

69. **Sennithulasi (*Stevia spp*)** leaves – A handful

Leaves are taken, dried and made it into a powder

Based on scientific reason

- Sweet glucoside is present in the entire plant with low calorific value of sugar.

70.	**Thandri -Belliric murobalan-*Terminalia bellirica*** Wood used for rough shafts, carts, and in coal mines as pit-props; also used for boards, packing-cases, boats, small crafts and catamarans and ploughs. Suitable for second quality slate frames. Fruits used for tanning. Also applied to rheumatic swellings and for eye-care. Bark diuretic and the gum demulcent and purgative. Irulas use the fruit for stomach ulcers, preventing white discharge and as a general health care medicine. Flowering Season: March to January **Ingredients** Thandri (*Terminalia bellirica*) leaves – A handful Kadukkai (*Terminalia chebula*) – A handful ,Amla powder (*Emblica officinalis*) – A teaspoonful ,Sukku (Dried ginger) **Preparation** Grind the leaves of Thandri and kadukkai and it is mixed with amla powder and sukku. **Dosage**: Take this medicine one time/day until it cures.	Based on scientific reason • Very old siddha medicine
71.	**Thumbai (*Leucas aspera*) Thumbe** Leaf extract Hand ful of Thumbai leaves are taken, cleaned and 10 ml juice is extracted which has to be given to person , who is affected from scorpion bite. Dosage: 10 ml / time/day.	Based on scientific reason • Flowers contain aromatic oil and seeds contain
72.	**Karunaikilangu *(Typhonium spp)*** It is a tuber crop it is used for the treatment of piles as well as constipation. It is given in the form of labium to the patients suffering from body temperature	Based on scientific reason • The tuber contains calcium oxalate.
73.	**Perunerinjal** Plant leaves and upper stem portion are kept in water for three to 4 hours and filter. The filtered water can be given to the women suffering from endometritis. It is having the capacity to cure endometritis, body heat and renal diseases	Based on scientific reason • Seeds are used for increasing the vigor and

ETHNOVETERINARY PRACTICES

All the 53 Ethno Veterinary Practices have been followed by the traditional veterinary practitioners based on their experiences and learned from their forefathers. The veterinary scientists informed that scientific rationality can be fixed only by conducting research. However, those practices have differed on the diseases of animals, which have been observed by many working on this line.

11

Rationality on Indigenous Technical Practices - Fisheries

FISHERIES

DSIR/TPDU/TIF 056 and E-28-GO-GOI Research Scheme on "Documentation and Preservation of Agricultural Traditional Knowledge through Farmer Participatory Approach"

S. No.	Documented ITK Practices	Rational
1.	**Suraa/shark mattu (Fishing Net)** It is used in Vilavancode taluk of Kanyakumari district. It is a bigger net used to catch suraa/shark fish. The main component of this mattu is a thoondai, which is made of iron. Thoondai measures about 10cm in length, has a sharp edge at one end and it is bent inwards. With this thoondai, saalai, kalavai, saemeen fishes are caught. For this, the fishermen go deep in to sea from 6 am to 6 pm. If the mattu gets struck on to any rocks the time may extend.	Based on scientific reason • Selective fishing gear. They will catch only above certain sizes. The hook size will suit only certain size of fish mouth.

Different types and their uses:
The 6 no.mattu is less in weight and used to catch nemmeen, sinna choorai, vaalai, ayilai where as no.8 is to catch ketta kembu, saarai, chiravu, kerai. The larger one of no.12 is used to catch suraa/shark.
Special features:
The mattu/nets are being made by the women in the fisherman family. The material used is silk and nylon. They make the mattu manually; do not use any machines for it. It also adds an additional employment to the fishing community.

2. **Maayakkaa (Fish trapping device)**
It is a trapping device created by the fishermen of kanyakumari district especially in the areas of kolachel, thengai patnam, eraviputhan thurai. It is made of iron and looks similar to fish. Its weight ranges from 150-200g.
Usage:
Mayakkaa is made as per the requirements of the fishermen, vary in size and weight. It is connected to one end of tangusam / role, blue in colour. Tangusam is made of nylon, one part of role is kept in hand and other left under the vellam –fishermen boat.
Benefits:
Maayakkaa invites by attracting the other fishes and acts as a prey to them. The fishes get trapped by this and struck on to the rope. With this nokkoli, saalai, kili, koduvaalai, ayilai fish are caught. Sometimes, the tangusam gets cut off when big fish are caught. To avoid this, experienced persons will get to know by the rope movement whether the fish is big or small. Immediately on seeing the movement of rope, pull out the tangusam by which they are freed off.

Based on scientific reason
- Traps offer shelter and food for fishes. The opening of traps will act as only way traffic device. The fish entered in to the traps can not come back.

3. **Muthu**
It is a fish trapping device used in Kanyakumari district. They are smaller in size when compared to maayakkaa. It is used in many numbers of 5-6for a single time. It is less weighty and weighs 25-100g.
Method of use
The muthu is tied on the rope and the net is used to catch fish. Between two muthu one gap is left. After tying, the net and the rope they are left in to the sea under the vellam. It is mainly used to catch, 'kanmai' fish. When bigger muthu is used 'periya neimeen' is trapped.

Based on scientific reason
- Traps offer shelter and food for fishes. The opening of traps will act as only way traffic device. The fish entered in to the traps can not come back.

4. **Soorai thumbai**

This is used when wind velocity is more and waves are fast so that the fisherman unable to go deep in to sea. It also looks similar to fish and traps soorai, kerai, nemmeen. Some other small fish are also caught. For this, the role used must be thick and strong.

Based on scientific reason

- Traps offer shelter and food for fishes. The opening of traps will act as only way traffic device. The fish entered in to the traps can not come back.

5. **Paamaram/paai maram**

Components:

- Paai-thaamaa paai
- Wooden pole- thadi
- Rope
- Small wooden boat

Structure:

The wooden boat at its bottom on which the wooden pole is kept and tied vertically. The rope is tied at an angle of 45^0 from the top. A triangular gap is created between the pole and the rope. Thaamaa paai is spread and tied in the gap.

Working principle:

The thaamaapaai gives protection when heavy winds occur. They move the boat in the direction the oar is swing. It mainly works with the oar and the wind blowing direction. With this up to 20kms fishing can be done.

Limitations:

During rainy days, the thaamaapaai gets completely wet and add more weight to the boat. To overcome this nowadays, fisherman use paai made of plastics.

Based on scientific reason

- Tied wooden logs will not sink out any situation. It can be used as life saving device also.

6. a **Catamaran**

Kattu Maram- Type I

The coastal fishing communities in Sirkali Taluk, Nagapattinam district (Tamil Nadu, India), have survived by their catamaran fishing for centuries. Catamaran is a light watercraft, which was named from the Tamil language words "kattu" (tie) and "maram" (wood, tree) –meaning "two trees tied together". It was invented by the "Paravas", an aristocratic fishing community in the southern coast of Tamil Nadu, south India.

"Catamaran" is a 10 feet boat made out of timber .Timber obtained from "Matha" tree is specially used for constructing it. This timber is transported from

Based on scientific reason

- Tied wooden logs will not sink out any situation. It can be used as life saving device also.

kerala where it is found abundantly. Generally, it is made by carpenters and bought by fishermen. The rate varies depending upon the size of the Catamaran .Normally, on an average one Catamaran costs about Rs. 3000/.Nearly 3-4 slightly curved wooden planks are tied together with rope, such that the entire structure forms a curve shape.

One end of the centre, timber is curved and lifted upwards to a height of nearly 1-1.5 feet. This facilitates the boat to adopt with wave movements, while roaring into the sea.

This kind of Catamaran can accommodate only one member. Where as by making an adjustment even3-4 persons can be accommodated i.e. when one or two more planks are added and tied together with a Catamaran more space is provided and so it enables more persons to go for fishing.

In Ramnad, Catamaran is made up wood of kalyana murungai (*Erythrina indica)* because it has good floating capacity in sea. The tree selected to make catamaran is cut down in "crescent moon" period. Then it is kept for 15 days. Thereafter it is shown in fire to remove outer barks. Kaatumaram is kept for 4-5 years based on usage. The larger woods are kept in outside, in the middle slightly bigger woods are kept in orderly manner. Finally these arranged woods are tied horizontally with wood and rope.

Cost effectiveness

Timber is the main component that involves more expenditure. On an average a Catamaran cost up to Rs.3000/ (one member boat).When compared to other boats (like motor boats made out of fiber) the cost of this kind of Catamaran is cheaper, affordable by an ordinary fisher man.

Feasibility and Relative advantage

Fishing by catamaran is a traditional way of harvesting marine life without damaging the marine environment and its biodiversity. Catamaran is highly preferable, because of its eco friendly usage. It is being rowed manually by using oar. Apart from that, entire structure is made out of wood. This ensures zero pollution of sea water. Since, the cost of construction is also cheaper, it is highly feasible and advantageous.

Constraints and Limitations

- Though it is cheaper and eco friendly, there are few constraints which limits its usage. It can be used only up to a distance of one mile (approximately) into the sea. So the distance covered is comparatively very less which effects in low fish catchment.

- It is difficult to rove the boat near marshy places, when the fishermen himself should get down and pull/ push the boat. It happens near to the sea shore.
- Fish catchment is 7-10 kg
- Using catamaran for fishing is also threatened by the mechanized boats and technological developments that have been used by commercial fishing operators over the last few decades. Marine capture fisheries have been stagnating in recent years in spite of an increased fishing effort –an indication that over-fishing by commercial mechanized boats is already taking a toll. This is threatening the survival of the poor fishing communities that use traditional catamaran.
- The number of boats in operation is estimated at about 70,000. It has been indicated that the small country craft do not operate beyond a few miles from the shore and spend much of their time in going to and from the fishing grounds. Consequently production per unit of effort is low. Mechanization of fishing operations would enable the fishermen to reach areas outside the range of the existing craft and also to fish for longer hours.
- It consumes more time in roving rather than fishing.
- During cyclonic wind period it is not used for fishing, because of heavy wind.

6. b **Catamaran- Type II**

It is a traditional fishing boat existing still today in the coastal areas kolachel, eravi puthanthurai of kanyakumari district. It is a U shaped, having 12-15 feet length at the middle portion. The middle portion is lengthier one, which is slightly raised at both ends.

Making of catamaran:

The wood used is taken from locally available trees (Kalyanamurungai). The wood is cut to required size and shape. For a single catamaran, 6pieces of wood with width of15cm, 12-15' length are taken and put it in to the sea by tying them to strong support preferably huge stones at sea shore. The process is continued for 3-5days. After 5days, they are taken out and tied together with rope to get U shape. Finally, kadiyaal and aniyam are kept and tied at front and back end

Based on scientific reason

- Tied wooden logs will not sink out any situation. It can be used as life saving device also.

respectively.Fishing can be done up to 20kms into the sea.

Limitation:

- Riskier than the karaimadi vellam
- Strong winds may lodge the boat in to sea.

Only 2-3persons can go at a single time, have to come back immediately on the next day itself.

6. c **Kattumaram- Type III**

Catamaran (Kattumaram) is traditionally used for fishing. It is made up of *Erythrina indica* because it has good floating capacity in sea. The tree is cut down in "crescent moon" period. It is kept for 15 days. Thereafter it is shown in fire to remove outer barks. Kaatumaram is kept for 4-5 years based on usage. The larger woods are placed in outside, in the middle slightly bigger woods are kept orderly manner, and finally tied horizontally with wood and rope. Fishing can be done in day hour's upto 7'0 clock using Kattumaram. Fisherman can catch 4 kg fishes by going 4 km range by using kattumaram. As the waves attack directly to fisher man under woods and also to the body, they can fish only for 2 hours.

Based on scientific reason

- Tied wooden logs will not sink out any situation. It can be used as life saving device also.

7. **Karaimadi vallam**

It is an advanced form of catamaran. This fish catching device still exists in coastal areas of Kolachel, Thengaipatnam of Kanyakumari district.

Structure and making of karaimadi vallam:

It is about 20'in length, has a bottom which is made up of various wooden planks. To the flat bottom is attached a shallow wooden planks 5'length and 1' breadth at all its sides. They are connected to each other with thick ropes. For this holes are made in the wooden planks. At the front and backend they are closed to form a sharp edge. Inside the vallam, there can be kept wooden planks to sit while fishing.

Benefits over catamaran:

Karaimadivallam has a space to store food in the boat. So, fishing can be done for a week and carried up to 50kms. up to10 can be taken in to the sea for fishing.

Limitations: It requires more initial investment than catamaran.

Based on scientific reason

- It is a banned fishing net. However many fisher man are still using it for fishing in near shore region.

8. **Nangooram/anchor**
It is made of iron with a single main piece to which 4 bent arrows like structures are joined. It weighs 5-10kg, 1'length. The bent arrows are 15cm in length. At the bottom of the main piece, a hole is present to which rope is tied. If the boat wants to stay in a place without moving, the fisherman put the nangooram/anchor in to the sea. It gets firmly held at a point and do not move. This is mainly used during natural disasters like storm, heavy rain with thunder and lightning. It can also be used during heavy waves or winds.

Based on scientific reason
- The fluke part of anchor will hold the sea bottom to take a vessel at sea.

9. **Traditional Fishing Basket**
It is designed by the farmer Th.Rajendran Agaraperunthottam in Sirkazhi taluk of Nagapattinam district
(i)Pari
It is an indigenous fishing basket which is made out of bamboo splits. Fresh bamboo splits are taken from the bamboo trees and cut into required length and width of 3feet and 2cm respectively. Pari has two openings. One is wide spaced top opening, the diameter of which is 1.5 feet and another one is bottom opening, whose 3/4 th portion is closed with a small hole. Width of the basket is gradually decreased from top to bottom, so it looks like a conical shaped basket. In four places, bamboo splits are tied with rope around the basket to obtain the conical shape. 30 cm from the top centre portion is blocked with small bamboo splits and two holes are provided with the diameter of 5cm for fish entry. After that it is divided into two halves. Y shaped stick is tied on the pari, which is buried in the ground in order to arrest movement of the basket along with running water. Stick is acting as an anchor for pari. Existence of pari in that is more than100 years which costs aboutRs.150.
Fishing Mechanism
Pari is used for fishing in the tail end of the running water like small canals. It is used for fishing during night time ie 7.00p.m-6.00a.m. Small hole in the bottom portion is blocked with paddy straw to prevent the fish to come out from the basket. It is kept on the tail end of the canal and anchored it. After that bund is made with muddy soil around two sides of the pari. It can be placed along the water flowing or running direction. Normally fish has the tendency to swim against the water flowing

Based on scientific reason
- Baskets will offer hiding place for the animals like octopus and lobsters.

direction. So the fish is jumped into the basket and it will enter into the bottom portion through two small holes. Trapped fishes are collected through the bottom opening.

(ii) Katcha

Katcha is a traditional fishing basket used for fishing in the rice fields. It is also a conical shaped bamboo basket and looks like a Pari. But there is no division in the inner side of the basket. Fresh bamboo splits are taken from the bamboo trees and cut into required length and width of 3feet and 2cm respectively. Katcha has only one opening i.e. mouth for fish entry and there is no bottom side opening for fish collection. After completion of fish capturing fishes are collected from the basket through the broad top opening (mouth). Existence of Katcha in that area is more than100 years which costs aboutRs.150.

(iii) Ootha

Ootha is also conical shaped basket made out of thin bamboo sticks. Length of the bamboo stick is 3 feet. Fifteen cm iron ring is taken and smooth bamboo sticks are closely tied around the ring with the help of nylon rope. To obtain conical shape bamboo sticks are tied in 3-4 places. To strengthen the Ootha thick notchi stick is attached in a circular manner. Ootha has two openings such as broad opening and small round opening (iron ring). Ootha is used to capture the fishes in running water and also stagnated water bodies. Ootha is placed randomly in the water bodies in such a manner that broad opening facing downward and small iron ring opening facing upward. Ootha is hold tightly by left hand and captured fishes are collected by right hand through the small iron ring opening. Two kilo gram of fishes can be captured in an hour. It costs about Rs 200/

Advantages

- Cost involved for making is very less.
- Less maintenance.
- Less weight and easy to handle.

Limitations

- It needs skilled persons for basket making.
- Fishing capacity is less.
- Only used to fish in the smaller areas.

10.	**Sorati (Crab collecting tool)** Sorati is used to collect the crabs from nest in the rice field. Two feet lengthy iron rod is inserted in the waste unused cycle peddle. Cycle peddle is used as handle. Iron rod is inserted into the crab nest and crabs will come out due to disturbance. Then crabs are collected manually for human consumption. One person can collect 40-50 crabs / day Cost required to make this Sorati is Rs. 15.	Based on scientific reason • Crabs will catch any thing external to its territory catch the iron rod.
11.	**Karuvadu Thayarithal (Dried Fish)** Excess unsold fresh fishes are used for making dried fish. Dried fish can be made out of all the species of fish. Fins and tail portion are cut by using knife and fish is split open. Inner portion is removed. Fish is washed without removing the scales (while drying the scales will fall off) and lined with knife, which facilitates better salt absorption. Salt is applied over the fleshy portion. It is dried in hot sunlight for a day. Next day the fish is washed and again salt is applied. Duration of drying varies from 3 days to 10 days, which depends upon the size of the fish. Dried fish is one of the major ingredients of the poultry feed. **Marketing** Captured fishes are sold in the local villages itself. Excess fishes are sold in the near by towns or far away places as ice preserved fishes. Price may vary. Normally price depends up on size and variety of the fish. Only 20 per cent of the sea fish catch is marketed as fresh fish and the remaining 80 per cent is either preserved as sun-dried fish or as salted fish or converted into fish meal and manure. Preserved fish is very largely used by the poor because of its cheapness.	Based on scientific reason • Salt act as a preserving agent to present the fisher for the longer duration. It would also give taste to fishes when stored long time. Salt also kill bacteria and prevent fungal growth
12.	**Dam fishing** **Traditional fishing in dam** This method of fishing involves traditional materials like bamboo splits, coir and net. A square dimensioned normally 7 ftx7ft or a rectangular dimensioned (10x7ft) net is tied on opposite two sides (Top & bottom) with bamboo splits that serve as a frame. Bamboo splits are framed such that one of the sides i.e., bottom side is bow shaped. The straight portion of the split is tied to the net and the bended (bow shaped) split is left without tying. On the opposite side i.e. top portion frame is tied	Based on experience • It also catches only the fishes more than the bamboo mesh size. Only used in fishing dam as well as shallow waters.

firmly with the net. On the four corners long coir ropes are tied.

Mode of operation

The net is suspended at the shutter outlet such that the bow frames a broad base at the bottom. It is kept as such for 20-25 minutes (Time varies from season to season). For a day nearly 15-20 times they suspend the net for fish catching. Each time when suspended, fish catch ranges from 1 kg to 2 kg.

Principle

While fishes are forced by the flushing water through the out let, it tries to jump backward. Thus it is caught in the net which is suspended at the outlet and the bow shape frame facilitates in withholding the fish that have fallen in the net.

Cost effectiveness

Since net, bamboo splits and the coir rope are locally available. They do not cost much. When compared to sea fishing, fish catch requirement cost is very cheap in this method.

Advantage

Necessary requirements are easily available. Making it as well as handling is easier.

References

1. Dewalt, B.R. (1994). "Using Indigenous Knowledge to Improve Agriculture and Natural Resource Management." Human Organization 53 (2). pp.123-131.
2. Emery, A.R. (1996), The Participation of Indigenous Peoples and Their Knowledge in Environmental Assessment and Development Planning (draft). Centre for Traditional Knowledge: Ottawa, Canada.
3. Giranand and Kamlesh kumar (1980) Folk beliefs Associated with Dry Farming. Indian Journal of Extension Education, 16 (3&4): 36-42.
4. Grenier, L. (1998). Working With Indigenous Knowledge: A Guide For Researchers. IDRC: Ottawa, Canada.
5. Gupta, Anil K. (1990). "Documenting Indigenous Farmers' Practices." ILEIA Newsletter, 6(2): 29-30.
6. IIRR (1996a), Recording and Using Indigenous Knowledge: A Manual. IIRR: Silang, Cavite, Philippines.
7. IIRR(1996). Recording and Using Indigenous Knowledge: A Manual. International Institute of Rural Reconstruction, Silang, Caviite, Philippines.
8. Kuhn, Thomas S. (1962) The Structure of Scientific Revolutions, Chicago and London: University of Chicago Press.
9. Kumar, M (2001). Organic Farming Practices. Num Vali Velanmai, 10 (1): 21-22.

10. Levi-Strauss, C. (1966) The Savage Mind, London: Weidenfeld and Nicholson
11. Massaquoi, J.G.M. (1993) "Indigenous Technology for Off Farm Rural Activites." Indig. Know. & Devt. Monitor, 1(3):2-3
12. Neela Mukherjee (1997) Participatory Appraisal of Natural Resources, Concept Publishing Company, New Delhi -59.
13. Neela Mukherjee (1997) Participatory Rural Appraisal and Questionnaire Survey, Concept Publishing Company, New Delhi -59.
14. Neela Mukherjee (1997) Participatory Rural Appraisal Methodology and Applications, Concept Publishing Company, New Delhi -59.
15. Sivanarayana, G (1993) Indigenous Technical Knowledge and Communication System. Unpublished Ph.D Thesis BHU, Varanasi.
16. Sundara Mari & T.T.Ranganathan (2003). Indigenous Agricultural Practices for Sustainable Farming. Published by Agro Bios (India).
17. Thrupp, L.A. (1989). "Legitimizing Local Knowledge: From Displacement to Empowerment for Third World People". Agriculture and Human Values. Summer Issue. Pp.13-24.
18. Warren,Michael D. and CashmanKristin,(1988), Indigenous Knowledge for Sustainable Agriculture and Rural Development, Gatekeeper Series No. SA 10, Sustainable Agriculture Programme, IIED, London.

Colour Plates

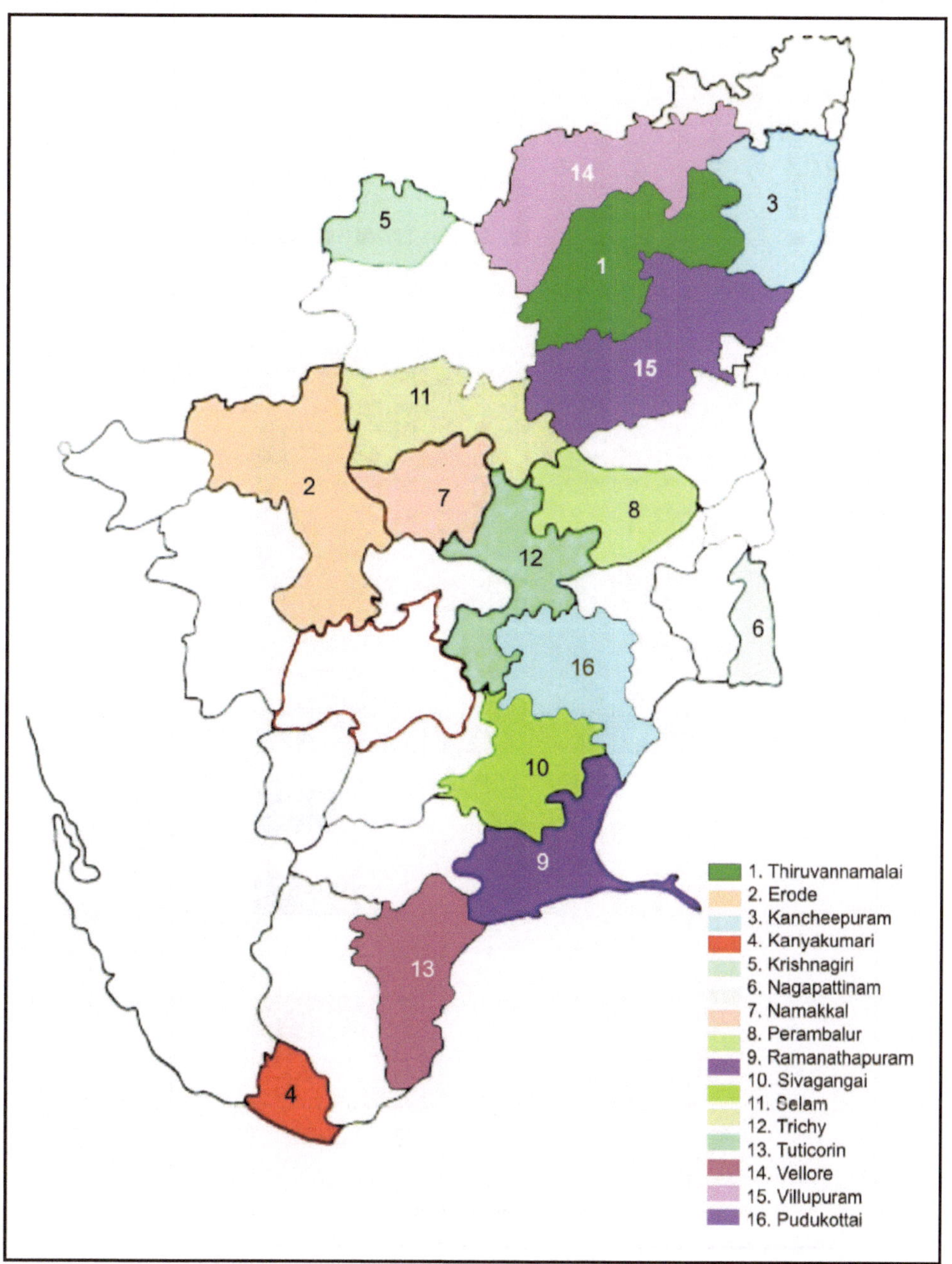

Fig. 1. Map Showing the study area in Tamil Nadu

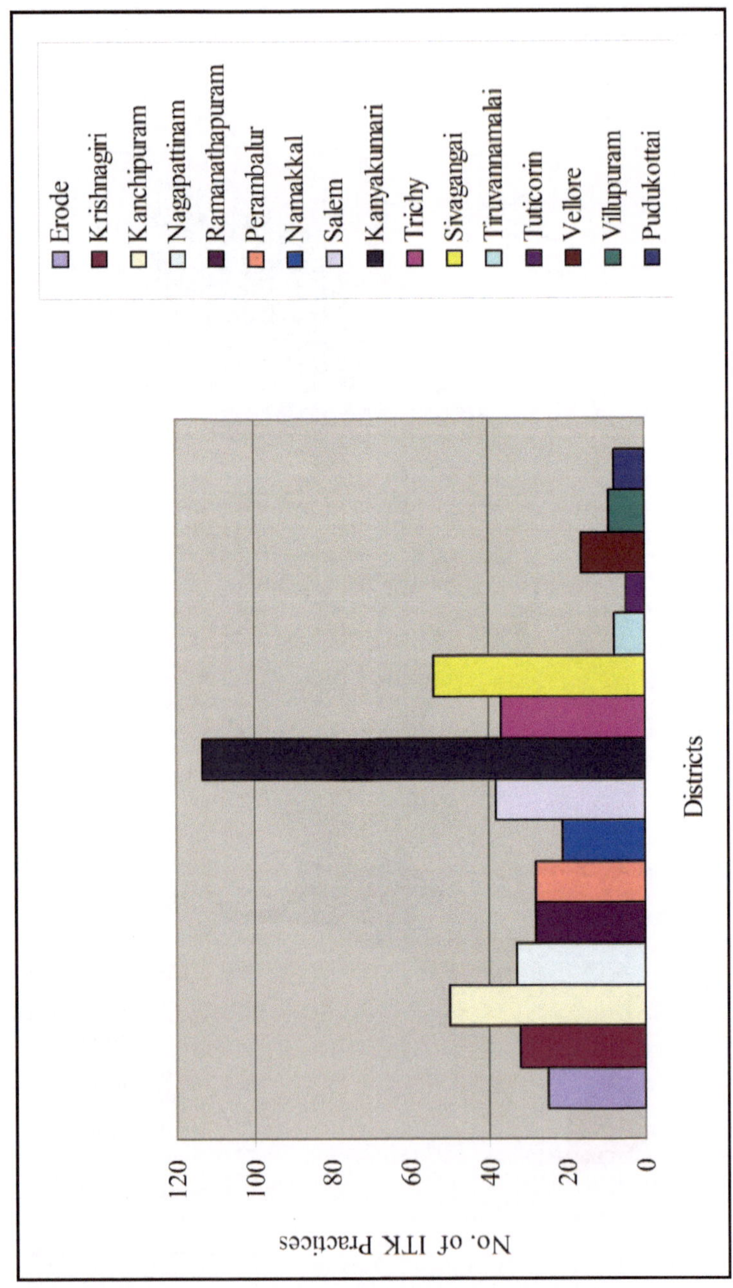

Fig. 2. District wise Documentation of ITK Practices

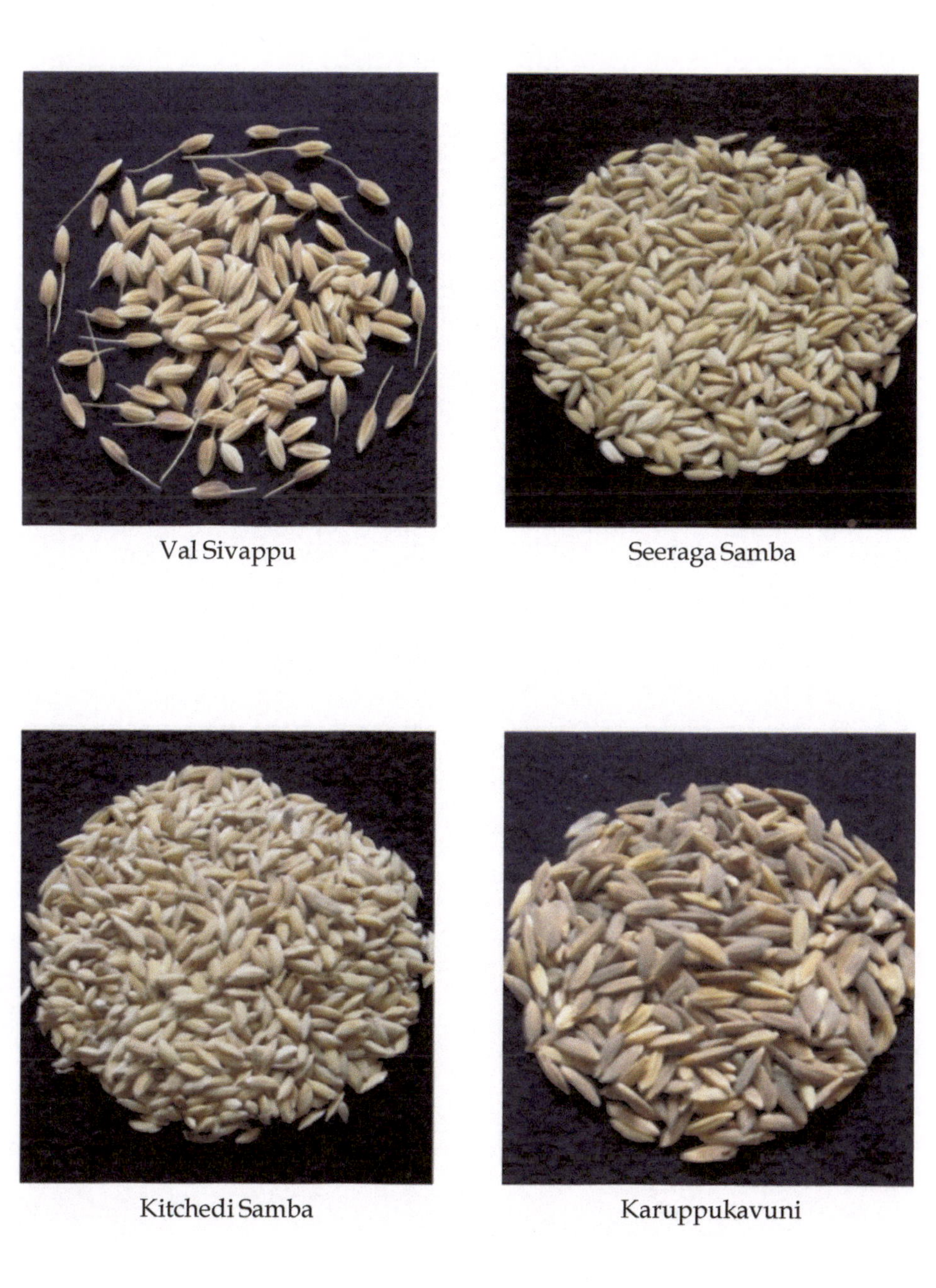

Val Sivappu

Seeraga Samba

Kitchedi Samba

Karuppukavuni

Fig. 3. Traditional Rice Varieties

Kuliparichan

Sivappu Chitraikar

Murungaikar

Sadaikkar

Fig. 4. Traditional Rice Varieties

Kuruvai Kalanchium

Norungan

Samba Mosanam

Kadaikazhuthan

Fig. 5. Traditional Rice Varieties

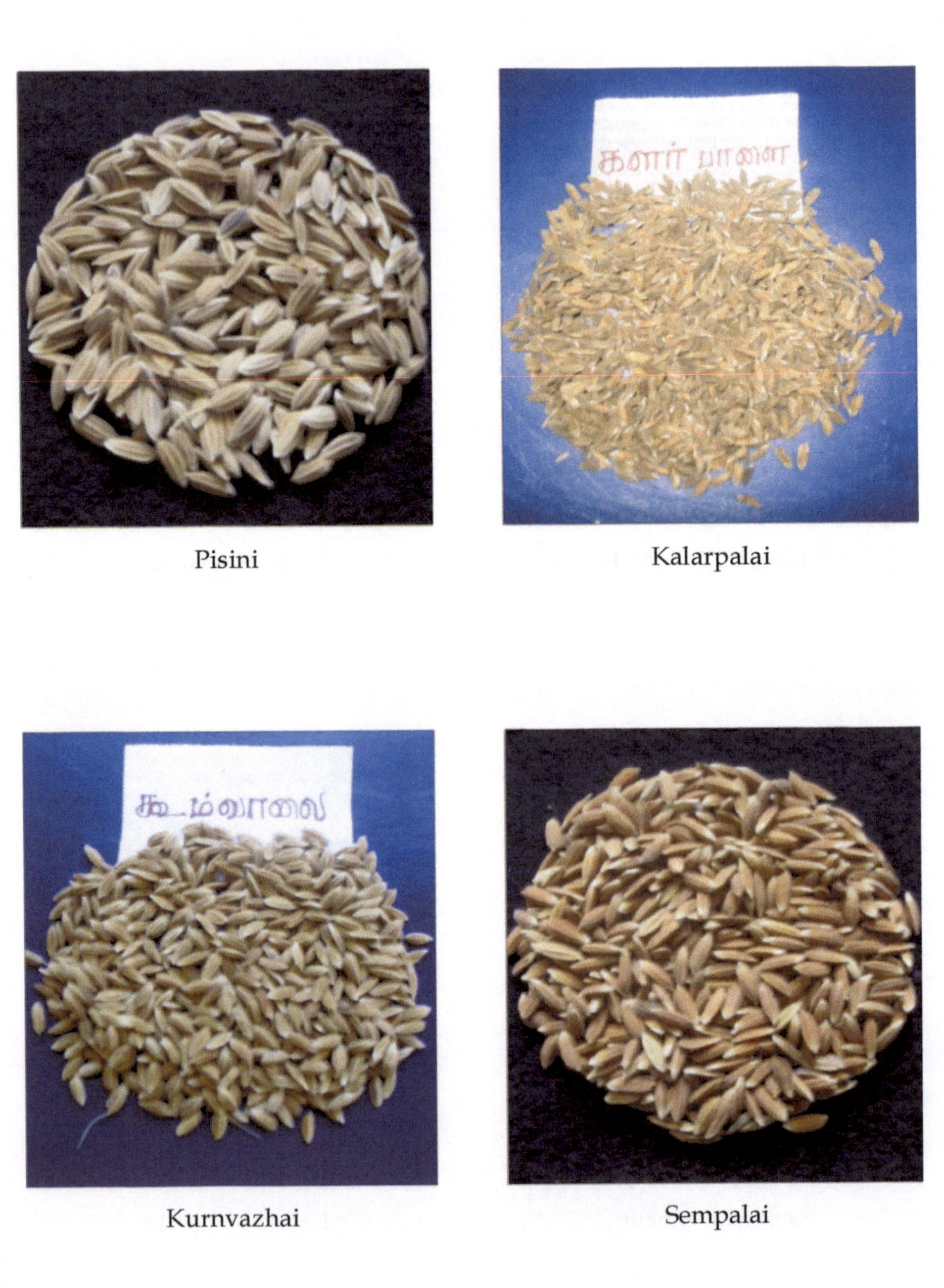

Fig. 6. Traditional Rice Varieties

Karikattai Ragi
Eleusine coracana

Kattucumbu
Pennisetum glacum

Thattai Varagu
Pespalum scrobiculatum

Kona Cholam
Sorghum bicolor

Fig. 7. Traditional Millet Varieties

Chittan Samai
Panicum sumatreuse

Sen Cholam
Sorghum bicolor

Red Gingelly
Sesamum indicum

Kothu Kadalai
Arachis hypogaea

Fig. 8. Traditional Millet and Oil Seed Varieties

Vazhuthunangai
Solanum melongena

Ramnad Gundu Milagai
Capsicum annum

Gandhari Chillies
Capsicum annum

Ther Kothavarai
Cyamopsis tetragonoloba

Fig. 9. Traditional Vegetable Varieties

Nattu Avarai
Lab lab purpureus

Betel Vine
Piper betel

Murungai
Moringa oleifera

Sweet Potato
Ipomoea batatas

Fig. 10. Traditional Varieties

Jathi Malli
Jasminum grandiflorum

Mullai
Jasminum auriculatum

Ramnad Gundu Malli
Jasminum sambac

Naadan
Ananas sativus

Fig. 11. Traditional Varieties

Ram Seetha
Annona reticulata

Chengavarkai
Mangifera indica

Chetti Ottu
Mangifera indica

Parukkan Pala
Atrocarpus heterophyllus

Fig. 12. Traditional Fruit Varieties

Varikkai Chakkai
Artocarpus heterophyllus

Koozhan Chakkai
Artocarpus heterophyllus

Sevvazhai
Musa sp

Nendran
Musa sp

Fig. 13. Traditional Fruit Varieties

Coconut + Banana

Rubber + Banana

Betel Vine + Agathi

Amla + Ramnad Jasmine

Castor + Red Gram + Cotton

Ground Nut + Red Gram

Fig. 14. Crop Management Practices

Pitcher Pot Irrigation

Stubble Mulching

Contour Trenching

Water Lifting Device

Green leaf Manuring

Pineapple Covering with Slip

Fig. 15. Crop Management Practices

Saalu Irrigation

Banana Bunch Covering

Rubber Cover Cropping

Shelter Belt in Banana

Tree Branch Panthal

Banana Tying with Ropes

Fig. 16. Crop Management Practices

Vermicomposting

Bird Perch

Rat Scaring

Ash Treated Seeds

Pest Repellent

Herbal Pest Repellent

Fig. 17. Crop Protection Measures

Paddy Storage Structure

Grains Storage Structure

Ragi Storage Pit

Bamboo See Box

Storage Mud Pot Container

Onion Storage Structure

Fig. 18. Storage Structures

Country Plough

Melur/Bose Plough

Koru Seed Driller

Seed Line Marker

Wooden Leveling Board

Spade

Fig. 19. Ploughing and Leveling Tools

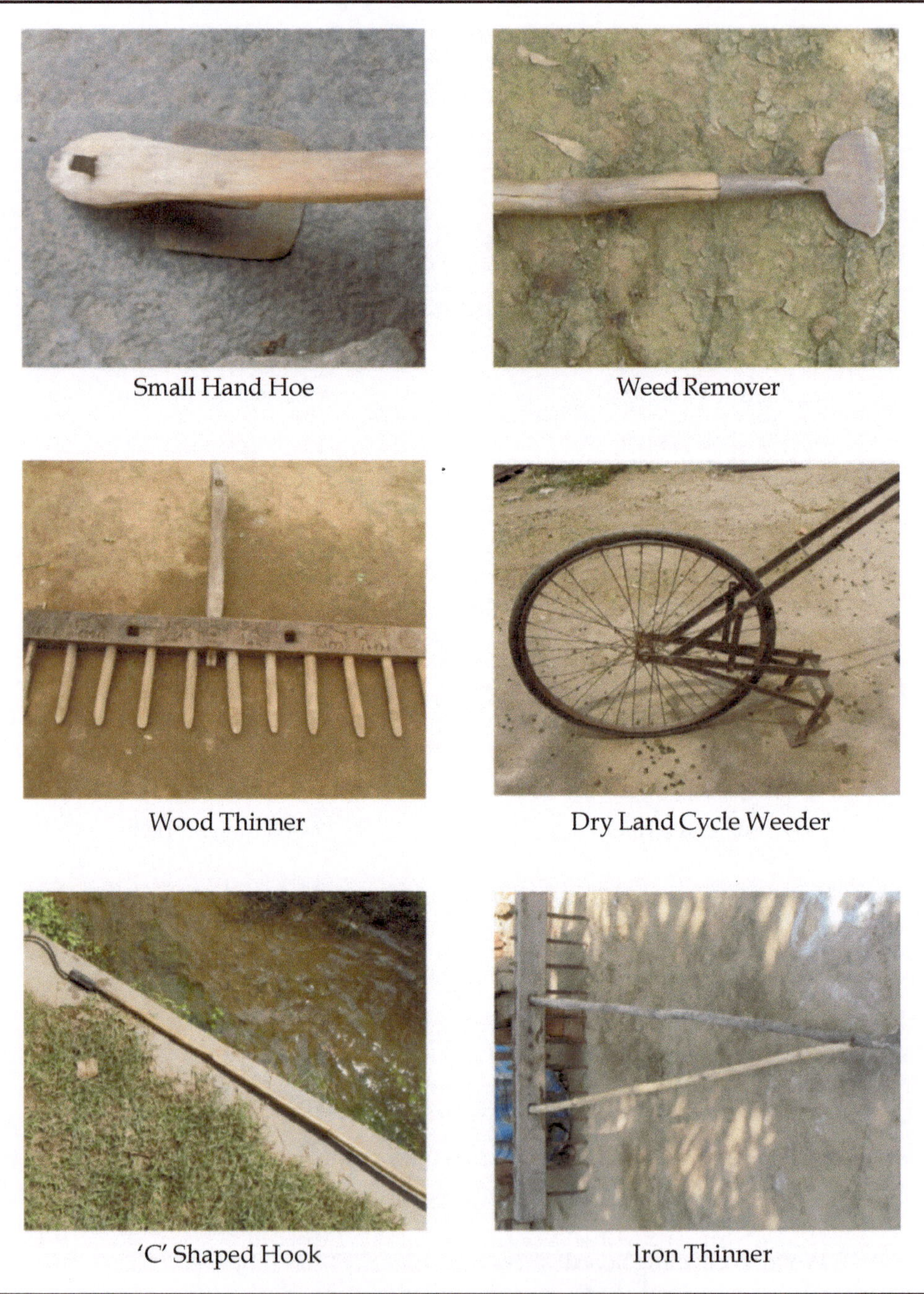

Small Hand Hoe

Weed Remover

Wood Thinner

Dry Land Cycle Weeder

'C' Shaped Hook

Iron Thinner

Fig. 20. Weeding and Thinning Tools

Iron Water Carrier

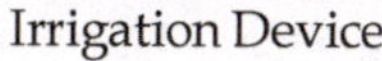

Irrigation Device

Measuring Equipments

Rat Trap

Weighing Balance

Palm Candy Preparator

Fig. 21. Measuring and irrigation tools

Sugar Cane Harvester

Pine Apple Harvester

Cutting Knife

Korai Grass Harvester

Betelvine Harvester

Sesbania Pruner

Fig. 22. Harvesting Tools

Rice Siever

Ground Nut Siever

Grains and Pulses Siever

Ground Nut Grader

Ragi Siever

Stone Grinder

Fig. 23. Sieveing and grading tools

Aanai Nerunji (Pedalium murex)

Adathoda (Adathoda vasica)

Nalla Vellai (Cleome gynandra)

Anaikathaalai (Agave americana)

Neer Mulli (Hygrophila auriculata)

Sudukadu Mallikai (Catharanthus roseus)

Fig. 24. Medicinal Plants

Mahuli Keerai (Celosia argentea)

Christmas Star (Euphorbia pulcherrima)

Andhimalligai (Mirabilis jalapa)

Marul (Sansevieria Zeylanica)

Omavalli (Coleus aromaticus)

Vallarai (Centella asiatica)

Fig. 25. Medicinal Plants

Sathissaranai (Trianthema decandra)

Karunthulasi (Ocimum basilicum)

Thaluthalai (Cerodendrum phlomides)

Eluthani Poondu (Launaea sarmentosa)

Palperuki (Euphorbia helerophylla)

Keelanelli (Phyllanthus amarus)

Fig. 26. Medicinal Plants

Seenthil kodi (Tinospora cordifolia)

Periyaanangai (Andrographis alata)

Gundumani (Abrus precatorius)

Vasambu (Acorus calamus)

Paruppu Keerai (Portulaca oleracea)

Sirukurinjan (Gymnema sylvestre)

Fig. 27. Medicinal Plants

Siriyanangai *(Andrographis paniculata)*

Senthalai *(Pandanus odoratissimus)*

Manjal Karisalankani *(Wedelia chinessis)*

Navakonchi *(Rauvolfia tetraphylla)*

Poonai Mesai Leaves *(Orthosiphon spiralis)*

Elumpotti Thalai *(Ormocarpum sennoides)*

Fig. 28. Medicinal Plants

Thalaisuruli *(Aristolachia indica)*

Syriyanangai *(Andrographis paniculata)*

Siruthumbai *(Leucas aspera)*

Periyanangai *(Andrographis alata)*

Sotrukatraalai *(Aloe vera)*

Nayurvi *(Achyranthus aspera)*

Fig. 29. Ethno Veterinary Medicinal Plants

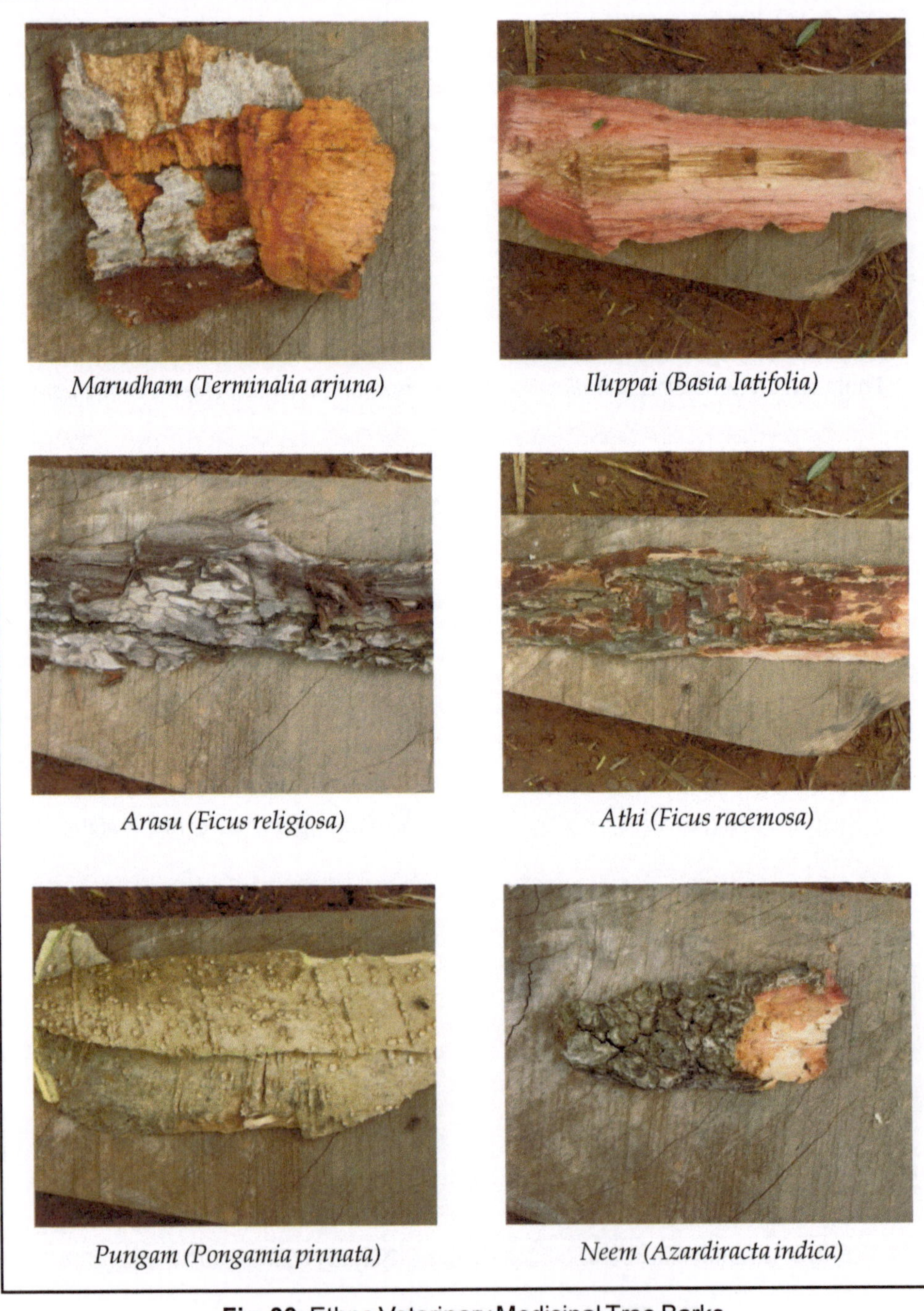

Marudham (Terminalia arjuna)

Iluppai (Basia Iatifolia)

Arasu (Ficus religiosa)

Athi (Ficus racemosa)

Pungam (Pongamia pinnata)

Neem (Azardiracta indica)

Fig. 30. Ethno Veterinary Medicinal Tree Barks

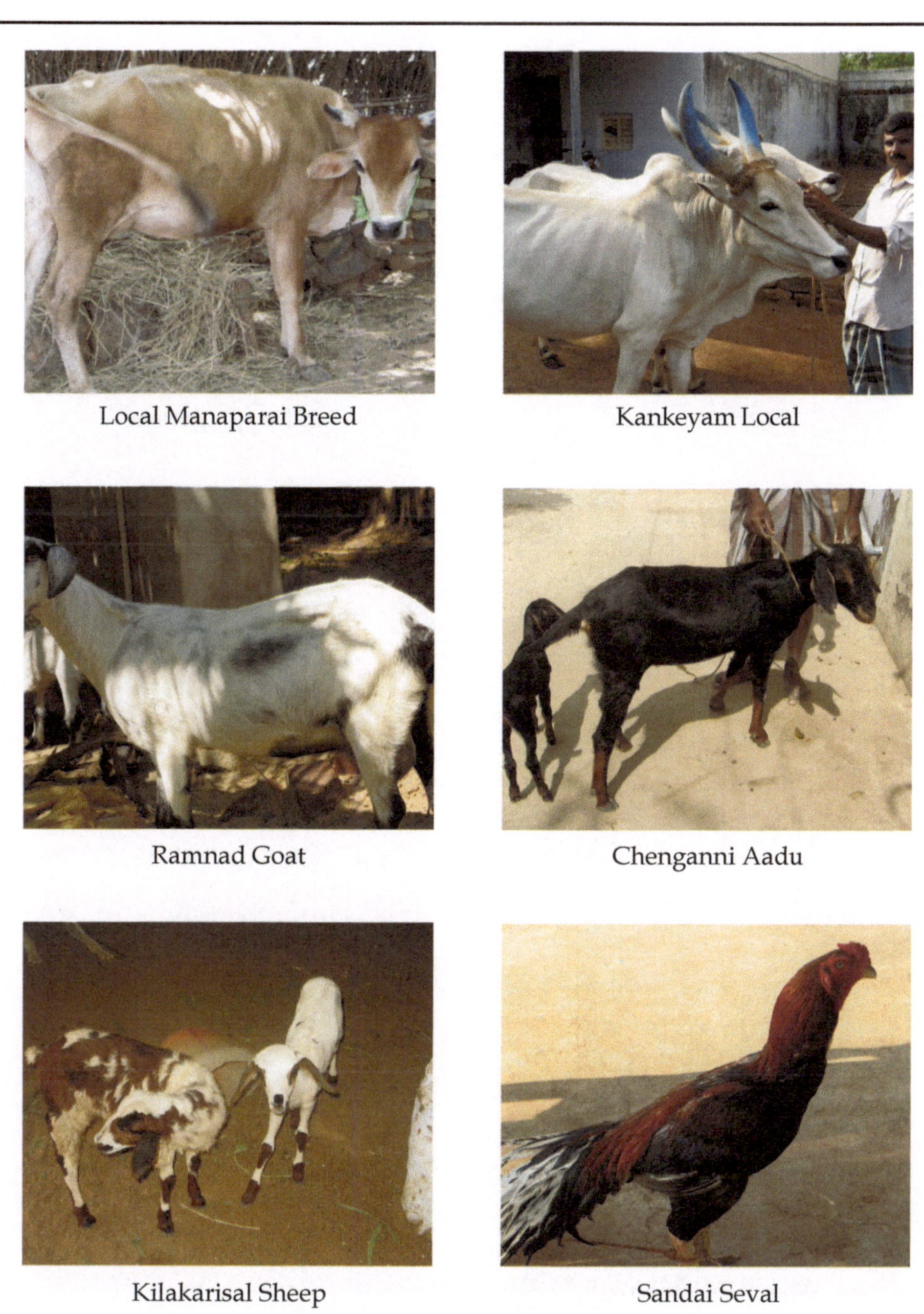

Local Manaparai Breed

Kankeyam Local

Ramnad Goat

Chenganni Aadu

Kilakarisal Sheep

Sandai Seval

Fig. 31. Local Breeds

Artificial Fish Bait

Fish Trapping Device

Fishing Net

Fishing Basket

Crab Collecting Tool

Fishing Boat

Fig. 32. Fishing Equipments

Traditional Chekku
Used for Oil Extraction

Traditional Oil Pestle-Ulakkai

Formation of Oil Cake
During Oil-extraction

Oil Outlet in Chekku
After Extraction

Fig. 33. Traditional Oil Extraction Chekku

Indigenous Tobacco
Curing Process

Tobacco Grading

Graded Tobacco
Grade 1, Grade 2, Grade 3

Removed Tobacco Sticks
During Grading

Fig. 34. Indigenous Tobacco Curing Process

Sugarcane Juice

Traditional Dust Removing Implements-mattipattai

Sugarcane Extracts Boiled for Jaggery Making

Traditional Jaggery Making Implement - Pagu Aath

Formation of Jaggery in Kopparai

Collection of Jaggery After Powedered Form

Fig. 35. Traditional Sugarcane Jaggery Making Process

Manapalli Aruval Pattarai

Coconut Oil Used
For Sharpening the Knief

Wooden Handle Prepared
For Implements

Different Types of Implements
Prepared in Manappalli

Fig. 36. Implements Preparation- Manappalli Aruval

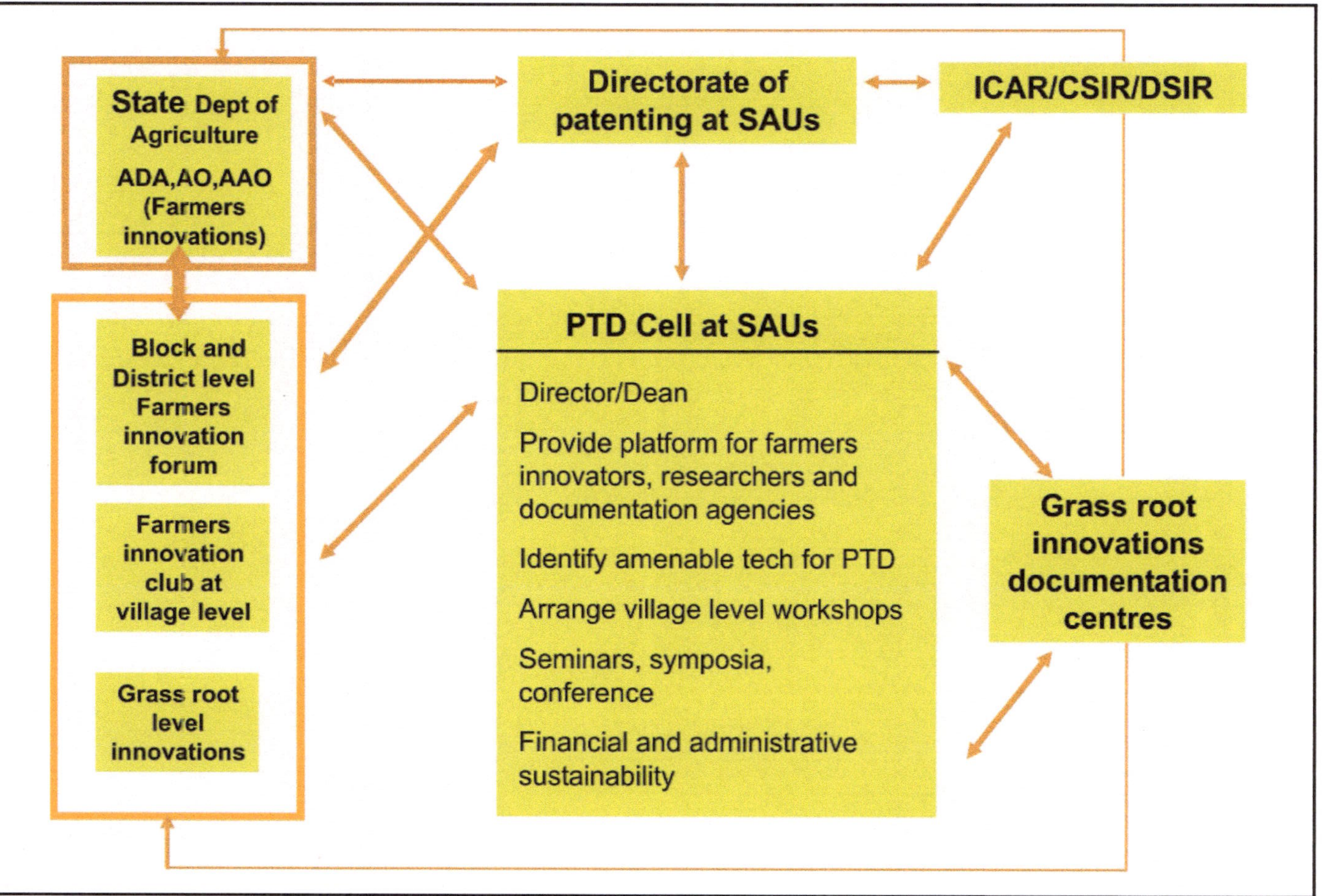

Fig. 37. Suggestive Model for Managing Farmers Innovations through PTD

Index

A

B

C

D

E

F

G

H

I

J

K

L

M

N

O

P

R

S

T

V

W

Y